Clean Energy Production Technologies

Series Editors
Neha Srivastava, Department of Chemical Engineering and Technology
IIT (BHU) Varanasi, Varanasi, Uttar Pradesh, India

P. K. Mishra, Department of Chemical Engineering and Technology
IIT (BHU) Varanasi, Varanasi, Uttar Pradesh, India

The consumption of fossil fuels has been continuously increasing around the globe and simultaneously becoming the primary cause of global warming as well as environmental pollution. Due to limited life span of fossil fuels and limited alternate energy options, energy crises is important concern faced by the world. Amidst these complex environmental and economic scenarios, renewable energy alternates such as biodiesel, hydrogen, wind, solar and bioenergy sources, which can produce energy with zero carbon residue are emerging as excellent clean energy source. For maximizing the efficiency and productivity of clean fuels via green & renewable methods, it's crucial to understand the configuration, sustainability and techno-economic feasibility of these promising energy alternates. The book series presents a comprehensive coverage combining the domains of exploring clean sources of energy and ensuring its production in an economical as well as ecologically feasible fashion. Series involves renowned experts and academicians as volume-editors and authors, from all the regions of the world. Series brings forth latest research, approaches and perspectives on clean energy production from both developed and developing parts of world under one umbrella. It is curated and developed by authoritative institutions and experts to serves global readership on this theme.

More information about this series at http://www.springer.com/series/16486

Neha Srivastava • Manish Srivastava
P. K. Mishra • Vijai Kumar Gupta
Editors

Substrate Analysis for Effective Biofuels Production

Springer

Editors
Neha Srivastava
Department of Chemical Engineering and Technology
IIT (BHU) Varanasi
Varanasi, Uttar Pradesh, India

P. K. Mishra
Department of Chemical Engineering and Technology
IIT (BHU) Varanasi
Varanasi, Uttar Pradesh, India

Manish Srivastava
Department of Physics and Astrophysics
University of Delhi
New Delhi, Delhi, India

Department of Chemical Engineering and Technology
IIT (BHU) Varanasi
Varanasi, Uttar Pradesh, India

Vijai Kumar Gupta
Department of Chemistry and Biotechnology, School of Science
Tallinn University of Technology
Tallinn, Estonia

ISSN 2662-6861 ISSN 2662-687X (electronic)
Clean Energy Production Technologies
ISBN 978-981-32-9606-0 ISBN 978-981-32-9607-7 (eBook)
https://doi.org/10.1007/978-981-32-9607-7

This Springer imprint is published by the registered company Springer Nature Singapore Pte Ltd.
The registered company address is: 152 Beach Road, #21-01/04 Gateway East, Singapore 189721, Singapore

Foreword

Never before has the need for clean renewable fuels been a greater focus, due to both the pollution caused by and the limited lifespan of fossil fuels. It is well-known that sourcing a viable cost-efficient and commercial alternative to fossil fuels is still challenging. In the series of renewable energy production, biofuels are the most promising. Production of biofuels from organic and cellulosic biomass waste grabs global attention due to its renewable, cost-effective and eco-friendly nature. Cellulosic feedstocks provide the raw material for biohydrogen, biogas, biomethane and biobutanol production, whereas wastes such as sewage, sludge and algal biomass provide the feedstocks for biodiesel and bio-oil production.

The major bottleneck in turning to mainstream commercialized biofuels has been the seemingly insurmountable costs involved. The costs of effective substrate, pretreatment, enzymes and associated bioconversion and fermentation technologies have retarded the route of biofuels to mainstream markets. In the past few years, modifications and improvements in biofuel technologies have been noted, and developments in substrate-based bioprocess technologies associated with biomass to biofuels contribute towards making this process more economically viable. Developing this green economically sustainable process demands focused attention on each individual parameter and subsequently considering the parameters collectively at a global scale.

Publication of the book entitled *Substrate Analysis for Effective Biofuels Production* is a notable effort in collating relevant information in the field of biofuels. I am writing this message with great joy and satisfaction as a researcher fascinated in the area of green fuel production. This book essentially holds ten chapters focusing on various biofuel production technologies with defined strategies to improve their production at an economic level. The book is focused on the effect of substrates and related parameters on different biofuel production technologies including biogas, biobutanol, bioethanol, biohydrogen and biodiesel. While addressing existing rollbacks in different biofuel production technologies, the book also discusses way-out technologies related with the production of different biofuels. It is my opinion that this book won't disappoint in providing a unique collection of practical information for scientists, researchers, teachers, students and industries that are interested in biofuel production technologies.

I appreciate the meticulous work of Dr. Neha Srivastava [IIT (BHU), Varanasi], Dr. Manish Srivastava [DU, Delhi], Prof. (Dr.) P. K. Mishra [IIT (BHU), Varanasi], and Dr. Vijai Kumar Gupta [TTU, Estonia] in effectively compiling this book. The effort made by the editors will certainly satisfy the knowledge gap in the field and fulfil the demands of scientists, researchers, teachers, students and industries. This book is an excellent resource for further research and advancement in the area of effective substrate selection in biofuel production processes. I am sure the readers will find this book as an informative source of relevant information in the field of biofuels, which also includes a reliable resource of supporting reference material.

Anthonia O'Donovan

Applied Biology and Biopharmaceutical Science,
School of Science and Computing
Galway-Mayo Institute of Technology,
Galway, Ireland
10.06.2019

Acknowledgements

We, the editors, are thankful to all the academicians and scientists whose contributions have enriched this volume. We also express our deep sense of gratitude to our parents whose blessings have always prompted us to pursue academic activities deeply. It is quite possible that in a work of this nature, some mistakes might have crept in text inadvertently, and for these, we owe undiluted responsibility. We are grateful to all authors for their contribution to present book. We are also thankful to Springer Nature for giving us this opportunity and the Department of Chemical Engineering and Technology, IIT (BHU) Varanasi, UP, India, for all technical support. We thank them from the core of our heart. Editor Manish Srivastava acknowledges the DST, Govt of India, for awarding the DST-INSPIRE Faculty Award [IFA13-MS-02] 2014 and also Science & Engineering Research Board for SERB-Research Scientist award 2019.

Contents

About the Editors

Neha Srivastava is currently working as Postdoctorate Fellow in the Department of Chemical Engineering and Technology, IIT (BHU) Varanasi, India. She has published 25 research articles in peer-reviewed journals with 3 patents and 1 technology transfer. She completed her PhD from the Department of Molecular and Cellular Engineering, SHIATS, India, in 2016 in the area of bioenergy. Furthermore, she has received 6 Young Scientist Awards. Presently, she is working on biofuel production (cellulase enzymes, production and enhancement, biohydrogen production from waste biomass, bioethanol production).

Manish Srivastava has worked as DST INSPIRE Faculty in the Department of Physics and Astrophysics, University of Delhi, India during June 2014 to June 2019. Currently he is working as SERB-Research Scientist in the Department of Chemical Engineering and Technology IIT (BHU) Varanasi, India. He has published 45 research articles in peer-reviewed journals, authored several book chapters and filed 1 patent. He worked as a Postdoctorate Fellow in the Department of BIN Fusion Technology, Chonbuk National University, from August 2012 to August 2013. He was an Assistant Professor in the Department of Physics, DIT School of Engineering, Greater Noida, from July 2011 to July 2012. He received his PhD in Physics from the Motilal Nehru National Institute of Technology, Allahabad, India, in 2011. Presently, he is working on the synthesis of graphene-based metal oxide hybrids and their applications as catalysts. His area of interest is the synthesis of nanostructured materials and their applications as catalyst for the development of electrode materials in energy storage, biosensors and biofuel production.

P. K. Mishra is currently Professor and Head in the Department of Chemical Engineering and Technology, Indian Institute of Technology (BHU), Varanasi, India. He obtained his PhD degree in Chemical Engineering from the Institute of Technology (Banaras Hindu University) in 1995. He has authored/coauthored over 60 technical papers published in reputed national/international journals and supervised more than 20 doctoral students. He has received several awards and honours and has five patents with one technology transfer. He is Fellow of the Institution of Engineers (India). He has received several awards and honours at national/international levels. He has also made significant contribution towards development of entrepreneurship ecosystem in the Eastern part of the country. He is Technology Business Incubator Coordinator at the institute and Member of the Executive Committee, NIESBUD, Ministry of Skill Development, Government of India.

Vijai Kumar Gupta, ERA Chair of Green Chemistry, Department of Chemistry and Biotechnology, School of Science, Tallinn University of Technology, Tallinn, Estonia, is one of the leading experts in the area of microbial biology and biotechnology. He is a Member of the International Subcommission on Trichoderma and Hypocrea, Austria, and of the International Society for Fungal Conservation, UK, and Secretary of the European Mycological Association. He is also the Fellow of the prestigious Linnean Society of London, UK; of the Indian Mycological Association; and of the Mycological Society of India. He has been honoured with several awards in his career, including Indian Young Scientist Award for his advanced research achievements in the field of fungal biology and biotechnology. He is the Editor of a few leading scientific journals of high repute and has many publications in his hands with h-index 21. He has edited many books for publishers of international renown such as CRC Press, Taylor and Francis, USA; Springer, USA; Elsevier Press, The Netherlands; Nova Science Publishers, USA; DE Gruyter, Germany; and CABI, UK.

Algal Biomass: Potential Renewable Feedstock for Biofuel Production

1

Archana Tiwari and Thomas Kiran Marella

Abstract

Algae possess immense potential to yield a myriad of valuable products which find wide application in waste water remediation, nutraceuticals, and aquaculture. The natural products from algae have attained significant attention owing to their extraordinary efficiency which makes them suitable for a plethora of application as biofuels, waste water remediation agents, therapeutics, food and feed, nutraceuticals, biocontrol agents, and aquaculture. Despite the huge potential of algae, studies are limited owing to the fact the difficulties in isolation and cultivation, the significant impact of nutrients, contaminants, and sessional variations. Waste waters tend to act as a source of nutrients, thereby facilitating the growth of algae, which in turn quench the excessive nutrients and heavy metals leading to the phycoremediation of waste waters. Algae can produce a plethora of biofuels including biodiesel, biogas, and bioethanol to name a few. This chapter elaborates the potential of algae biomass as a renewable feedstock for biofuel production and their further utilisation as a promising source of nutraceuticals and high-value products, which can be a sustainable solution making the best out of waste for a better global environment and economy.

Keywords

Algae · Biofuel · Nutraceuticals · Phycoremediation · Wastewater treatment

A. Tiwari (✉)
Diatom Research Laboratory, Amity Institute of Biotechnology, Amity University, Noida, India

T. K. Marella
International Crops Research Institute for the Semi-Arid Tropics, Patancheru, Hyderabad, Telangana, India

N. Srivastava et al. (eds.), *Substrate Analysis for Effective Biofuels Production*, Clean Energy Production Technologies, https://doi.org/10.1007/978-981-32-9607-7_1

1.1 Introduction

The problem of global environmental pollution and rapid depletion of nonrenewable energy sources call for exploring alternate energy sources. Worldwide, researchers are exploring options for eco-friendly, renewable fuels. Algae can serve as a promising option as the most appropriate energy generating organisms as they are efficient in consuming the atmospheric carbon dioxide, which would limit the carbon dioxide emissions. They have been widely explored as the source of a plethora of biofuels and bioproducts (Laurens et al. 2017; Show et al. 2017; Halim et al. 2012; Dong et al. 2016; Li et al. 2017a, b). The acumen of algae in biofuel production coupled with long-chain polyunsaturated fatty acids (PUFAs) was reported to be encouraging (Harwood and Guschina 2009) and can grow promptly doubling algal biomass by 24 h, while in the exponential growth phase, the doubling time is nearly 3.5 h. Algae are abundant in oil content, which can exceed 80% of dry biomass weight (Spolaore et al. 2006).

There exists a variation in the presence of oil, lipids, hydrocarbons, etc. in different algal species, for example, *Botryococcus braunii* yields copious hydrocarbons, particularly oils in triterpene form, which are nearly 30–40% of its dry weight. The main hydrocarbons include *Botryococcus* oils: botryococcenes, alkadienes, and alkatrienes (Banerjee et al. 2002). The major product of *Botryococcus braunii* is quite similar to the fossil fuel compounds and these valuable bio-oils from photosynthetic algae can yield a wide range of fuels like diesel, gasoline, and jet fuel which is in fact the extreme demand of the present time (Xu et al. 2006).

Biofuels are the need of today. Algae are wonderful tiny factories that can convert biomass to bioenergy resolving the problems related to the exhaustion of fossil fuels and global increase in pollution. In addition to biofuels, algae are also exclusive reservoirs of a plethora of extraordinary products ranging from nutraceuticals to antimicrobial products, UV protectants, high-value products, and aquaculture (Fig. 1.1). Focusing on better growing and harvesting methods will help in early commercialisation. Emphasis on screening new algal strains, techniques to improve production of algal biofuels, and improvisation in downstream processing will open new doors in the area of algal biofuels. Biofuels can attribute towards conservation of our nonrenewable sources of energy and environment, a way towards better future.

Algae-derived biofuel generation needs sufficient innovative tool considerations for the extensive exploitation of biotechnological techniques, greater culture condition optimisation, new strains and consortium, and above all understanding fundamental mechanisms (Rodionova et al. 2017). The production of hydrogen from algae has been well documented, yet the commercial implementation requires further investigations. Amongst the major biofuels envisaged till date, the algae biodiesel has established the most focus and perhaps is the only algal biofuel venture which has been upscale to the pilot-scale and full-scale utilisation, though algae-based bioethanol production requires proper attention and adequate consideration.

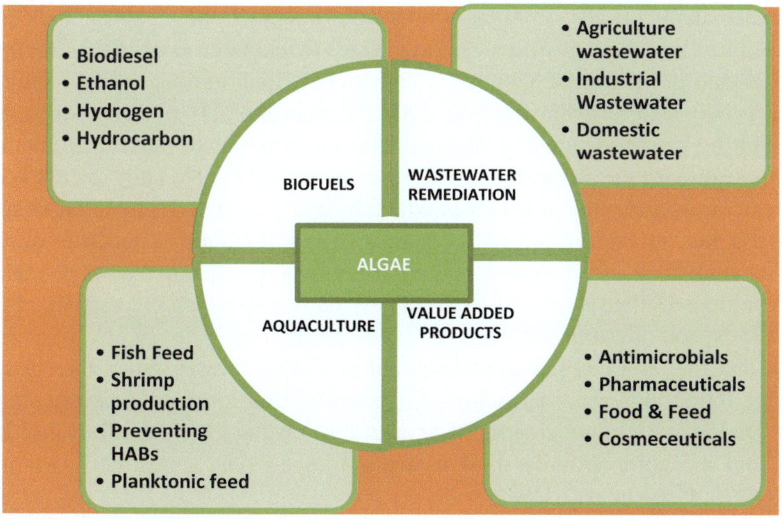

Fig. 1.1 Application of algae

1.2 Algal Biomass: A Unique Reservoir

Algae possess a special photosynthetic system which makes them ideal organisms for biofuel production with distinctive properties, and scientific reports have indicated the biofuel production efficiency in many algal species ranging from cyanobacteria, green algae, diatoms, to macroalgae (Archana and Anjana 2012; Marella et al. 2018). Algae are photosynthetic organisms inhabiting varied habitats and performing a plethora of ecological roles which are irreplaceable and commendable for the environment. They are efficient photosynthetic autotrophic systems that can convert atmospheric carbon dioxide utilising the sunlight into energy so proficiently that they can exponentially increase their algal biomass by twofold. The process of oxygenic photosynthesis takes place in almost all algae, and actually most of the knowledge about the photosynthetic process was for the first time elucidated in *Chlorella*, which is green alga. Similar to the higher plants in algae, photosynthesis also involves two reactions – light reactions and dark reactions (or Calvin cycle). Carbon dioxide is fixed by the oceanic primary producers during photosynthesis. The photosynthetic pigment systems which are the pioneers of the solar energy entrapment are also more of similar in nature compared to the higher plants. The algal pigments include chlorophyll, phycocyanin, phycoerythrin, carotenoids, fucoxanthin, etc. depending upon the algae class. The efficiency of the algal photosynthetic system is imparted by the pigments present in the diverse classes of algae and helps the algae to inhabit the range of habitants (Archana 2014).

The algal biomass is considered as the fastest-growing biomass on the planet and omits oxygen which is one of the survival sources for the living beings on the earth (Chakrabarti et al. 2018; Cheng et al. 2006). They are the potential source of oils and feedstocks (Chu 2017). Algae produce a number of chemicals such as chemicals used for the synthesis of resins and plastics (Mekonnen et al. 2013; Wang et al. 2016). Algae produce much higher concentrated lipid substances, and they have antibacterial properties (Satoru et al. 2007; Minhas et al. 2016) under stress conditions. During silicon deficiency, they can change activities of those enzymes which are responsible for the accumulation of lipids (Roessler 1988). They release secondary metabolites and other substances which have cytotoxic effects on other species of microalgae, invertebrates, and cell lines of mammalians (Ruocco et al. 2018; Dhanker et al. 2015). Red algae produce carotenoids which help to cure retinal damage in animals (Sathasivam and Ki 2018). Algae became even the priority research source for space missions for the sustainable production of feedstock, recycling of carbon dioxide, and oxygen replenishment on future long-distance interplanetary missions (https://www.nasa.gov/mission_pages/station/research/experiments_category). Diatoms are key elements of aquaculture (Wang 2015; Clifford and Kevan 2014) and are commercially used as feeds to secondary consumers (Mann and Droop 1996; Chen et al. 2018).

Several investigations are being conducting for the optimum utilisation of microalgae for the production of desired compounds (Gerardo et al. 2015; Chew et al. 2017) and make them as hot candidates for biofuel production (Saranya et al. 2018), as a food supplement for human and animals (Chew et al. 2017), to cure the problem of waste water treatment (Marella et al. 2018, 2019), their aquaculture importance (Li et al. 2017a, b), natural insecticidal and industrial uses of their fossilised form (Korunić et al. 2016), and many more.

When we compare the capability of oil production per acre in algae, it is nearly 15 times more compared to other biofuel-producing plants. The ability of algae to process sugar from sources rich in cellulose like wood chip or grass is incredible and definitely more proficient compared to other microorganisms. The processing of cellulosic biomass is challenging for microorganisms as the presence of lignin in such biomass inhibits the microorganism and the biomass has to be processed prior to removal of lignin. But the forbearance of the algae to lignin makes it possible to avoid the process of lignin removal from cellulosic biomass, thereby making the process less cumbersome, convenient, and cost-effective. The advantages of algae as source of different biofuels are highlighted below:

- Algae consume atmospheric carbon dioxide, thereby leading to the mitigation of greenhouse gas.
- The growth requirements of algae are minimal in nature, and they can be readily cultivated in the areas which are not fit for the plant growth making the land arable for agriculture. They can grow in salty water, freshwater, or waste waters.
- The algal photosynthetic machinery produces the bio-oils utilising solar power, carbon dioxide, and water. The biofuels originating from the algal oil have similar molecular structures to the existing fossil-based fuels like petroleum and diesel making them more suitable for use in the current automobile engines without many changes.

- The biofuel production per acre of algae is more compared to the other sources of biofuel generation. The productivity of algae-based biofuels is more as they can be rapidly cultured requiring less time.

1.3 Algae-Derived Biofuels

Biofuels in the present context have seen the first two initial generations. The first-generation biofuels originate from the crop-based plants, thereby committing with the agriculture and the food demands of the population, while the second-generation biofuels are derived from the residues of the forest and agriculture and feedstocks which are not food based in nature. There were several limitations associated with these two generations of biofuels, and the third-generation biofuels aptly addressed the lacuna with the previous generations of biofuels. The biofuels originating from the algae are quite diverse in their nature as algae produce copious amounts of biomolecules like carbohydrates, lipids, and proteins, and these can be readily sort out into biofuels like biodiesel, ethanol, hydrocarbons, lipids, hydrogen, and high-value substances with wide applications (Fig. 1.2).

1.3.1 Algal Biomass

The algal biomass in dry form can be converted directly to yield electricity and heat. The technique of high temperature can be used to generate the fuel gas from dry algal biomass; also high pressure can generate fuel oil from the dry algal biomass since these processes require the dry algal biomass; there are many apprehensions associated with their use to generate the biofuels. The algal biomass drying demands

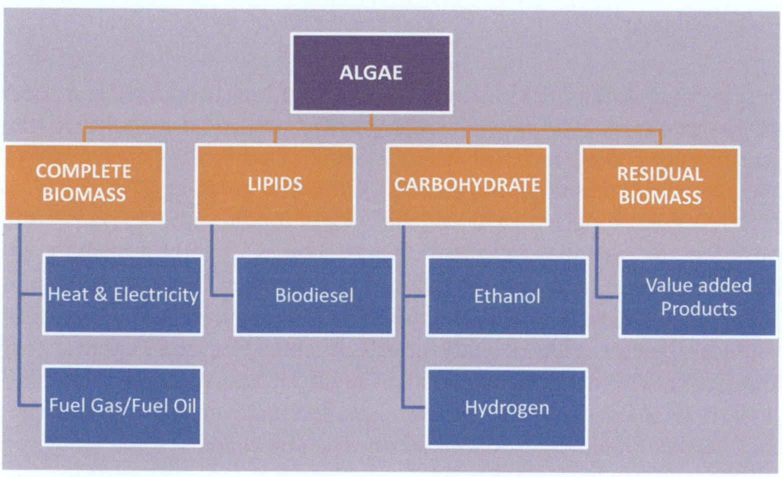

Fig. 1.2 Diverse fuels derived from microalgae

a good amount of energy, which is quite negative in part of the energy balance and investment costs of essential equipment (Wijffels 2007) though the sustainable approaches advocate the application of solar energy in order to dry algal biomass.

Another option to dry algal biomass is through the process of liquefaction which can take place through the application of both heat and chemicals and involves treatment at a very high pressure and temperature (Banerjee et al. 2002; Tsukahara and Sawayama 2005), but this technique is quite naïve and in the incipient phase (Meuleman 2007). The process of anaerobic digestion can be used as an alternate method to process algal biomass, and through this method biogas can be produced from the wet stream wherein the energy requirements are relatively less compared to the thermal process. The biogas contains nearly 55–75% methane which can be converted to generate electricity and heat (de Mes et al. 2003).

1.3.2 Lipids

The lipids comprise the key constituents of algae, and these lipids find application as a liquid fuel as straight vegetable oil (SVO) in the adapted engines, while the free fatty acids and triglycerides can be transformed into biodiesel by the action of the catalysts which can be homogeneous or heterogeneous in nature. A number of algal strains have been extensively documented to produce lipids (Li et al. 2017a, b). A comparative analysis between the straight vegetable oil (SVO) and algal oil indicates that the algal oil is basically unsaturated, thus less suitable for automobile engines which are sensitive. Intricate investigation is much essential for finding more novel algal species capable of yielding good amount of lipids, and improving the lipid production through innovative methods is certainly essential for prompt commercialisation so that the algal lipids can see the light of the day.

1.3.3 Biodiesel

Biodiesel is a non-polluting eco-friendly substitute fuel resulting from renewable sources (Hossain et al. 2008) which is at par in fuel performance to the existing fossil fuels like petrol and diesel with lesser amount of particulate matter and sulphur release (Miao and Wu 2006; Scragg et al. 2002). The green algae *Chlorella protothecoides* has been reported as a suitable genus for the production of biodiesel because it has the potential to utilise diverse sources of carbon, namely, acetate, glucose, and glycerol, and can produce good amount of biomass and lipids (Xu et al. 2006). In algae, the biodiesel and glycerol are produced by the transesterification of triacylglycerols (TAGs) with the aid of an acid or alkali catalyst (Johnson and Wen 2009). The transesterification of algal oil and processing of fatty acid methyl esters (FAME) lead to the biodiesel production from algae, and this process requires a catalyst of either an acid or alkali. The key benefits of the biodiesel from algae oil are included in Fig. 1.3.

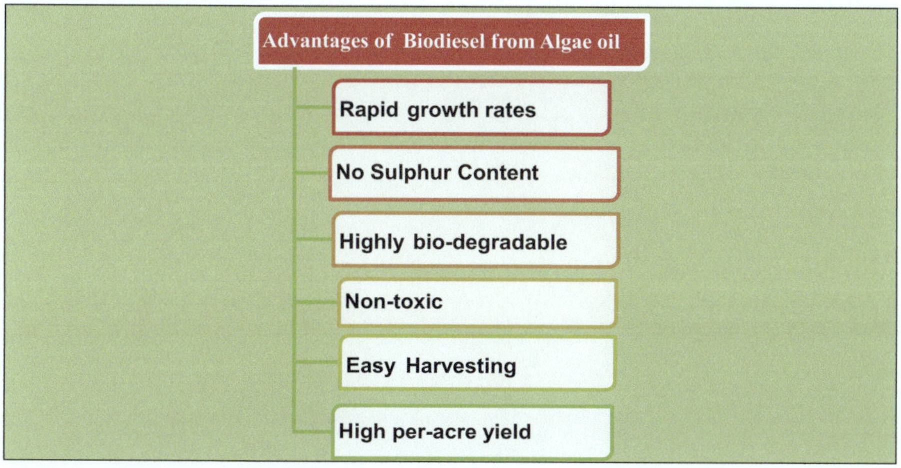

Fig. 1.3 Advantages of biodiesel

Biodiesel production can be mediated through two routes:

1. Direct transesterification of biomass from algae (Lewis et al. 2000)
2. Two-step process wherein the lipids are extracted, collected, and transesterified (Johnson and Wen 2009)

Both these biodiesel production techniques need the extraction of lipids utilising the amalgamations of different alcohols and solvents like chloroform/methanol, hexane/isopropanol, or petroleum ethers and methanol (Johnson and Wen 2009; Mulbry et al. 2009). Amongst the two production methods, the direct method is profitable, since it conglomerates the lipid extraction and transesterification into one process, thus saving more time compared to the two-step extraction-transesterification process (Fig. 1.4).

Though the biodiesel is a suitable alternative to the nonrenewable petroleum diesel, there are several bottlenecks associated with the biomass-based production, which need to be efficiently addressed through technological interventions in order to eradicate the current fossil-based diesel crunch. More exploratory studies are indeed required which investigate new strains or novel consortium with a much higher rate of growth and oil yield. In this direction, Ali et al. (2017) have advocated a new chemical-based method for increasing the algal biomass and production of lipid and the fatty acid composition alteration in the green algae *Chlorella protothecoides* employing exogenous bioactive molecules – phytohormones and antioxidants.

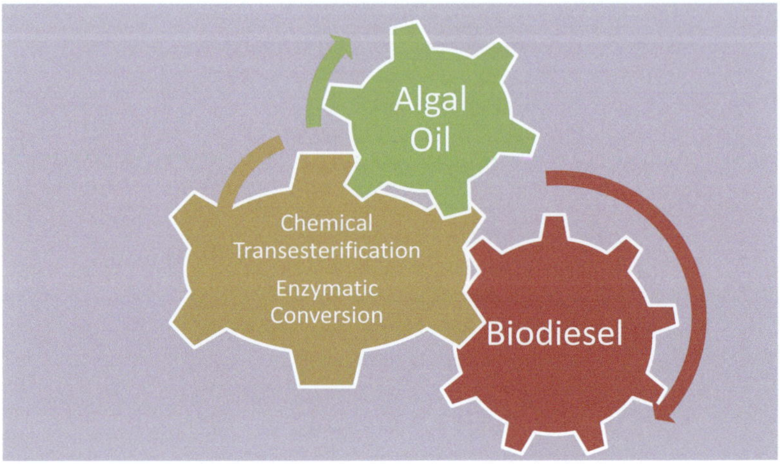

Fig. 1.4 Origin of biodiesel from algal oil

1.3.4 Hydrocarbons

Algae can produce hydrocarbons, and similar to petroleum, algal hydrocarbons can be converted to fuels like diesel, gasoline, and kerosene. A renowned alga, *Botryococcus braunii*, is extremely popular owing to its potential to yield hydrocarbons (Hillen et al. 1982). In comparison to other algal species which generally contain nearly 1% of hydrocarbons, *B. braunii* possesses nearly 20–60% dry weight which can reach up to 80% (Wijffels 2006). On the basis of the algal strains, the hydrocarbons are either C30 to C37 alkenes or C23 to C33 odd-numbered alkenes (RangaRao and Ravishankar 2007). The algal hydrocarbons are majorly gathered on the outer surface of the cell, which makes the process of their extraction relatively simpler and convenient (Wijffels 2006).

1.3.5 Hydrogen

Algae produce hydrogen, which has great potential as a future fuel owing to numerous advantages (Archana and Anjana 2012). Hydrogen as fuel has immense prospects as it liberates only water and no toxic emissions are there, which makes it environment savvy but he major associated lacuna is the non-availability of a sustainable system of hydrogen production. At present, the hydrogen gas is formed by the fossil fuel steam reformation process and the water electrolysis, but these processes are electricity consuming, making the hydrogen yield not at all economic. Biologically, hydrogen can be produced by a range of resources like the bio-oil steam reformation (Wang et al. 2007), dark and photofermentation of organic materials (Kapdan and Kargi 2006) and algae-mediated photolysis of water (Ran et al. 2006). However, algae can produce hydrogen straight from sunlight and water, but only under the condition wherein oxygen is not present in the vicinity.

In a nutshell it can be stated that the conditions required for hydrogen production are either expensive or difficult, making the production very challenging, so making the production commercially possible, exhaustive investigations are needed. Exploring new algal species producing high yield of hydrogen needs to be envisaged for an advanced understanding of the enhanced hydrogen production mechanism, and use of modern biological tools like genetic engineering are indeed needed (Ashutosh et al. 2007; Archana et al. 2019).

1.3.6 Ethanol

Bioethanol is unique as it can be an excellent substitute to the existing fossil fuel – petrol. Though the ethanol production from algae is quite promising, at present it is in the incipient research stage and a lot of attention is required for the functionalisation of the large-scale system of its production. In algae the fermentation of the polysaccharides like starch, cellular, and cellulose leads to the production of ethanol. Algae are rich in carbohydrate content, for example, in some species the carbohydrate content can rise up to 70% under optimum conditions, which makes them suitable organism for the bioethanol production. Algal strains like *Chlorella, Scenedesmus, Chlamydomonas,* and *Dunaliella* produce a good amount of starch, and hence they are very suitable candidates for the production of bioethanol. The cell wall of algae comprises of outer and inner cell wall layer, wherein the outer cell wall is trilaminar and is rich in polysaccharides such as agar, alginate, and pectin and the inner part of the cell wall comprises of mainly hemicellulose and cellulose. Though the abundance of cellulose and starch in algae, which can be further fermented to ethanol, makes them ideal for bioethanol production, elaborative exploratory research is needed to transform the preliminary research into concrete large-scale commercially profitable process. Table 1.1 highlights the biofuels from algae and their yield (Cuellar-Bermudez et al. 2014).

1.3.7 High-Value Products

The sustainable approaches towards algal biofuels necessitate the demand of valuable co-products to make the whole process economically pragmatic. Rapid commercial implementation of third-generation biofuels from algae requires innovative initiatives on algal energy generation. Alga culture can turn out to be a promising option as algae are excellent feed and generate yields with a greater commercial value compared to energy.

Algae have magnificent metabolism, and they produce an array of secondary metabolites which find vivid application as food, food supplements, food colourants and nutraceuticals, fish feed, shrimp and shellfish, and omega-3-fatty acids like DHA, EPA etc. (Grima Molina et al. 2003; Reith 2004; Li et al. 2017a, b).

Table 1.1 Microalgal biofuels

Species	Product	Yield
Chlamydomonas reinhardtii (CC124)	**Biohydrogen**	102 mL/1.2 L
		0.58 mL/hL
		0.30 mol/m^2
		0.6 mL/L h
Chlamydomonas reinhardtii (Dang 137C mt+)		175 mL/L
		4.5 mmol/L
		71 mL/L
Chlorella vulgaris MSU 01		26 ml/0.5 L
Scenedesmus obliquus		3.6 ml/μgChl a
Platymonas subcordiformis		11,720 nL/h
		7.20 mL /h
		0.339 mL/hL
Dunaliella tertiolecta	**Bio-oil**	43.8%, 34 MJ/Kg
		42.6%, 37.8 MJ/Kg
		25.8%, 30.74 MJ/Kg
Chlorella protothecoides		52%
		57.9%
Chlorella sp.		28.6%
Chlorella vulgaris		35.83%
Nannochloropsis sp.		31.1%
Chlorella vulgaris	**Biogas**	0.63–0.79 LCH_4/gVS
Dunaliella salina		0.68 LCH_4 /gVS
Euglena graciis		0.53 LCH_4 /gVS
Scenedesmus		140 LCH_4 /KgVS
Scenedesmus (biogas from lipid-free biomass)		212 LCH_4 /KgVS
Scenedesmus (biogas from amino acid-free biomass)		272 LCH_4 /KgVS
Scenedesmus obliquus		0.59–0.69 LCH_4/gVS
Botryococcus braunii	**Lipid content for biodiesel**	25–75%
Chlorella sp.		28–32%
Chlorella vulgaris		56%
Crypthecodinium cohnii		20%
Monallan thussalina		20–70%
Nannochlorisis sp.		20–35%
Nannochloropsis sp.		31–68%
Neochloris oleoabundans		35–54%
Nitzschia sp.		45–47%
Scenedesmus dimorphus		6–40%
Scenedesmus obliquus		11–55%
Schizochytrium sp.		77%

(continued)

Table 1.1 (continued)

Species	Product	Yield
Chlorella pyrenoidosa	**Carbohydrate content for bioethanol**	26%
Chlorella vulgaris		12–17%
Dunaliellasalina		32%
Scenedesmus obliquus		10–17
Porphyridium cruentum		40–57%
Euglena gracilis		14–18%

Cuellar-Bermudez et al. (2014)

1.4 Cultivation of Algae for Biofuels

Algae can produce a diverse range of biofuels and till date many algal species are reported to yield biofuels. The process of cultivation of algae has substantial benefits as algae have autotrophic ability to utilise sunlight to yield sugar, via photosynthesis, so it is relatively convenient to cultivate them in open ponds, closed system like photobioreactors, and hybrid system. Table 1.2 highlights the advantages and disadvantages of the various cultivation systems.

The selection criterion for algal-based biofuels includes:

1. Easy cultivation and rapid growth on minimal requirements
2. High content of oil and good productivity
3. Good ability of harvesting with easy purification
4. Contamination resistance and endurance
5. Tolerance towards high concentration of oxygen and temperature extremes
6. Adaptation to the altered culture conditions encountered in the growth ponds

1.4.1 The Open System

The open system is perhaps the oldest and the simplest method meant for algal cultivation. In the open system the natural resources are utilised for the algal growth. The energy is obtained via sunlight, which converts the atmospheric carbon dioxide to sugars. There are a couple of options available with open ponds with algae cultivation, which include raceway pond, circular pond, and unstirred pond. There are several advantages and disadvantages associated with the open system. The open systems are cost-effective and low maintenance demanding systems, but they are directly under the influence of environmental factors, and in countries where weather is not very suitable for algal cultivation, it is a difficult system.

1.4.1.1 Raceway Ponds

The raceway ponds can be built as an individual as well as after joining more than one raceway ponds together having depth of 15 and 30 cm. Channels may be made in concrete, plastics, and compacted earth. Paddlewheel, pumps, and airlifts are

Table 1.2 Different types of cultivation methods with advantages and disadvantages

Cultivation system	Advantages	Disadvantages
Open system	Low construction and maintenance cost	More prone to different kinds of contamination such as bacterial, virus, fungal, and other microalgal species including invasive ones
	Easy to scale up	
	Possibility to integrate with waste water	Low productivity – limited photosynthetic activity of algal species due to poor mixing and water and gas transfer
		Difficult to control cultivation parameters such as seasonal temperature, light, temperature, pH, and nutrient
		High harvesting cost: due to mostly small-sized algae and poor algal cell
		High evaporation of water especially in tropical or desert areas
		High volume to surface ratio
Closed system	Higher cell biomass production and cell density due to better mixing and maximised photosynthetic capabilities which directly reduced algae harvesting and drying costs	Prohibitively high construction cost at large scales
	Less contamination risk due to less exposure to the environment	Lowest volume to surface ratio
	Possible contamination from bacteria and fungi, even from other algae species also	
	Better control of cultivation parameters such as light, temperature, pH, and nutrient adjustment	
Immobilised system	Easy harvesting of algal biomass because algal cells are enclosed in small spaces (e.g. beads) or attached to solid carriers	Unavailability of appropriate cost-effective and durable supportive matrix for higher algal biomass growth
		Can be utilised for selected species only
		Having low lipid content
	Improved productivity – can improve nutrient/gas transfer and prevent light shielding	Difficult to scale up outdoor immobilised systems
	Main purpose is to remove nutrients not to produce biomass and oil production	

used to drive water continuously. However, paddlewheel is mostly popular for generating water velocity in comparison to others. The raceway ponds are more safe and cheaper in comparison to other open pond systems for commercial algal production. Because mixing is done in an open pond, the system is more efficient in capturing carbon dioxide and sunlight. Additionally, the water-derived movement continuously prevents the settling down and deposition of algae and helps make the operation more successful. The construction cost maintenance cost of these ponds are very low. Many companies around the world are using this type of system for large-scale production of food supplements or β–carotene. The algal productivity is varying with environmental factors such as light duration and intensity, temperature, and salinity.

1.4.1.2 Circular Ponds

Circular ponds have depth of 30–70 cm and, normally, having a diameter of 45 m. It has a central pivoted agitator. These ponds are being used at large scale for b carotene production and higher annual algal biomass production in Taiwan and Japan. These ponds are basically used commercially for β-carotene production. The most commonly used algae in this production is *Chlorella* sp. The integrated use of algal biomass production (such as *Chlorella* sp. and *Oscillatoria* sp.) and waste water treatment is conducted in circular ponds for a long time (Sheehan et al. 1998).

1.4.1.3 Unstirred Ponds

Unstirred ponds are the natural uncovered body. These ponds can be as large as the lake. These ponds are economically more efficient and need less technical support in comparison to other commercial production ponds. Generally, the depth of these ponds is not more than 50 cm deep. The unstirred ponds have been used for carotene production in few parts of Australia by some companies. However, these kinds of ponds are more prone to contaminations such as growth of invasive microalgal growth and bacterial, protozoan, and fungal infections. Similarly, in another study done by Lee in 1997, they noted approximately 32 tons of microalgal biomass production per year from natural lakes in Southeast Asia.

1.4.2 The Closed System

The closed systems have controlled set of conditions suited for the maximum growth of algae for enhanced biofuel production. The dependence on the climatic conditions is not a constraint in closed system. The chances of contamination are reduced in the closed systems compared to the open systems (Posten and Schaub 2009; Yeang 2008).

There are multiple options of photobioreactors employed in algal cultivation ranging from flat-panel to horizontal systems. The photobioreactor (PBR) systems are assembled to culture algae in tubes, bags, and transparent materials wherein they are not secluded from the direct influence of the environment (Lehr and Posten 2009; Shen et al. 2009). For nutrient supply either waste water is used or supplements are to be provided for algal growth. The Biomass productivity is much higher because the algae are exposed to abundant sunlight (Chisti 2008a).

The various culture systems have been highlighted in Fig. 1.5.

The tubular photobioreactors (TBPR) are specialised as they possess more ratios in terms of surface-to-volume (Shen et al. 2009). The following are the different types of TBRs:

- The helical tubular PBRs
- The fence-like tubular PBRs
- The horizontal tubular PBRs

All these types of TPBRs establish and standardise flow rate to avert the algal biomass process of sedimentation (Chisti 2008a). In these systems there is production of oxygen in high amounts due to dense algal growth, and this high oxygenic conditions lead to photosynthesis and algal cell damage owing to photo-oxidation (Chisti 2008a). Since the removal of dissolved oxygen is not possible, it is suggested to degas the reactor vessel (Chisti 2007). The optimum temperature can be maintained by the cooling or heating of the vessel coupled with use of agitators. Tubular PBRs are highly productive systems due to their large surface-area-to-volume area. They have been used to produce a significant amount of biomass algal productivities. The world's largest PBRs are vertical PBRs which is found in Germany which is arrayed in a greenhouse. The algal biomass production is about 35 to 41 gm2 per day. In another study, the highest productivity of 47 gm2 per day was noted in tubular PBRs with south–north orientation. A horizontal airlift tubular PBR with a size of 0.2 m2 was recorded with the average biomass production of 25.2 $gm^{-2} d^{-1}$ in the case of *Phaeodactylum tricornutum*.

Flat-panel photobioreactors (FPBR) are made up of transparent substances and are thin rectangular vessels with unique angles for utmost solar power utilisation. Though they have much solar light exposure, the algal productivity is reported to be similar to that of the tubular photobioreactors (Shen et al. (2009). The advantage of flat-panel photobioreactor is the complete direct light illumination available throughout the reactor (Hu et al. 1996). The inclined and vertical alignment are the

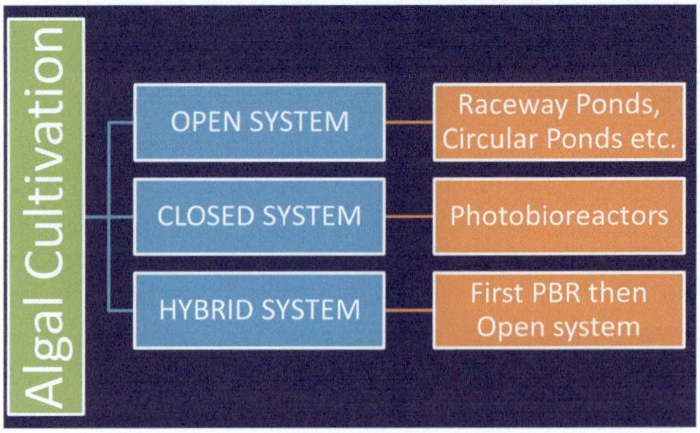

Fig. 1.5 Systems for algal cultivation

two most commonly used plate PBRs. These are flat panels having a transparent rectangular container whose light path generally lies between 1 and 30 cm (Hu et al. 1996). A seasonal variation in the production of *Synechocystis aquatilis* is recorded by Zhang et al. They successfully cultivated this algal species under 11.2 ± 2.1 MJ m^{-2} d^{-1}, an average irradiation. They have calculated an average algal biomass production of 30gm^{-2} d^{-1} with a cell concentration of 1–2 gL^{-1}. In another study, Hu and Sommerfeld have used the vertical plate PBRs, and the biomass productivity and have achieved 12.5 gm^{-2} d^{-1} cell density of about 7 g L^{-1} in Arizona.

Novel photobioreactor systems are innovative algal cultivation systems called the Offshore Membrane Enclosure for Growing Algae (OMEGA) which has been designed by the National Aeronautics and Space Administration (NASA) and consists of bags that are semipermeable in nature and float in waste water (Trent 2009).

1.4.3 Immobilised Culture Systems

It is the least studied algal culture system amongst all algal culture methods. In these systems, the unialgal cultures are immobilised in a polymeric matrix, or attached algal communities grow in shallow, artificial streams or on surfaces of rotating biological contactors. These systems can be classified as enclosure and non-enclosure methods.

1.4.3.1 Enclosure Methods

These methods are mostly used by the industries. These methods use the polymeric matrix for the algal culture, which helps the algae to attach to the substrates. However, these methods have been explored at a very limited scale for the algal production. However, it is extensively utilised for bacterial, yeast, and enzyme technology for industrial purposes for algal cultivation. The algal production is mostly utilised for waste water treatment through this method, because this system is good for species control and has high efficiency of waste removal, etc. (Robinson 1986). Some researchers have indicated this method as more efficient compared to free cell production systems. For instance, the algae *Chlamydomonas reinhardtii* cells are immobilised in Ca alginate. Due to highly costly polymeric fibre, it has limited use in large-scale cultivation of algae.

1.4.3.2 Non-enclosure Methods

Non-enclosure methods are similar to the enclosure method because this method is also applied to waste water treatment mostly. However, the difference between enclosure and non-enclosure is to grow algae on solid support without enclosure. Algal Turf Scrubber is one of the examples of the non-enclosure methods (ATS). The main component of this system is the solid support for growth and algal harvesting as well as wave surge for agitation. According to Kebede-Westhead et al., the algae removed nitrogen and phosphate from waste water efficiently and have very higher biomass productivity with mean biomass productivities between 7.1 and 9.4 g m^{-2} d^{-1}.The algal cells allow to grow on polymer matrix and on gel beads.

1.4.4 The Hybrid System

These systems amalgamate the features of both open and closed systems. Such systems are well articulated to eliminate the limitations of both open and closed systems of algal cultivation. In the hybrid cultivation system, the algal cultivation is done in controlled closed systems with optimum growth factors and later on the algal cultures are shifted to the open system of cultivation thereby facilitating enhanced algal growth and biofuel production. This system holds substantial features and is an outstanding cultivation technique for promoting biofuel production via algae (Archana and Kiran 2018).

1.5 Algal Harvesting Techniques

The water content is removed from algal cells through a process known as harvest. The algal harvesting is a downstream processing. There are so many downstream techniques available which are utilised for algal harvesting. However, due to higher cost, no harvesting technique is fully efficient for commercial large-scale harvesting till date. The basic steps which are involved in algal harvesting are screening, coagulation, flocculation, flotation, sedimentation, filtration, and centrifugation (Brennan and Owende 2010). However, other techniques are electroflotation, electrophoresis, and ultrasound which are less important compared to the above-mentioned general steps (Heasman et al. 2003). Generally, a particular type of harvesting technique should be applied on the basis of algal cell size and cell density with lower operating cost to achieve desirable yield of biomass. Further, the applied technique must have little influence on the further processes (Kim et al. 2013). Basically, harvesting technique can be divided into two types: first, based on the bulk separation of algal biomass using sedimentation process and, second, filtering of the biomass and preparing slurry through the process of filtration and centrifugation.

1.5.1 Centrifugation

The mixture of different-sized particles is separated on the basis of density through spinning in the centrifugation process. The centrifugation works on the principle of centripetal force. Although it is the most widely accepted and efficient technique for harvesting, it is costly when applied at large-scale production (Sim et al. 1988). In another study, Heasman et al. (2003) based on the harvesting efficiency trial, 90–100% harvesting efficiency was recorded via centrifugation. However, high gravitational force and shear stresses which are applied during the centrifuge process can damage the algal cell structure.

1.5.2 Gravity Sedimentation

In sedimentation procedure, the particles settled down in the bottom out of the fluid in which they are mixed. Gravity is one of the forces which are responsible for the sedimentation. In gravity sedimentation harvesting procedure, the heavier algal particles are settled down to the bottom against the barrier. A sample having higher cell density and large-sized algae are efficiently harvested through this procedure (Edzwald 1993) in comparison to lower cell density, but this procedure is effected by the density and radius of the microalgal cells, other factors are algal particle settling time, specific gravity suspension, and gravity of particles.

1.5.3 Filtration

This procedure is applied to separate solid matter from fluids or gases through different kinds of mechanical, physical, and biological processes by proving a medium through which only liquid can pass (Brennan and Owende 2010). For this procedure the filters with specific pore size are used (Chen et al. 2011; Mohn 1980). The filtration process can be of several types such as microfiltration, dead-end filtration, vacuum filtration, ultrafiltration, and tangential flow filtration (TFF). Out of all filtration process, the least energy is consumed in the pressure filtration and TFF that is why they are better for harvesting than others. However, overall, the filtration is an expensive process because in it we need to change filters and membrane frequently.

1.5.4 Microstrainers

It is a closed cylindrical structure having a cylindrical screen within. It is commonly used in sewage water for separating the microalgae and other plankton species from water. This system is rotated by applying the centrifugal force which pushes the larger particles away from the screen in turn preventing the clogging and maintaining the continuality of the procedure (Chen et al. 2011). It is made of a rotating drum in which a stainless steel micromesh is fabricated and partially dipped in the water. A range of mesh size such as 15–64 μm is used so that maximum species of algae and plankton can be trapped. The microstrainers have a simple function and manufacturing capacity. The filtration ratios are high under low investment and low energy consumption. The disadvantages with microstrainers are partial removal of solid particles. Another thing is that if the alga is in higher concentration than moderate, it can block the screen; due to this the small-sized algae can be driven away from the screen very easily.

1.5.5 Electrophoresis

In electrophoresis process the charged particles are separated on the basis of charge and size. In an electrical field, the charged algae shift out in a growth medium of the electrophoretic tray. The advantage of this process is that it separates the algae without addition of any chemical. Another advantage of this technique is it is economically and environmental feasible with very less consumption of energy. A significant number of algal species have been separated by the electroflotation method having very less contamination of solids. A complex effect of algal cell and liquid is recorded during the electrophoresis which pressurises the fluid for movement (Pearsall et al. 2011). In spite of the above-mentioned techniques, ultrasonication is also used as a harvesting technique. The sound energy is used for vibrating the matter for disturbing the algal cells in a sample (Li et al. 2011). The cost in algal cultivation and harvesting influences the overall cost of the algal biorefinery process. The full potential use of the diatom as a holistic approach at large scale is a challenge due to limited reliable techniques. Therefore, for the commercial and industrial use of diatoms, we need to explore more environmental and economically feasible techniques (Yang et al. 2009).

Wang and Seibert (2017) used 80% of the daily produced biomass and left weight 20% for further reseeding the tank units. Thereafter, concentrated diatoms can be pasted, frozen, and dried, and dry mass is determined. The average annual yield calculated by Wang in 2017 was found to be close to 132 MT of dry diatoms per hectare. Another study done by Huntley et al. calculated 75 MT ha^{-1} year^{-1} yield for the diatom *Staurosira*. Out of different kinds of harvesting techniques, coagulation and flocculation, centrifugation, and filtration were found to be more effective for biorefinery purpose (Singh and Patidar 2018).

1.5.6 Lipid Extraction Techniques

Although diatom is abundant in nature, still it is a big challenge to maintain the contamination-free culture of diatoms. Few diatoms have been successfully cultivated in many countries. *Chaetoceros* species is being cultivated commercially in Taiwan, in Hawaii, and in many more countries (Syvertsen 2001). Wang and Seibert (2017) have issued a patent, and their findings have suggested a high yield of diatoms per unit surface area. Their finding suggested that Chaetoceros are less prone to contamination compared to animals. The reference has selected some species and manipulated the nutrient content; thereafter the increased concentration of accumulated lipids was calculated. It is noticed that for higher triacylglycerol production, starved algal biomass is more useful than well feed. The concentration of the different lipids varies within the species with varying culture conditions and cultivation methods.

The laser-based technique further leads to metabolite separation and partial purification steps such as ultrafiltration and nanofiltration (Pohl et al. 1968). Extremely short electric field pulses – with small temperature rise – stimulates pore creation and disruption on the basis of osmotic difference. Richard Nuccitelli of Bio Electro Med Corp. used the pulsed electric fields (PEFs, 30 kv/cm), which make nanopores in liquid membrane and also send shock waves through water to break cells open (Nuccitelli et al. 2010). Centrifugation and freezing use solvents and extraction of lipids. Elimination of cell breakage requirement is a better approach for lowering the cost of a biocrude production process. Other processes used for biofuel and biocrude production are hydrothermal liquefaction (HTL) process and foam fractionation. According to Rossignol et al. (1999), the higher pressure quick release (HPQR) for cell disruption is competitive to conventional laboratory-scale techniques such as sonication or shear-based systems. Lipid accumulation can be greater in heterotrophic cultivation in comparison to autotrophic culture. When similar experimental conditions were provided to autotrophic and heterotrophic algal culture, approximately 55% higher growth of lipids were noted in case of heterotrophic culture (Mohammad Mirzaie et al. 2015), which clearly shows higher biomass production in heterotrophic algal culture. Better carbon fixation is directly proportional to increase in higher biomass production. Thus some properties play a pivotal role in algae biomass production and enhance their role in fuel production.

1.6 Understanding the Factors Influencing Algal Growth for Biofuel Production

The algal growth depends on various factors which profoundly impact the cultivation of algae exclusively for the purpose of biofuel production. These factors can be categorised as physical factors, chemical factors, or biological factors, and each factor plays a crucial role in algal metabolism and hence growth. The factors can also be segregated as intrinsic factors including the strain of algae, the metabolic potential of algae, the light harvesting efficacy, endurance and adaptability, and extrinsic environmental factors like temperature, light, pH, and nutrient availability. Elucidating the key factors responsible for optimum growth of algae would lead to high yield of biofuels derived from algae. The major factors influencing the algal growth have been discussed.

1.6.1 Light

The entire process of photosynthesis and thus the production of algal biomass are exclusively dependent on the presence of light, which is the major energy driving force for the autotrophic machinery. The presence of light, duration of light, and the intensity of light are the important parameters which need adequate consideration if the algal growth is to be induced optimally for utmost growth and biofuel production (Melis and Happe 2001; Chisti 2007; Gao et al. 2007).

1.6.2 Temperature

Temperature has a significant impact on algal growth. The algae have growth seasons in the temperate and subtropical areas in the world. At optimum temperature the algae attain maximum growth and hence more biofuel productivity. In open algal cultivation system, the temperature is a critical issue, while in closed systems the temperature is maintained with the help of cooling jackets or heating coils attached to the bioreactors (Chisti 2007; Chini Zittelli et al. 1999).

1.6.3 Carbon Dioxide

Algae require the presence of carbon dioxide to carry out the process of photosynthesis and there is no possibility of growth in the absence of carbon dioxide. In algal growth scarcity of carbon dioxide can turn out to be an inhibitory factor leading to reduced algal productivity. Enhanced incorporation of carbon dioxide can be done through air bubbling in water because the suspension of carbon dioxide from the natural air into water is not sufficient. Alternate options for carbon dioxide include flue gas from boiler or pure form of carbon dioxide or growing calcium carbonate-rich algae capable for additional carbon dioxide sequestration (Brown 1996; Doucha et al. 2005; Moheimani and Borowitzka 2007).

1.6.4 Nutrients and Other Factors

The algae require nutrients like nitrogen and phosphorous to grow in addition to the presence of carbon dioxide and light. There are many macronutrients like carbon, hydrogen, oxygen, phosphorous, and nitrogen essential for the algal growth and metabolism. The trace elements needed for algal growth include iron, manganese, cobalt, zinc, copper, and nickel. There are many options of nutrient supply for algal growth such as fertilizers, waste waters from fishery, piggery waste, agriculture waste, and domestic waste. The utilisation of waste water for algal growth is an excellent sustainable approach leading to the exorbitant remediation of waste waters, thus reducing the impact of eutrophication for a clean environment (Braun and Reith 1993; Chisti 2008b).

1.7 Concerns Related to Commercially Available Fossil Fuels

There are many concerns with nonrenewable resources, which are responsible for the increasing need to switch towards renewable energy resources such as climate change, external cost of fossil fuels and nuclear fuel use, geopolitical instability, and inadequate fossil fuel reserves (Boyle 2012). That is why to tackle the above issues, a worldwide big network is created. This network is scheduled to work for social, economic, and environmental protection by using renewable energy resources. However, to what extent the renewable energy resources may be successful in terms of desirable productivity, with reasonable cost, is still a challenge (Gifuni et al. 2018).

There are several governmental and nongovernmental organisations which are promoting the use of renewable energy resources to overcome the constraints related to fossil fuels. The increasing demand of energy forces us to think about alternate and more reliable resources. Natural gas has been the fuel of the last quarter of the twentieth century. Natural gas has higher energy efficiency and less environmental impact in comparison to fossil fuels.

Methane, being a main component of fossil fuels, is responsible for carbon dioxide emission. During combustion methane shows a higher amount of emission (Heede and Oreskes 2016). However, coal is a major carbon dioxide producer compared to oil and methane. Conventional energy sources are not only responsible for greenhouse gas emission but also responsible for sulphur dioxide emission which is the main component of acid rain (Li et al. 2017a, b). Thus, renewable energy resources decrease greenhouse gas emission; on the other hand, they reduce the impact of acidic rain too. Therefore, they may protect our environment indirectly by reducing health-related problems of living beings particularly of human beings. India imports 80% oil from other parts of the world such as from OPEC. It is the third biggest oil importer worldwide. This also causes India to face huge trade deficits, a vital factor of economic crisis here (https://www.opec.org/opec_web/en/4567.htm). The anticipated results say that by 2040, India's oil demand will increase by more than 150% in future.

1.7.1 Remedies to Tackle the Issues Related to the Commercially Available Fuels

To solve the above-mentioned issues, some remedies are provided such as public awareness, supporting research and development (R and D), setting targets, etc. to spread publicity and education about alternate resources of energy such as biofuels and many others, making consumers aware with latest technologies so that they trust and support alternate resources. It is necessary to promote R and D with reliable and more scientific technologies to meet the demand for fossil fuels through renewable energy sources such as biofuel uses in which microalgae especially diatoms seem a promising tool (Hildebrand et al. 2012). The government can promote renewable energy resources by setting targets. India has set a target that renewable energy must provide 25% of India's energy by 2022. However, it depends up to what extent its implementation is done at reality level. Even international unions and international conferences also provide targets to individual countries to overcome the present scenario, arising due to the use of commercially produced energy. Countries like Denmark and Germany have taken pledge to switch to 100% and 65% renewable energy by 2050 and 2030, respectively.

Undoubtedly, the renewable energy resources are the demand of time. However, we need to have a reservoir of commercially produced fuels for the purpose of security. We should think about potential desirable output and cost-effective technologies. This way the transition from nonrenewable resources to renewable sources can be fruitful. In IEA's (https://www.iea.org/newsroom/news/2019/february/iea-becomes-facilitator-of-biofuture-platform.html) blue map scenario, the world primary energy demand continues to grow and renewable energy by 2050 contributes almost 40% of primary energy supplies. It is necessary to promote integrated biomass for renewable energy production (Boyle 2012).

1.8 Innovative Strategies Towards Enhanced Biofuel Productivity

The major lacuna related to biofuels from algae lies in the large-scale production on industrial parameter with better biomass harvesting of algae coupled with economic downstream processing techniques yielding biofuels and valuable products. Envisaging novel strategies to channelise algal biorefinery approach towards enhanced algal fuel production would include methods like novel strains and consortium, nutrient enrichment, efficient cultivation and harvesting, etc. (Fig. 1.6).

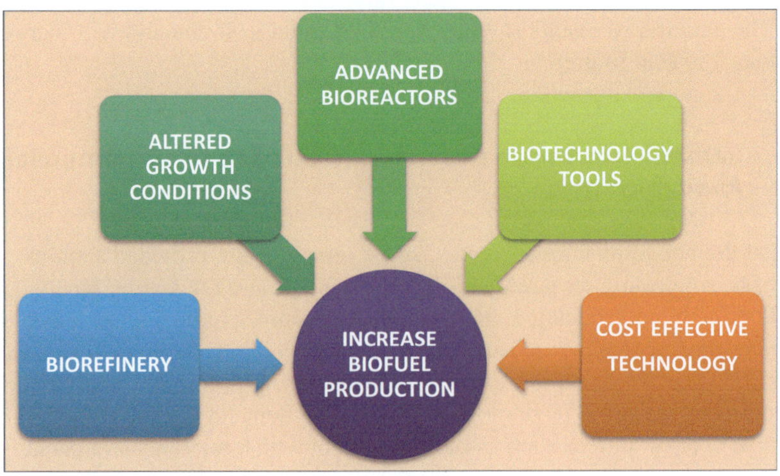

Fig. 1.6 Approaches to enhance biofuel production

The economic viability of algal biofuel can be increased manifold by incorporation of valuable co-product approach, which comprises consecutively algal cultivation, biofuel production, and bioactive compound extractions from residual biomass. The application of valuable co-product approach via the integrated biorefinery approach can lead to the significant increment in making the algal biofuel production highly cost-effective (Li et al. 2008).

Advanced biotechnological tools can be employed to increase algal biofuel production. Advanced biotechnology tools like metabolomics engineering and synthetic biology hold enormous prospective for enhanced biofuel production. Triggering of lipid production can also be attained by imposing stress on microalgae. Improvising downstream processing techniques are also required in this regard to attain an economic productivity on a sustainable platform.

1.8.1 Algal Biorefinery: A Holistic Approach

Biorefinery is defined as the fabrication of an extensive array of substances and biofuels from biomass by the assimilation of bioprocessing and suitable environment-friendly tools in an economical and pragmatically viable way (Li et al. 2008; Chisti 2007). Algae produce an array of bioactive molecules which find a wide range of application as therapeutics, health supplements, colourants, antioxidatives, vitamins, and omega 3 fatty acids. Incorporation of these co-products with the biofuel production can make the entire production cost-effective; thus this co-product scheme is a promising option (Fig. 1.7).

1.8.2 Growth Optimisation for Enhanced Productivity

The growth of algae is dependent of multiple factors and the cultivation can be stimulated by altering the growth conditions. There are a plethora of reports on the effects of different environmental and intrinsic factors on the biofuel production efficacy of algae. It has been reported that algae fed with a high amount of carbon dioxide and organic substances can yield up to 40% more oil (www.sciencedaily.com). A complete understanding of all the parameters that can actually increase the biofuel yield in algae is another important aspect which needs adequate consideration and elaborative studies. Elucidating the optimum factors for algal growth can also lead towards enhanced biofuel productivity. For example, in green algae *Chlorella vulgaris* ferrous chloride supplements during the late-exponential growth phase resulted in more total lipid content and dry cell weight (Liu et al. 2008).

1.8.3 Efficient Design of Cultivation Systems

The cultivation systems used for biomass growth is one of the pioneer factors impacting the yield of the biofuel production. The cultivation options include open

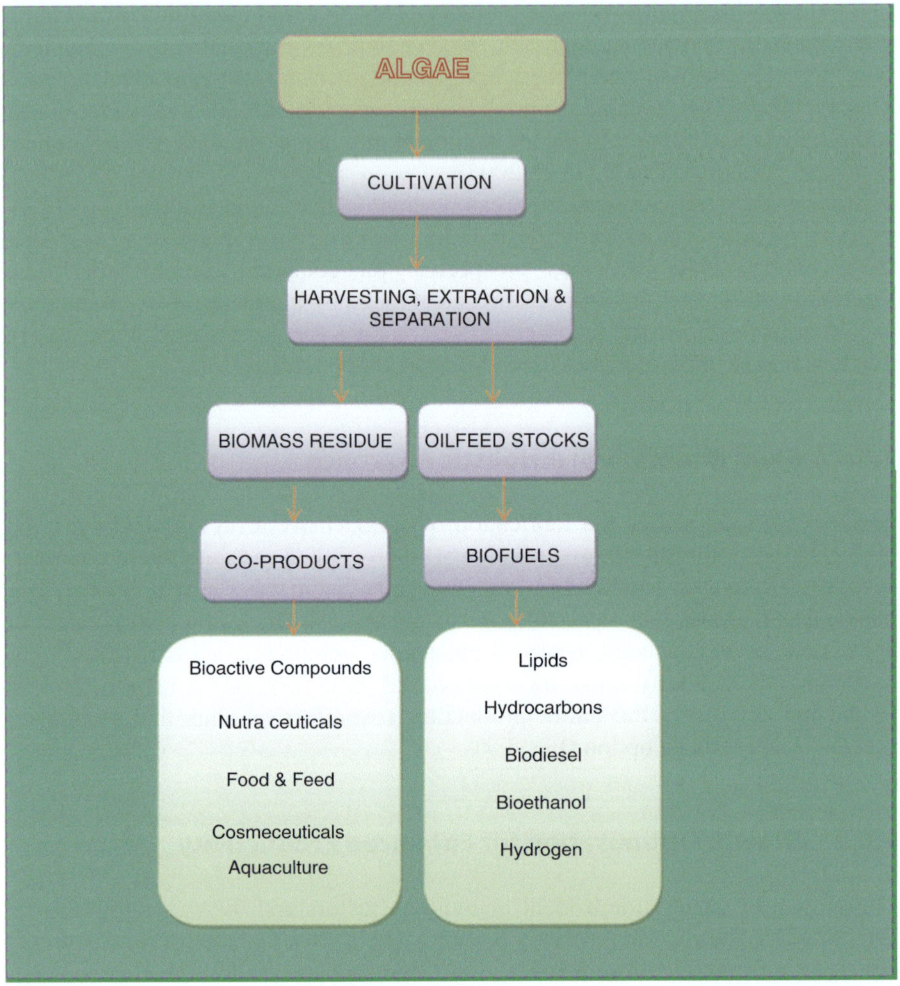

Fig. 1.7 Biorefinery approach to algae

systems, closed systems, and hybrid systems (Janssen et al. 2003; Huntley and Redalje 2007). Each cultivation system which can be used for the growth of algae has its own limitations and advantages. On the global scale the open and the closed systems have been critically evaluated. Designing the cultivation system with recent technological advances is the demand of the hour and requires attention so that the biofuel production from algae can be increased manifold.

1.8.4 Altering the Antenna Size

There exists a competition between photosynthetic organisms and algae for light to out show the photosynthesis process in the natural habitat, so if the receptors of the chlorophyll are the specialised antenna part of chlorophyll which captures the light has cutting-edge seizure efficiency to procure extra light than the normal photosynthesis under ideal lighting circumstances (Melis and Happe 2001). Researchers have successfully reduced the size of chlorophyll antenna, thereby making more proficient usage at high light intensity (Mitra and Melis 2008).

1.8.5 Induced Stress-Triggered Lipid Production

The induction of stress has been correlated with enhanced accumulation of enormous lipids which are intracellular in nature in many algal strains. The cellular mechanism which stimulates the increase in lipid production in algae when they encounter stress is subject of investigation and requires much elucidation so that the stress induction can further be enhanced for more yields. Application tools like synthetic biology and metabolomics engineering can also aid in triggering the algae to yield more lipids during major phases of the algal growth cycle (Sheehan et al. 1998).

1.8.6 Cost-Effective Biomass Harvesting and Processing

The costs behind the efficient harvesting and biomass processing are important constrains and need to be aptly envisaged for rapid biofuel commercialisation. The technological selection of suitable biomass harvesting is determined by the characteristics of algal strains, biomass productivity, and the employed cultivation process. Though there are many technological methods in use for algal biomass harvesting which primarily include flocculation, centrifugation, filtration, drying, etc., few techniques like flocculation are relatively cheap, while uses of centrifugation, filtration, or ultrafiltration are expensive. The drying of algal biomass is yet another significant consideration, and the available alternatives include sun drying which is an inexpensive option but time- and space-consuming and not unanimously suitable, while the shelf drying at low pressure is a better option but less efficient too. Advanced efficient algal drying techniques include freeze-drying, drum drying, fluidised bed drying, and dehydration technologies. The algal harvesting and drying techniques are to be chosen in such a way that they are economic but at the same time not compromising the product and co-product quality.

1.9 Bioenergy Policies: India and Beyond

The recent concerns on fossil fuel depletion and environmental pollution have triggered the demand for alternate energy sources. In this aspect biofuels efficiently address these challenges with removal of bottlenecks. The Kyoto and Paris Protocols emphasise on the practice of renewable and eco safe fuels to substitute the existing nonrenewable fuel sources like petrol and diesel (United Nations 2016). Many countries around the world have put forth policies, mandate, and objectives for bioenergy. In developed countries like the United States and the European Union, a large number of biofuel programmes have been implemented keeping the objectives of lesser emissions and less import of existing fossil fuels (Ramos et al. 2016).

Moreover, the International Risk Governance Council (IRGC) opines that the impetus behind these programmes needs to be envisaged pragmatically on the global platform. The contributions of biofuels though appear less, but the impact on the environment can be more and significant with intervention of new approaches. The recommendations of the International Risk Governance Council (IRGC 2008) are as follows:

- Minimise any negative impact of bioenergy production on resources of water.
- Stimulate more viable agricultural processes for the production of food and fuel.
- Optimum utilisation of waste for bioenergy production.
- Development and adequate risk assessment approaches like lifecycle assessments and Environmental Impact Assessments (EIAs) and their local application.
- Adoption of internationally acclaimed standards and conditions for certification that would be documented under the international trade rules.

The IRGC also endorsed that the designers of the policies related to biofuels should formulate the regulations which are transparent and market-friendly on the global platform. The policymakers of the government worldwide need to envisage methods for biofuel production based on minimum resources concomitantly utilising the wastes. These holistic approaches towards sustainability will pave path for a better future.

1.10 Conclusion and Future Prospects

The algal-based products find wide applications ranging from biofuels to aqua feed, pharmaceutical products, waste water remediation, and generation of different valuable products towards a sustainable world. The cultivation of algae is relatively cost-effective, easy, and eco-friendly in nature as that is the reason behind their exhaustive exploration as biofuel reservoirs. The algal biorefinery approach involves a holistic strategy towards the complete utilisation of the complete algal biomass in a plethora of avenues with multiple applications as nutraceuticals, waste water management, cosmeceuticals, food, and feed. The integrated system incorporates the algal farming concomitant with the production of biofuels and other high-value products to establish the economics of the algal production process.

To make the process robust and practically functional, there are several angles which need to be exclusively estimated and well elucidated. On the front foot the well-designed exploratory envision is essential to establish fast-growing algal species with greater adaptability and much higher biomass productivity, thereby leading to the enhanced biofuel yield and other valuables from the residual biomass. The design and the choice of the cultivation system is equally significant as it proportionally impacts the cultivation and thereby productivity. The cultivation system whether open, closed, or hybrids has its own limitations and advantages. The system which suits the rapid growth and efficient cultivation based on the optimum requirement of the organism should be selected. The geographical interface also comes into picture in this arena, wherein we optimise the cultivation system based on the prevalent climatic conditions since the physical conditions like light, temperature, pH, etc. have profound influence on the algal cultivation and the maintenance of culture conditions is yet another important consideration.

At present research is relatively young and requires wide elucidations on the multifaceted parameters. The downstream processing is very crucial to accomplish better-quality product with utmost purity. The optimisation is essential in the post-cultivation system as yield purity with minimal cost is a challenging task. The technical interface and the economical paraphernalia are significant and probably a holistic or integrated synchrony is much essential in this case. The recent biotechnological approaches can be beneficial in the strain improvisation; better biomass processing including harvesting, purification, and drying; and residue biomass valorisation. The entire system has to be articulated in a pattern so that yield, the production cost, and the environmental safety are well considered integrating an excellent amalgamation of the biological, technological, and economical perspectives.

Sustainability urges to explore new sources of energy to conserve the environment. Focusing on better growing and harvesting methods will help in early commercialisation. Emphasis on screening new algal strains, techniques to improve production of algal biofuels, and improvisation in downstream processing will open new doors in the area of algal biofuels. Biofuels can attribute towards conservation of our nonrenewable sources of energy and environment, a way towards a better future.

Acknowledgement The authors would like to thank the Department of Biotechnology, Ministry of Science and Technology, India, for research funding (BT/PR15650/AAQ/3/815/2016).

References

Ali CH, Qureshi AS, Mbadinga SM, Liu JF, Yang SZ, Mu BZ (2017) Biodiesel production from waste cooking oil using onsite produced purified lipase from Pseudomonas aeruginosa FW_SH-1: central composite design approach. Renew Energy 109:93–100

Archana T (2014) Cyanobacteria: nature, potentials and applications. Astral International Publishing House, New Delhi

Archana T, Anjana P (2012) Cyanobacterial hydrogen production – a step towards clean environment. Int J Hydrog Energy 37:139–150

Archana T, Kiran T (2018) Biofuels from microalgae. In: Advances in biofuels and bioenergy. https://doi.org/10.5772/intechopen

Archana T, Kiran T, Pandey A (2019) Algal photobiohydrogen production. In: Konur O (ed) Bioenergy and biofuels. CRC Press, Boca Raton

Ashutosh P, Archana P, Priya S, Anjana P (2007) Using reverse micelle as microreactor for hydrogen production by coupled system of Nostoc/P4 and Anabaena/P4. World J Microbiol Biotechnol 23:269–274

Banerjee A, Sharma R, Chisti Y, Banerjee UC (2002) *Botryococcus braunii*: A renewable source of hydrocarbons and other chemicals. Crit Rev Biotechnol 2002(22):245–279

Boyle G (ed) (2012) Renewable energy: power for a sustainable future, 3rd edn. Oxford University Press/Open University, Oxford

Braun A, Reith J (1993) Algen in de Nederlandseenergiehuishouding, in opdracht van het programma, EnergiewinninguitAfval en Biomassa (EWAB) van Novem. Utrecht, Braun Consultants

Brennan L, Owende P (2010) Biofuels from microalgae–a review of technologies for production, processing, and extractions of biofuels and co-products. Renew Sust Energ Rev 14:557–577. https://doi.org/10.1016/j.rser.2009.10.009

Brown K (1996) The utility of remote sensing technology for carbon sequestration. Winrock international, 1611 N. Kent St., Suite 600, Arlington, VA 22209, USA

Chakrabarti T, Krishnamurthi K, Devi SS, Fulke BA (2018) CO2 sequestration by microalgae: advances and perspectives. Published by OMICS Group eBooks, Foster City

Chen CY, Yeh KL, Aisyah R, Lee DJ, Chang JS (2011) Cultivation, photobioreactor design and harvesting of microalgae for biodiesel production: a critical review. Bioresour Technol 102:71–81. https://doi.org/10.1016/j.biortech.2010.06.159

Chen Z, Wang G, Zeng C, Wu L (2018) Comparative study on the effects of two diatoms as diets on planktonic calanoid and benthic harpacticoid copepods. Ecol Integr Physiol 329:140. https://doi.org/10.1002/jez.2215

Cheng L-H, Lin Z, Chen H, Gao C (2006) Carbon dioxide removal from air by microalgae cultured in a membrane-photobioreactor. Sep Purif Technol 50(3):324–329

Chew KW, Yap JY, Show PL, Suan NH, Juan JC, Ling TC, Lee DJ, Chang JS (2017) Microalgae biorefinery: high value products perspectives. Bioresour Technol 229:53–62. (in place of Muhammad et al., 2018)

Chini Zittelli G, Lavista F, Bastianini A, Rodolfi L, Vincenzini M, Tredici MR (1999) Production of eicosapentaenoic acid by *Nannochloropsis* sp cultures in outdoor tubular photobioreactors. J Biotechnol 70(1–3):299–312

Chisti Y (2007) Biodiesel from microalgae. Biotechnol Adv 25(3):294–306

Chisti Y (2008a) Biodiesel from microalgae beats bioethanol. Trends Biotechnol 26(3):126–131

Chisti Y (2008b) Response to Reijnders: do biofuels from micromicroalgae beat biofuels from terrestrial plants. Trends Biotechnol 26(7):351–352

Chu WL (2017) Strategies to enhance production of microalgal biomass and lipids for biofuel feedstock. Eur J Phycol 52:419–437

Clifford RM, Kevan LM (2014) Microalgae (Diatom) production – the aquaculture and biofuel nexus. Oceans-St. John's, IEEE2014, pp 1–10

Cuellar-Bermudez SP, Romero-Ogawa MA, Rittmann BE, Parra-Saldivar R (2014) Algae biofuels production processes, carbon dioxide fixation and biorefinery concept. J Pet Environ Biotechnol 5:185. https://doi.org/10.4172/2157-7463.1000185

de Mes TZD, Stams AJM, Reith JH, Zeeman G (2003) Methane production by anaerobic digestion of wastewater and solid wastes. In: Biomethane & bio-hydrogen: status and perspectives of biological methane and hydrogen production. Dutch Biological Hydrogen Foundation, The Hague, pp 58–102. isbn:9090171657

Dhanker R, Molinero JC, Kumar R, Tseng LC, Ianora A, Hwang JS (2015) Responses of the estuarine copepod Pseudodiaptomus annandalei to diatom polyunsaturated aldehydes: reproduction, survival and postembryonic development. Harmful Algae 43:74–81

Dong T, Knoshaug EP, Davis R, Laurens LML, Van Wychen S, Pienkos PT, Nagle N (2016) Combined algal processing: A novel integrated biorefinery process to produce algal biofuels and bioproducts. Algal Res 19:316–323

Doucha J, Straka F, Livansky K (2005) Utilization of flue gas for cultivation of microalgae (Chlorella sp.) in an outdoor thin-layer photobioreactor. J Appl Phycol 17:403–412

Edzwald JK (1993) Coagulation in, drinking water treatment: particles, organics and coagulants. Water Sci Technol 27(11):21–35

Gao KS, Wu YP, Li G, Wu HY, Villafane VE, Helbling EW (2007) Solar UV radiation drives CO_2 fixation in marine phytoplankton: A double-edged sword. Plant Physiol 144(1):54–59

Gerardo ML et al (2015) Harvesting of microalgae within a biorefinery approach: a review of the developments and case studies from pilot-plants. Algal Res 11:248–262. (in place of Plaza et al., 2009)

Gifuni I et al (2018) New ultra-flat photobioreactor for intensive microalgal production: the effect of light irradiance. Algal Res 34:134–142

Grima Molina E, Belarbi EH, Acién Fernández FG, Robles Medina A, Chisti Y (2003) Recovery of microalgal biomass and metabolites: process options and economics. Biotechnol Adv 20:491–515. https://doi.org/10.1016/S0734-9750(02)00050-2

Halim R, Danquah MK, Webley PA (2012) Extraction of oil from microalgae for biodiesel production: a review. Biotechnol Adv 30(3):709–732

Harwood JL, Guschina IA (2009) The versatility of microalgae and their lipid metabolism. Biochimie 91(6):679–684

Heasman M, Diemar J, O'Connor W, Sushames T, Foulkes L (2003) Development of extended shelf-life microalgae concentrate diets harvested by centrifugation for bivalve molluscs – a summary. Aquac Res 31:637–659. https://doi.org/10.1046/j.1365-2109.2000.00492.x

Heede R, Oreskes N (2016) Potential emissions of CO2 and methane from proved reserves of fossil fuels: an alternative analysis. Glob Environ Chang 36:12–20

Hildebrand M, Davis AK, Smith SR, Traller JC, Abbriano R (2012) The place of diatoms in the biofuels industry. Biofuels 3(2):221–240

Hillen LW, Pollard G, Wake LV, White N (1982) Hydrocracking of the oils of *Botryococcus-Braunii* to transport fuels. Biotechnol Bioeng 24(1):193–205

Hossain A, Salleh A, Boyce A, Chowdhury P, Naqiuddin M (2008) Biodiesel fuel production from microalgae as renewable energy. Am J Biochem Biotechnol 4:250–254

https://www.iea.org/newsroom/news/2019/february/iea-becomes-facilitator-of-biofuture-platform.html

https://www.opec.org/opec_web/en/4567.htm

Hu Q, Guterman H, Richmond A (1996) A flat inclined modular Photobioreactor for outdoor mass cultivation of photoautotrophs. Biotechnol Bioeng 51(1):51–60

Huntley ME, Redalje DG (2007) CO2 mitigation and renewable oil from photo- synthetic microbes: a new appraisal. Mitig Adapt Strat Glob Change 12:573–608

IRGC (2008) Risk governance guidelines for bioenergy policies. http://www.irgc.org/IMG/pdf/IRGC_Bioenergy-Exec_Sum.pdf

Janssen M, Tramper J, Mur LR, Wijffels RH (2003) Enclosed outdoor photo- bioreactors: light regime, photosynthetic efficiency, scale-up, and future prospects. Biotechnol Bioeng 81:193–210

Johnson M, Wen Z (2009) Production of biodiesel fuel from the microalga *Schizochytrium limacinum* by direct transesterification of algal biomass. Energy Fuel 23(10):294–306

Kapdan IK, Kargi F (2006) Bio-hydrogen production from waste materials. Enzym Microb Technol 38:569–582

Kim J, Yoo G, Lee H, Lim J, Kim K, Kim CW et al (2013) Methods of downstream processing for the production of biodiesel from microalgae. Biotechnol Adv 31:862. https://doi.org/10.1016/j.biotechadv.2013.04.006

Korunić Z, Rozman V, Liška A, Lucić P (2016) A review of natural insecticides based on diatomaceous earths. Agriculture 22:10–18

Laurens LML, Chen-Glasser M, McMillan JD (2017) A perspective on renewable bioenergy from photosynthetic algae as feedstock for biofuels and bioproducts. Algal Res 24:261–264

Lehr F, Posten C (2009) Closed photo-bioreactors as tools for biofuel production. Curr Opin Biotechnol 20(3):280–285

Lewis T, Nichols PD, McMeekin TA (2000) Evaluation of extraction methods for recovery of fatty acids from lipid-producing micro heterotrophs. J Microbiol Methods 43(2):107–116

Li YQ, Horsman M, Wang B, Wu N, Lan CQ (2008) Effects of nitrogen sources on cell growth and lipid accu-mulation of green alga *Neochloris oleoabundans*. Appl Microbiol Biotechnol 81(4):629–636

Li Y, Chen YF, Chen P, Min M, Zhou W, Martinez B et al (2011) Characterization of a microalga Chlorella sp. well adapted to highly concentrated municipal wastewater for nutrient removal and biodiesel production. Bioresour Technol 102:5138–5144. https://doi.org/10.1016/j.biortech.2011.01.091

Li X-l, Thomas Kiran M, Tao L, Li R, Tiwari A, Li G (2017a) Optimization of growth conditions and fatty acid analysis for three freshwater diatom isolates. Phycol Res 65(3):177–187. https://doi.org/10.1111/pre.12174

Li X-l, Marella TK, Tao L, Tiwari A, Li G (2017b) A novel growth method for diatom algae in aquaculture wastewater for natural food development and nutrient removal. Water Sci Technol 75(12):2777–2783. https://doi.org/10.2166/wst.2017.156

Liu ZY, Wang GC, Zhou BC (2008) Effect of iron on growth and lipid accumulation in Chlorella vulgaris. Bioresour Technol 99(11):4717–4722

Mann DG, Droop SJM (1996) Biodiversity, biogeography and conservation of diatoms. Hydrobiologia 336:19–32

Marella TK, Parine NR, Tiwari A (2018) Potential of diatom consortium developed by nutrient enrichment for biodiesel production and simultaneous nutrient removal from waste water. Saudi J Biol Sci 25:704–770

Marella TK, Datta A, Patil MD, Dixit S, Tiwari A (2019) Biodiesel production through algal cultivation in urban wastewater using algal floway. Bioresour Technol 280:222–228

Mekonnen T, Mussone P, Khalil H, Bressler D (2013) Progress in bio-based plastics and plasticizing modifications. J Mater Chem A 1:13379–13398

Melis A, Happe T (2001) Hydrogen production. Green microalgae as a source of energy. Plant Physiol 127(3):740–748

Meuleman B (2007) Personal communication on biomass conversion technology

Miao XL, Wu QY (2006) Biodiesel production from heterotrophic microalgal oil. Bioresour Technol 97(6):841–846

Minhas AK et al (2016) A review on the assessment of stress conditions for simultaneous production of microalgal lipids and carotenoids. Front Microbiol 7:1–19

Mitra M, Melis A (2008) Optical properties of microalgae for enhanced biofuels production. Opt Express 16(26):21807–21820

Mohammad Mirzaie MA, Kalbasi M, Mousavi SM, Ghobadian B (2015) Investigation of mixotrophic, heterotrophic, and autotrophic growth of Chlorella vulgaris under agricultural waste medium. Prep Biochem Biotechnol 46(2):150–156

Moheimani NR, Borowitzka MA (2007) Limits to productivity of the alga *Pleurochrysiscarterae* (Haptophyta) grown in outdoor raceway ponds. Biotechnol Bioeng 96(1):27–36

Mohn FH (1980) Experiences and strategies in the recovery of biomass from mass cultures of microalgae. Algal Biomass 1980:471–547

Mulbry W, Kondrad S, Buyer J, Luthria D (2009) Optimization of an oil extraction process for microalgae from the treatment of manure effluent. J Am Oil Chem Soc 86:909–915

Nuccitelli R, Tran K, Sheikh S, Athos B, Kreis M, Nuccitelli P (2010) Optimized nanosecond pulsed electric field therapy can cause murine malignant melanomas to self-destruct with a single treatment. Int J Cancer 127:1727–1736

Pearsall RV, Connelly RL, Fountain ME, Hearn CS, Werst MD, Hebner RE et al (2011) Electrically dewatering microalgae. IEEE Trans Dielectr Electr Insul 18:1578–1583. https://doi.org/10.1109/TDEI.2011.6032827

Pohl P, Wagner H, Passig MVT (1968) Inhaltsstoffe von algen-II: Über die unterschiedliche fettsäurezusammensetzung von salz- und süßwasseralgen. Phytochemistry 7:1565

Posten C, Schaub G (2009) Microalgae and terrestrial biomass as source for fuels—A process view. J Biotechnol 142(1):64–69

Ramos J-L, Valdivia M, García-Lorente F, Segura A (2016) Benefits and perspectives on the use of biofuels. Microb Biotechnol 9(4):436–440

Ran CQ, Chen ZA, Zhang W, Yu XJ, Jin MF (2006) Characterization of photo- biological hydrogen production by several marine green microalgae. Wuhan Ligong Daxue Xuebao 28(Suppl 2):258–263

RangaRao A, Ravishankar GA (2007) Influence of CO_2 on growth and hydrocarbon production in *Botryococcus braunii*. J Microbiol Biotechnol 17(3):414–419

Reith JH (2004) Duurzame co-productie van fijnchemicaliën en energieuit micro-algen: openbaareindrapport E.E.T. project K99005/398510-1010. Petten, Energieonderzoek Centrum Nederland

Robinson CD (1986) Some factors influencing sedimentation. Ind Eng Chem 18:869–871. https://doi.org/10.1021/ie50200a036

Rodionova MV, Poudyal RS, Tiwari I, Voloshin RA, Zharmukhamedov SK, Nam HG, Zayadan BK, Bruce BD, Hou HJM, Allakhverdiev SI (2017) Biofuel production: challenges and opportunities. Int J Hydrog Energy 42(12):8450–8461

Roessler PG (1988) Changes in the activities of various lipid and carbohydrate biosynthetic enzymes in the diatom *Cyclotella cryptica* in response to silicon deficiency. Arch Biochem Biophys 267:521–528

Rossignol N, Moan R, Jaouen P, Robert JM, Quemeneur F (1999) Continuous high-pressure disruption of marine diatom Haslea ostrearia. Assessment by laser diffraction particle sizer. Biotechnol Tech 13:909–913

Ruocco N, Costantini S, Zupo V, Lauritano C, Caramiello D, Ianora A, Budillon A, Romano G, Nuzzo G, D'Ippolito G, Fontana A, Costantini M (2018) Toxigenic effects of two benthic diatoms upon grazing activity of the sea urchin: morphological, metabolomic and de novo transcriptomic analysis. Sci Rep 8:5622

Saranya G, Subashchandran MD, Praksah M, Ramachandra TV (2018) Prioritization of prospective third-generation biofuel diatom strains. Energy Ecol Environ 3:338–354

Sathasivam R, Ki J-S (2018) A review of the biological activities of microalgal carotenoids and their potential use in healthcare and cosmetic industries. Mar Drugs 16:26. https://doi.org/10.3390/md16010026

Satoru T, Hirata JA, Laws E (2007) Silicate deficiency and lipid synthesis of marine diatoms. (*Chaetoceros gracilis*; Hantzschia sp. ; Cyclotella sp). J Phycol 23:260–267

Scragg A, Illman A, Carden A, Shales S (2002) Growth of microalgae with increased calorific values in a tubular bioreactor. Biomass Bioenergy 23(1):67–73

Sheehan J, Dunahay T, Benemann J Roessler P (1998) Look back at the U.S. Department of energy's aquatic species program: biodiesel from microalgae; close-out report: size, 325 pages

Shen Y, Yuan W, Pei Z, Wu Q, Mao E (2009) Microalgae mass production methods. Trans ASABE 52:1275–1287

Show PL, Tang MSY, Nagarajan D, Ling TC, Ooi C-W, Chang J-S (2017) A holistic approach to managing microalgae for biofuel applications. Int J Mol Sci 2017(18):215–249

Sim TS, Goh A, Becker EW (1988) Comparison of centrifugation, dissolved air flotation and drum filtration techniques for harvesting sewage-grown algae. Biomass 16:51–62. https://doi.org/10.1016/0144-4565(88)90015-7

Singh G, Patidar SK (2018) Microalgae harvesting techniques: a review. J Environ Manag 217:499–508

Spolaore P, Joannis-Cassan C, Duran E, Isambert A (2006) Commercial applications of microalgae. J Biosci Bioeng 101:87–96

Syvertsen KE (2001) Optimizing fatty acid production in diatom Chaetoceros spp. by modifying growth environment. In: Biosystems engineering 2001, University of Hawaii at Minoa, Honolulu

Trent J (2009) Wind sea microalgae. In: Trent J (ed) International workshop on offshore microalgae cultivation, Maribo, Denmark, April 20–22; Baltic Sea Solutions and STP Productions, Lolland, Denmark, p 274

Tsukahara K, Sawayama S (2005) Liquid fuel production using microalgae. J Jpn Pet Inst 48(5):251–259

United Nations (2016) Framework convention on climate change. FCCC/CP/2015/10. URL http://unfccc.int/resource/docs

Wang J-K (2015) An absorbent and method of application for the treatment of heavy metal waste water. Chinene Patent, 201410072325

Wang J-K, Seibert M (2017) Prospects for commercial production of diatoms. Biotechnol Biofuels 10:16. https://doi.org/10.1186/s13068-017-0699-y

Wang Z, Pan Y, Dong T, Zhu X, Kan T, Yuan L, Torimoto Y, Sadakata M, Li Q (2007) Production of hydrogen from catalytic steam reforming of bio- oil using C12A7-O-based catalysts. Appl Catal A 320:24–34

Wang K, Mandal A, Ayton E, Hunt† R, Zeller MA, Sharma S (2016) Modification of protein rich algal-biomass to form bioplastics and odor removal. In: Protein byproducts. https://doi.org/10.1016/B978-0-12-802391-4.00006-9

Wijffels R (2006) Energie via microbiologie: Status en toekomstperspectiefvoor Nederland. Utrecht, Senter Novem

Wijffels R (2007) Presentation. Microalgae for production of energy. www.sciencedaily.com

Xu H, Miao XL, Wu QY (2006) High quality biodiesel production from a microalga *Chlorella protothecoides* by heterotrophic growth in fermenters. J Biotechnol 126(4):499–507

Yang Z, Kong F, Yang Z, Zhang M, Yu Y, Qian S (2009) Benefits and costs of the grazer induced colony formation in *Microcystis aeruginosa*. Ann Limnol Int J Limnol 45:203–208. https://doi.org/10.1051/limn/2009020

Yeang K (2008) Biofuel from microalgae. Archit Des 78(3):118–119

Algal Butanol Production

2

Enosh Phillips

Abstract

The energy of the sun converted to chemical energy by photosynthetic plants drives the life on earth. Energy has become an important aspect of the development of human civilization. Presently 80% of the energy demand is fulfilled by the fossil fuels which are non-renewable and has a reserve up to a few years. Moreover, the combustion of fossil fuels has increased the concentration of greenhouse gases effecting the climate and ecosystem. Renewable sources of energy like biofuels are promising alternatives to reduce the dependence on fossil fuels. Butanol is one such biofuel which has shown to blend in with the present fuel like petroleum, fulfilling the energy demands of transportation and industries. Butanol was initially being produced by the crop plants hence threatening the food security. Microalgae, an easily grown photosynthetic organism, has shown its capacity to enhance and increase the production of butanol without affecting the crop fields as well as the ecology. It can be grown in large quantities in a small cost-efficient manner and then can be digested by *Clostridium* sp. to produce butanol through the ABE method. Apart from *Clostridium*, *E.coli* has shown its competency in genetically modifying it to concentrate on butanol production.

Keywords

Butanol · Microalgae · Fossil fuels · Clostridium · Biofuel

E. Phillips (✉)
Department of Biotechnology, St. Aloysius College (Autonomous), Jabalpur, India

N. Srivastava et al. (eds.), *Substrate Analysis for Effective Biofuels Production*,
Clean Energy Production Technologies, https://doi.org/10.1007/978-981-32-9607-7_2

2.1 Energy Demand of the World

The flow of energy on earth begins from the sun; that is why earth is called as a solar-powered system. The helium fission reaction which begins in one of the biggest stars of the Milky Way triggers the process of photosynthesis that absorbs and harnesses the energy residing in the incoming photons into chemical energy. This phenomenon drives the life on earth. Therefore, energy is one of the basic components for everyday living (Lindsey 2009; Afework et al. 2019). For the continuous upgrade in the life of human species, energy plays a vital role. The history records in itself that with every new milestone achieved in a step towards civilization development, the demand for energy has increased simultaneously. This speaks that development of human civilization cannot be achieved without increasing the production of energy. The energy demands of the world increase linearly with the increase in population to meet its needs (Alam et al. 2007).

At present the energy demand of the world is dependent on non-renewable fossil fuel resources (Shafiee and Topal 2009). Ever since their discovery, 80% of energy demand is fulfilled by three different forms of fossil fuels, that is, coal, petroleum or gasoline and natural gas. The dependency on renewable resources and nuclear power for energy production is just 20% (Asif and Muneer 2007). As of 2004, the human population has surpassed 1.3 billion, and the economic growth globally stands near the average of 8%. This summons for higher industrial productions especially in countries like China and India which holds the major part of the world population to meet the needs of the people. Increasing industrial production requires increase in energy supply. China alone stands as the second largest consumer of energy using an energy equivalent amount of coal that is about 1.7 billion tons of coal, the USA being the first using more than this amount to meet its energy demand to satisfy the need of the third largest group of people. It consumes alone about 24% of the total world energy production. Figure 2.1 clearly represents the dependence of USA about 87% on non-renewable sources and only about 13% over renewable

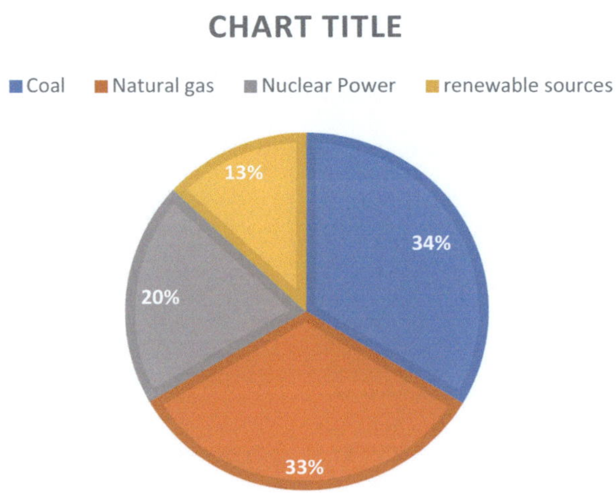

Fig. 2.1 A graphical representation of different types of fuels used for the production of electricity in the USA in the year 2015 (Source: NAS)

sources for meeting its electricity demand. On the other hand, India, the second most populated land mass just next to China, is the sixth largest consumer of energy, about 3.5% of world energy production. These data clearly suggests that there is an imbalance in the energy consumption between the developed and developing parts of the world, forcing the developing countries to feed less on energy when their population is more and seeking more energy to developed countries where amenities are already in surplus (Asif and Muneer 2007; Ragauskas et al. 2008; Crompton and Wu 2005).

The energy is consumed by three sectors of a nation:

- Industry
- Transport
- Others (agriculture, service sector and residential)

The connection between the consumption of energy, social development in relation to the economic status of the society and growth in population of a country puts up a serious question on the dependability on the non-renewable sources for energy. The phenomena of globalization, higher living standards and communication all point towards higher energy consumption, and current global policies over energy production will bring economic instability and unsustainable development and living standards (Lombard et al. 2008). Empirical studies on these relationships for many regions of the world suggest a nexus between economy, energy and emissions. Such a nexus points out that economic growth adheres to energy requirements or increase in energy consumption requires a strong economic support (Pao and Fu 2013). Here in Table 2.1 is the data on energy consumption in different parts of the continents:

The Table 2.1 clearly indicates that the top economies of the world whether developed or developing consumes more energy than produced by them and world at large have balanced the production and consumption at the cost of less consumption by lower economies. This suggests the unequal growth or growth opportunities to all humans of the world.

Table 2.1 Energy production and consumption by the world and top 10 developed and developing economies

S. no.	Country	Energy production (in BTU[a])	Energy consumption (in BTU[a])
1.	World	570.1	569.5
2.	China	114.2	137.1
3.	The USA	88.2	97.5
4.	Russia	57.8	31.5
5.	India	15.9	28.4
6.	Brazil	11.1	12.8
7.	France	5.3	10.0
8.	Germany	4.9	13.4
9.	Japan	1.8	19.4
10.	South Korea	1.7	12.0

Source: International Energy Statistics (2015)
[a]*BTU* British thermal unit

2.2 Non-renewable Fuels and Their Efficiency

Fossil-based energy sources that are primarily in use:

- Coal
- Natural gas
- Petroleum

The major offset of energy production in developed and developing countries is *coal*-based energy. It is an organic and inorganic sedimentary rock which is full of carbon, hydrogen and oxygen with trace amounts of sulphur and oxygen. As of the year 2001 the world coal reserve is 1083 billion short tons estimated to serve as energy source for approximately 230 years as detailed by the Energy Administration Information, USA. Coal is present around the world including Antarctica. In India the coal reserve is 93,000 million short tons. Coal is being used for energy purposes since 1000BC by the Chinese. But the impact and usage of it as the primary source was boosted only after the Industrial Revolution (Longwell et al. 1995; Miller 2005). The current efficiency of coal-based power plants is 36–45% (Buskies 1996; Rosen 2001). Coal is basically of three types: anthracite, bituminous and lignite. On ignition they produce CO_2, SO_2 and NO_x. They are also known to contain trace amount of natural radioactive elements like uranium, thorium, radon, etc. which is released in the atmosphere on burning coal (Baigh and Yousaf 2017).

On the other hand, another type of non-renewable fuel that is being used for energy production is *natural gas*. This gas is formed in the deepest part of the earth's crust from fossils. Mainly it is composed of hydrocarbon like methane (CH_4) which is the primary type of natural gas found in abundance. Apart from this, it also contains natural gas liquids and other non-hydrocarbon products like carbon dioxide (EIA 2018). The hydrates of natural gas (methane trapped in ice like crystals of water), which are solid compounds, are an alternative source of energy. There are so far 220 reserves that have been discovered. The growing energy demands of the future look upon the promises of this fossil reserve, whose physical properties like crystalline structure and efficiently compressing the gas make it handy for energy demands (Solan 2003; Makogon 2010). It is considered as a clean fuel due to less exhaust emission and is found to be favourable to the engines. But it possesses the disadvantage of low thermal efficiency that arises from slow burning ability and poor lean-burning ability. These factors decrease the efficiency of the engine. The traditional method optimized for improving on this issue causes the emission of NOx (Huang et al. 2006).

Petroleum, which is also a fossil fuel, is formed from the dead remains of marine organisms under the beds of the earth from a period of millions of years. It is found beneath the beds of ocean and seas over the vast area of land mass where ancient seas were located. It is found in gas, liquid and nearly solid form and may occur in combinations. It is being utilized in many industrial process and running engines. Operation of petroleum produces large amount of toxics in the environment which includes greenhouse gases that are responsible for the greenhouse effect (Morse and Turgeon 2018; Caudle and McLeroy 2019).

2.3 Biofuel: An Alternative Source of Energy

It is a well-known fact that the world stock remains of coal, petroleum and natural gas is limited as they are obtained from the fossilized stores buried deep inside the earth. It took millions of years to convert the dead remains of the organic life to be converted into such carbon components that can be broken to give energy for today's everyday work. We may reach global peak position in oil production by 2035 as the reports from various studies indicate. These resources are limited and may get extinguished in the coming future. Therefore, there is a need for opting an alternative and low-priced energy source with comparatively less emission of pollutants. These characters are met when efforts are put to produce energy from biomass. Biomass are the dead organics of plant matter which can be utilized to produce energy. The developed countries are moving towards developing such methods and techniques for efficient fabrication of energy from organic waste (biomass), which is termed as biofuel, in such a manner that it is becoming competitive and cost-efficient with fossil fuels (Demirbas 2008). It is a good alternative for reducing the emission of greenhouse gases like CO_2 (Durre 2007), even though the term biofuel is referred to as the fuel utilized for energy production in transport systems which is manufactured from the biomass. However, according to ASTM (American Society for Testing and Materials), it is defined as "a fuel composed of mono alkyl esters of long chain fatty acids derived from vegetable oils or animal fats" (Escobar et al. 2009; ASTM D6751 2018).

Biofuel is obtained from biomass via thermochemical and biological routes and have the following desirable attributes which are much needed in the present scenario:

- Does not produce any sort of undesirable exhaust, in low values
- To be available locally
- Easy access
- Should be maintainable
- Dependable fuel

These features attract the world market to enforce much research in developing technologies for the enhancement of the processes for the competent production of biofuels so as to decrease the dependability over fossil fuels (Vasudevan et al. 2005). Biofuel includes:

1. Bioethanol – gasoline equivalent
2. Biodiesel – diesel equivalent (Kapasi et al. 2010)

Table 2.2 Data that shows the approaches made in developing better technologies for biofuel production

Biofuel generation	Raw material for biofuel	Technology used	Limitations
First generation	It is obtained from food crops having good value of glucose like sugarcane. These materials have higher octane rating	Enzymation	Food crops diverted from market to industries for biofuel production hence threatening food security
Second generation	It includes cellulosic material, switchgrass, waste biomass, cornstalks, wheat stalks and wood	It utilized liquid technology to produce biofuel from solid biomass. Exactly how ruminant animals digest the grass they have eaten	Use large land mass for the production of crops that make more biomass for fuel production, hence decreasing the balanced crop yield
Third generation	Algae are used as the third-generation biofuel raw material with much ongoing research	Biochemical, thermochemical and chemical	Unstable fuel as compared to first- and second-generation fuels due to high level of unsaturation
Fourth generation	Photobiological and solar biofuels	Ongoing research to convert solar energy into fuel. Basically synthetic biology tools	Efficient technology for better usage of materials used for fuel energy production

Biofuels showed their presence in meeting the needs of human civilization since the late 1800s, preferably in those days for cooking and heating. The urge for more energy and increased dependability over limited fossil fuels accompanied by the commercialization and technology development have paved the way for the extensive use of biofuels in the European nations. They have succeeded in blending as an alternative for petroleum. At present there are four generations of biofuels produced so far, as described in Table 2.2:

The breakthroughs that are achieved so long and is evident from the data in Table 2.2 have been promising in developing paths for the economies which are earning carbon neutrality and affordable biofuel production that bears both features of quantity and quality (Awudu and Zhang 2012; Behera et al. 2015; Tao and Aden 2009; Aro 2016).

2.3.1 Biofuel Efficiency

Due to the fact that fossil fuel-based energy production has shown negative impact over the environment, the need to depend more on an alternative source of energy which is eco-friendly in nature is in the uproar. Biofuels in this context have shown its candidature as the best alternative to replace or decrease the dependability over fossil fuels like petroleum for energy production. An alternative source should not only be environmental-friendly but also cost competitive and can be produced in such abundance that it can come across the energy requirements of the increasing population, as it is the energy production that drives the way for standard living of human civilization. Biofuels have in themselves the capacities to answer all these questions. Therefore, there is a need for a thorough analysis of biofuels.

For the production of bioenergy from biofuel, there is a requirement of energy input which is growing crops and converting them for the production of biofuel (Hill et al. 2006).

Among the scientific community it is agreeable that the life cycle assessment (LCA) method could be adopted to assess the validity of biofuel as replacement without having an impact upon the life on the planet. It is evident that petroleum, natural gas and coal are fossil fuels that are not derived from the components that are involved in the sustenance of life on earth but rather derived from the dead organic parts which are buried millions of years ago. On the contrary biofuel is derived from the biomass that has to be produced from crops that can cause a competitive behaviour in the market to use land resources for biofuel production unit or for food requirement. Therefore LCA has been conducted to analyse all such queries (Cherubini et al. 2009; Davis et al. 2009). From the studies of LCA, it has become evident that the balance of energy and greenhouse gas emission depends much upon a number of factors:

1. Feed stock resource
2. Technique used for conversion of biomass to bioenergy
3. Technology involved in deriving energy from biofuel
4. Reference energy with which bioenergy is compared

It is evident by the studies conducted by researchers by applying LCA that there is a net reduction in the discharge of greenhouse gases when equated with that of fossil fuel energy. This is mainly found when biofuel was the chief fuel for transportation (Kim and Dale 2005; von Blottnitz and Curran 2007). LCA has been conducted to see the impact of biofuel on certain environmental aspects like inoculation of air with pollutants, acidification of terrestrial elements like water sources, eutrophication, ozone depletion, land use etc. The studies indicate that biofuel may have a negative impact on many of these environmental issues. For instance, for increased production of bioenergy there is a requirement for more amount of biomass which calls for use of pesticides and fertilizers for increased crop production. The use of fertilizers and pesticides poses harm of contaminating water bodies and soil quality. So biofuel on one hand can decrease the emission of green-house gas but may pose negative impact on other natural phenomenon (Cherubini et al. 2009). Moreover, studies have indicated that till 2006 about 5% of the world energy requirement was fulfilled by biomass energy. But the present scenario says that it is insufficient to compete with fossil energy and if the biomass energy is pushed beyond this, it would definitely impact on the food security by reducing it and may bring about change in climate, as it is evident that more dependence on biomass energy would ask for more of land resources be utilized for biomass production which may bring

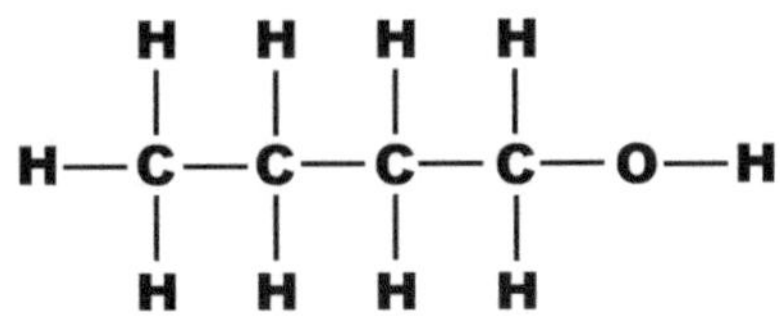

Fig. 2.2 Molecular structure of n-butane

down the land being utilized for crop production and may decrease the forest area which will be destroyed for availing land for biomass production (Field et al. 2007).

Studies like these will always be helpful in developing technologies that are much clean as that of present to minimize the negative effect on the environment overall and not just on one aspect (Cherubini et al. 2009).

2.3.2 Butanol as a Biofuel

Butanol or butyl alcohol or n-butanol is a tetra carbon chain with an alcoholic group (Fig. 2.2 details the butanol structure) having a molecular formula C_4H_9OH. It has a boiling point of 118 ° C. The commercial production of butanol dates back to World War I where it was being produced for lacquer industry. Butanol is utilized in paint, polymers and plastics which has seen a market value of $5 million, it is likely that the production of butanol will increase at a rate of 3.2% pa (Ward and Singh 2002; Green 2011).

It has shown its advantages as a biofuel in terms of internal combustion engines. Studies are indicative that it is a better biofuel when compared with ethanol with 30% more energy efficient and can be produced from various biomass. It has the properties to overcome the drawbacks of biodiesel (Qureshi 2009; Jin et al. 2011). Butanol can blend in with gasoline in a much better way as it has similarity with it due to long-chain structure. For instance, David Ramey drove his car with 100% butanol without doing any changes to the car engine. Although it has a low octane number as compared to ethanol, its pollutant emission is very well lower than ethanol (Szulczyk 2010). It contains 22% oxygen (see Fig. 2.2) that suggests that the combustion efficiency is more if butanol is used as a fuel than ethanol and decreases the possibility of CO emission, thereby beneficial for the environment. This gives butanol a new suffix bio-butanol (Qureshi et al. 2010).

These all features urges for the production of bio-butanol for pushing the need for biofuel.

2.4 Production of Bio-Butanol

Butanol or bio-butanol is being produced at a large scale by fermentative process. Microbes have shown their potentiality in the bioconversion of agricultural crops like corn and other biomass to butanol (Qureshi et al. 2008). Butanol is produced commonly by a process known as ABE process, that is, acetone, butanol and ethanol

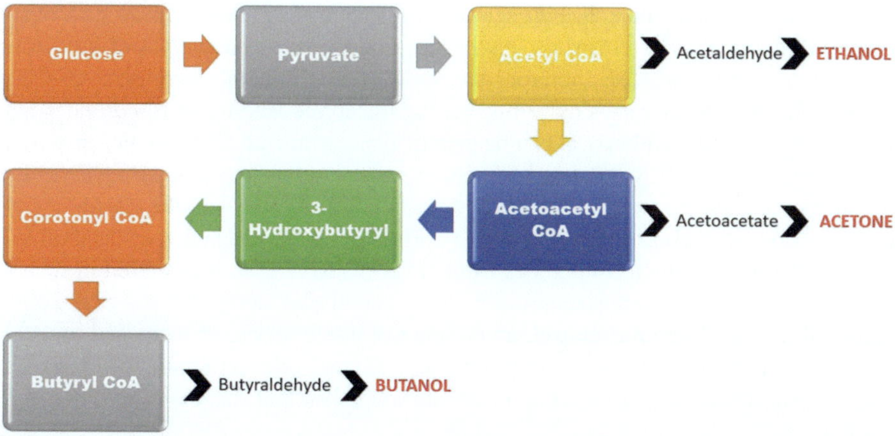

Fig. 2.3 Metabolism of glucose to butanol. The coloured boxes show the major path followed by fermenting bacteria in converting glucose to butyryl-CoA. The short arrows show the pathway acquired in converting the intermediate substrates produced during the metabolism to ethanol, acetone and butanol. This is seen in ABE fermentation and is a natural phenomenon

production process. Figure 2.3 shows how butanol is produced by ABE process. It is being synthesized by the fermentative ability of clostridia which is a rod-shaped, gram-positive bacteria. It is so commonly used because of its ability to consume an extensive variety of substrates for conversion. Among many clostridia, *C. acetobutylicum* is commonly used for industrial production of butanol (Monot et al. 1982; Ezeji et al. 2007; Lee et al. 2008). The metabolic pathway of clostridia first converts the glucose present in the substrate to pyruvate by the process known as glycolysis. The pyruvate so synthesized is then converted to acetyl-CoA. On the one hand, this substrate can be utilized to produce ethanol by a different metabolic path, whereas for butanol production acetyl-CoA in a stepwise manner is first converted to butyryl-CoA which is metabolized to butanol, as can be seen in Fig. 2.3 (Lee et al. 2008).

In one of the studies conducted by Nasib Qureshi, Badal C. Saha and Michael A. Cotta in 2007, the wheat straw biomass was used for the preparation of butanol by the mediation of *C. beijerinckii.* The spores of this microbe were first transferred to a medium known as cooked meat medium (abbreviated as CMM) and incubated at 35 °C for 16–18 h. In the meantime, the wheat straw collected from the farmland was grounded to fine particles. Wheat straw have high values of cellulose and hemicellulose and low values for lignin. It is added to dilute H_2SO_4 and autoclaved and cooled. The substrate was then inoculated with an enzyme mixture of cellulase, β-glucosidase and xylanase and kept at 45° for 72 h with continuous agitation. After this the entire mixture is filtered out and the filtrate was then inoculated with the developing culture. The fermentation is carried out at 35 °C till the ABE production ceases. In such a process, the concentration of butanol is always more than the concentrations of acetone and ethanol. Apart from these three, acetic acid in low quantity is also present in the mixture (Qureshi et al. 2007).

Other than *Clostridium*, *E. coli* at several instances has been employed to overcome the drawbacks of clostridia. As in the production of butanol from clostridia contains a mixture of ABE therefore the yield of sole butanol decreases. Also the growth of clostridia is very slow. However the metabolic activity of *E. coli* does not proceed for butanol synthesis, it can be genetically engineered to do so. For instance, in a study conducted by Atsumi with fellow researchers, *E. coli* when genetically modified was able to metabolize glucose to butanol. *E. coli* has the capability of breaking down 6-carbon hexose sugar to 3-carbon pyruvate which is then catabolized to acetyl-CoA. By transferring genes from clostridia to *E. coli*, it is altered in such a way that n-butanol is synthesized from *E. coli* (Atsumi et al. 2008; Neilson et al. 2009; Xue et al. 2013). Butanol is in itself an inhibitor of its synthesis, on increased concentration. Attempts have been made to engineer butanol-tolerant strains of clostridia (Lin and Blaschek 1983; Bowels and Ellefson 1985; Isar and Rangaswamy 2012).

2.5 Butanol from Algae

Due to the problem imposed by the use of agricultural biomass-based butanol production in terms of food security and climatic aspects, a search for new biomass was required. This search seems to lead to a known photosynthetic organism which is of eukaryotic origin and belongs to the family of protists – algae. As it is evident that with the advent of the twenty-first century, biofuels have been a promising source of energy while eradicating the environmental concerns created by fossil fuel-based energy system. Biofuel production on the other hand depends largely on agriculture biomass. Therefore a serious concern was that a society based on carbon will have an imperative fuel and food sustainability (Subhadra and George 2010). It means that it becomes quite difficult to have food and fuel security by and by. So, the biorefineries were tasked to find an alternate source for biofuel production, in which algae is seen as potential candidate. Studies have indicated that algae grown in small ponds, which take comparatively less land area than agriculture land, can produce enough biomass to be utilized in the production of biofuel, i.e. butanol. It finds its ecological preferences in terms of lowering the release of anthropogenic pollutant and requires less water subsidies (Smith et al. 2010).

Moreover, microalgae (also known as microscopic algae as can be seen in Fig. 2.4, which are unicellular algae present in freshwater and marine system) which are the most used form of algae for butanol production have desirable components and features. For instance, it is found that they grow rapidly with high mass value in a small pond and have attributes like CO_2 fixation and accumulation in high quantity of lipid and carbohydrates. To convert biomass to butanol, it is required the presence of fermentable sugars. Microalgae contain cellulose and starch (found in plastids) with no values of lignin or hemicellulose. This is beneficial as it lowers down the production cost since there is no procedure involved for removing lignin and hemicellulose as they are undesirable components that are basically found in agricultural biomass (Chen et al. 2013). It is evident from the studies conducted by researchers that more amount of butanol can be produced from microalgae when compared with the production of the same from crop biomass. Cheng and his associates

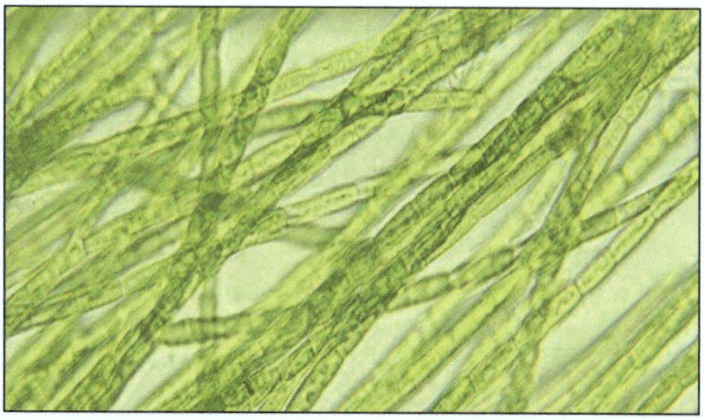

Fig. 2.4 A microscopic look at the microalgae grown by landfill technique to be used for making biofuels and bioplastics. (Credit: Gen3Bio)

demonstrated from their batch experiments in which they evaluated the potentiality of microalgae in butanol production taking *C. acetobutylicum* for fermentation. Their results have shown that when glucose is taken as sole carbon source, then butanol is produced at a concentration of 0.2 g per gram of glucose, and this may be enhanced by adding butyrate. However, *C. acetobutylicum* can produce 3.86 g L^{-1} of butanol when the substrate was microalgae with 1/3 of carbohydrate still remaining in unused form (Efremenko et al. 2012; Cheng et al. 2015).

2.5.1 Benefits of Microalgae

The first- and second-generation biofuels, although equally eco-friendly in terms of reducing CO_2 emission, are not ecologically competent as they develop threats for food security and climatic changes. The third-generation biofuel utilizes microalgae, which can be produced in large quantity in a given small area used for the production of biofuel. It is the most sustainable answer for the replacement of fossil fuel by a natural renewable source. Microalgae have a wide range of benefits to be used for butanol production:

- Due to high rate of photosynthesis, it gives certain benefits like:
  - Lipid accumulation
  - Carbon sequestration
  - Oxygen production
  - Nitrogen cycle
- Remover of greenhouse gases
- In waste water treatment
- Pollution control

Microalgae can be grown over in any environment like in freshwater, waste water and marine environment. Certain strains have also been identified and isolated from high-temperature regions. From among the 80,000+, many of which can be grown in the presence of light anywhere has the potential to produce chemical energy from photosynthesis which is beneficial for butanol. It is evident from many studies that microalgae produce more oil than feedstock (Maity et al. 2014).

2.5.2 Cultivation of Microalgae for Butanol Production

Studies have shown that microalgae have the capacity to trap energy from the sun, called photon energy, and convert it into chemical energy in the form of metabolites. There have been attempts since World War II for the production of microalgae at that time as a supplement for protein (Chaumont 1993). They are a fast-growing species that can grow up to 10 meters in length. It is seen that marine algae grow at a higher pace than algae that are grown terrestrially. Microalgae are carbohydrate-rich and sustainable and uses no agricultural land. It is not a major part of the animal food chain. This removes the concern of the role of feedstock on climatic change as it has to be grown on water bodies, so the utilization of agricultural land reduces considerably giving enough place for growing crops. Microalgae is not only a promising feedstock for butanol production but is extensively used in waste water treatment (Kraan 2013).

One of the approaches in microalgae cultivation is cultivating it on effluent water as part of waste water treatment. It has been worked upon worldwide as it is the most cost-effective method of cultivation. However, they are sensitive to a number of components present in waste water like toxic substances, pollutants and imbalanced nutrient profiles, identification of such species which shows tolerance to all these components to develop into a biomass having high carbohydrate and lipid content is in the mind. Two such species have been identified and worked upon, *Chlorella* sp. and *Scenedesmus* sp. These are found to be tolerant to different types of waste water and hence can be grown in many treatment plants. Algal biomass can be cultivated over POME (palm oil mill effluent) which is an effective way for its development. Research works have indicated that when microalgae are grown on 75% POME, they have shown to develop the highest mass of the algae. In another study it was seen that high concentration of POME was inhibiting the growth of the species like that of *Chlorella* but addition of synthetic nutrient in 20% POME encouraged the growth at the highest level. There it can be assumed that POME is a successful way of cultivating the growth of the algae and supplanting with synthetic nutrient can enhance their growth. POME is also efficient in removing COD from the effluents through the mediation of microalgae (Selmani et al. 2013; Cheah et al. 2016).

Municipal solid wastes are known to be treated by natural bioreactors like *landfills*. Landfills are known from a very long time to be used for treatment of solid wastes especially in urban areas. This draws attention for using landfill for the production of microalgae. In landfill leachate, a mixture of algae has been developed as the studies have reported and can be seen in Fig. 2.4 and it is evident that toxic components are responsible for inhibiting the growth of algae. It has been reported that algae can be enhanced in landfill leachate by adjusting the pH to around 7.0 due to which the toxicity is reduced thus increasing algal biomass production. Studies have indicated that landfills have to be pretreated for the growth of microalgae (Borowitzka 1999; Cheah et al. 2016).

Microalgae can also be grown while doing sewage treatment (Park et al. 2011; Raouf et al. 2012; Cai et al. 2013).

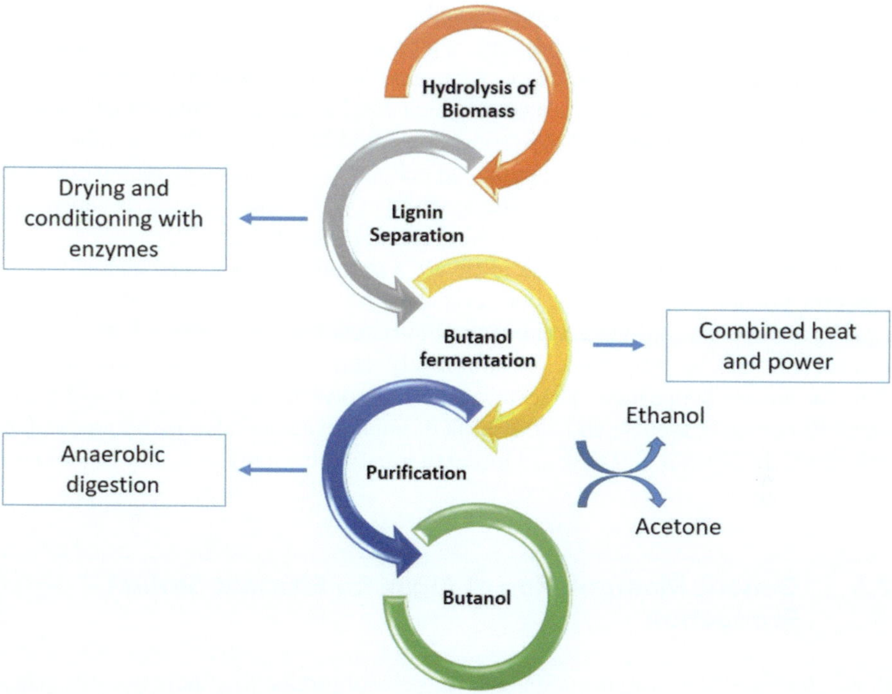

Fig. 2.5 A generalized view of butanol production employed

2.5.3 Making of Butanol

Butanol is produced from algae after harvesting it on any of the above known methods. In one of the methods used by Potts et al., as depicted in Fig. 2.5, microalgae naturally occurring in Jamaican bay were collected. Now this growth took place in the presence of phosphates, nitrates and carbon dioxide which were mixed into the water over which the algae have developed. These algae were then harvested from the pond and dried and subjected to mechanical grinding. After it is being ground to fine form, the algae were introduced to acid hydrolysis. This is done in view that algal biomass is comprised primarily of cellulose which is non-fermentable by fermenting bacteria like *Clostridium.* Acid hydrolysis converts this polymer into fermentable sugars so that the fermenting bacteria can act upon these broken stretches. Now the biomass is known as algal sugar biomass. The hydrolysate after acid hydrolysis consisted of 15.2 g/L of sugar which shows that algae contain a good amount of carbohydrate for the synthesis of butanol. The acetone, butanol and ethanol (ABE) extraction method has been used in this fermentation for the synthesis of butanol. An estimation of 4 g/L of butanol was produced (Lakaniemi et al. 2012; Potts et al. 2011).

Ellis et al. made an attempt in a similar way in which butanol was produced by ABE method. Here the waste water algae were used as the raw material for carbon source for the metabolic activities of the *Clostridium* sp. Algae were grown over the waste water lagoon. The growth was enough at a high rate proving that algae can be a renewable source of biomass for butanol production. About 10% of algae was subjected for the production of butanol. After acid hydrolysis and ABE fermentation, about 2.74 g/L of butanol was produced which was increased on supplementing with 1% glucose. It was found that the production could be increased on providing glucose as well as enzymes like xylanase and cellulase. The results of such work concluded that butanol production can be increased to 160% when glucose is added and to 250% when xylanase and cellulase were added (Ellis et al. 2012).

These results are indicative that algae are a promising renewable source of biomass for butanol production. It does contain a huge amount of carbohydrate that could be converted into large volumes of bio-butanol provided the polymer carbohydrate present in the source is completely converted to fermentable sugars. Approaches are being made at several levels to solve these issues as it will bring great difference in the use and cultivation of microalgae and also in the production of butanol thus boosting its use as biofuel to satisfy the energy demands of the budding population.

2.6 Genetic Manipulation of Algae for Efficient Biofuel Production

In the shadow of the use of microalgae for butanol production, there are still many barriers that are needed to be removed in clearing the path towards the mass usage of algae economically for biofuel production. Genetic engineering will be a handy tool in overcoming the issues of productivity optimization and identifying and developing strains that contain the traits that are favourable for biofuel production. Out of the 40,000 identified species of microalgae, only 3000 algae have shown the potentiality for the production of biofuel. It has been suggested that a manipulation to enhance the photon conversion efficiencies will reduce the land mass required for the production of the algae and will also reduce the cost of fuel production. Therefore, expressed sequence tag (EST) databases have been established which contain the data of genomes from nuclear, mitochondria and chloroplast. Such a database provides the window through which a manipulation can be introduced in the genome to enhance the ability of the algae to contain the traits of biofuel. As a result of such constructions have facilitated the genetic transformation of more than 30 strains successfully (Radakovits et al. 2010).

2.7 Conclusion

Solar photons are captured by greenery of the earth and subsequently transferred to other parts of the living systems which in the timeline turn to fossil of different forms, the present-day energy source for various activities. Although fossil fuels are efficient, they pose a threat to the environment due to the emission of greenhouse gases, petroleum being the major contributor. Biofuel hence came out to be a good alternative, being eco-friendly and utilizing organic waste for its production. LCA studies have revealed that use of butanol gradually decreases greenhouse gases, but there is negative impact on the environment on the issue of harvesting microalgae. Even then microalgae have become a prominent client in butanol production that could replace the use of petrol and diesel in the very future. Butanol blends in with fossil fuel so the requirement of a new technique for its combustion is not obligatory. Microalgae is an abundant source of the raw substrate required for butanol production. *Clostridium* sp. has been widely used so far for butanol production. Moreover genetic interference in the cellulose producing genes of microalgae would greatly enhance the production by increasing the concentration of cellulose in the organisms, much more studies and assessment are required for the use butanol as an alternative fuel and completely remove fossil fuel use. Moreover, with increasing concerns on global warming and climatic and pollution issues, use of butanol is much more the call of nature, a natural product to save nature.

Acknowledgement I would like to thank the God almighty in the first place for giving the wisdom to write this chapter. Indebted to Dr. Neha Shrivastava for the opportunity given and all the help, suggestions and advices without which it would have not been possible to do this. I am thankful to my family who always stood by me during the entire work.

References

Afework B, Hanania J, Llyod E, Meyer R, Sheardown A, Stenhouse K, Yyeland B, Donev J (2019) Earths energy flow. https://energyeducation.ca/encyclopedia/Earth's_energy_flow. Accessed 11 Jan 2019

Alam S, Fatima A, Butt MS (2007) Sustainable development in Pakistan in the context of energy consumption demand and environmental degradation. J Asian Econ 18:825–837

Aro ME (2016) From first generation biofuels to advanced solar fuels. Ambio 45(1):24–31

Asif M, Muneer T (2007) Energy supply, its demand and security issues for developed and emerging economies. Renew Sust Energ Rev 11:1388–1413

ASTM D6751-18 (2018) Standard specification for biodiesel fuel blend stock (B100) for middle distillate fuels, ASTM International, West Conshohocken, www.astm.org

Atsumi S, Cann AF, Connor MR, Shen CR, Smith KM, Brynildsen MP, Chou KJY, Hanai T, Liao JC (2008) Metabolic engineering of E. coli for 1-butanol production. Metab Eng 10:305–311

Awudu I, Zhang J (2012) Uncertainties and sustainability concepts in biofuel supply chain managements: a review. Renew Sust Energ Rev 16:1359–1368

Baigh SK, Yousaf M (2017) Coal fired power plants: emission problems and controlling techniques. J Earth Sci Clim Chang 8(7):1–9

Behera S, Singh R, Arora R, Sharma NK, Shukla M, Kumar S (2015) Scope of Algae as third generation biofuels. Front Bioeng Biotechnol 2(90):1–13

Borowitzka MA (1999) Commercial production of microalgae: pond, tanks and fermenters. Prog Ind Microbiol 35:33–321

Bowles LK, Ellefson WL (1985) Effects of butanol on *Clostridium acetobutylicum*. Appl Environ Microbiol 50(5):1165–1170

Buskies U (1996) The efficiency of coal-fired combine-cycle powerplants. Appl Therm Eng 16(2):959–974

Cai T, Park SY, Li Y (2013) Nutrient recovery from waste water stream by microalgae: Status and prospects. Renew Sust Energ Rev 19:360–369

Caudle BH, McLeroy PG (2019) Petroleum production. https://www.britannica.com/technology/petroleum-production. Accessed on 12 Jan 2019

Chaumont D (1993) Biotechnology of algal biomass production: a review of systems for outdoor mass culture. J Appl Phycol 5:593–604

Cheah YW, Ling TC, Show LP, Juan JC, Chang JS, Lee DJ (2016) Cultivation in wastewater for energy: a microalgae platform. Appl Energy 179:609–625

Chen CY, Zhao XQ, Yen HW, Ho S, Cheng CL, Lee DJ, Bai FW, Chang JS (2013) Microalgae based carbohydrates for biofuel production. Biochem Eng J 78:1–10

Cheng HH, Whang LM, Chan KC, Chung MC, Wu SH, Liu CP, Tien SY, Chen SY, Chang JS, Lee WJ (2015) Biological butanol production from microalgae based biodiesel residues by *clostridium acetobutylicum*. Bioresour Technol 184:379–385

Cherubini F, Bird ND, Cowie A, Jugmeier G, Schlamadinger B, Gallasch SW (2009) Energy and green-house gas based LCA of biofuel and bioenergy systems: key issues, ranges and recommendations. Resour Conserv Recycl 53:434–447

Crompton P, Wu N (2005) Energy consumption in China: past trends and future directions. Energy Econ 27:195–208

Davis SC, Teixeira KJA, DeLucia EH (2009) Life cycle analysis and the ecology of biofuels. Trends Plant Sci 14(3):140–146

Demirbas A (2008) Biofuel sources, biofuel policy, biofuel economy and global biofuel projections. Energy Convers Manag 49:2106–2116

Durre P (2007) Butanol: an attractive fuel. Biotechnol J 2(12):1525–1534

Efremenko EN, Nikolskaya AB, Lyagin IV, Senko OV, Makhlis TA, Stepanov NA, Maslova OV, Mamedova F, Varfolomeev SD (2012) Production of biofuels from pretreated microalgae biomass by anaerobic fermentation with immobilized *Clostridium acetobutylicum* cells. Bioresour Technol 114:342–348

EIA (2018) Natural gas explained. https://www.eia.gov/energyexplained/natural-gas/. Accessed on 13 Jan 2019

Ellis JT, Hengge NN, Sims RC, Miller CD (2012) Acetone, butanol and ethanol production from waste water algae. Bioresour Technol 111:491–495

Escobar JC, Lora ES, Venturini OJ, Yanez EE, Castillo EF, Almazan O (2009) Biofuels: environment, technology and food security. Renew Sust Energ Rev 13:1275–1287

Ezeji TC, Qureshi N, Blaschek HP (2007) Butanol production from agricultural residues: impact of degradation products on *Clostridium beijerinckii* growth and butanol fermentation. Biotechnol Bioeng 97:1–8

Field CB, Campbell JE, Lobel DB (2007) Biomass energy: the scale of the potential resource. Trends Ecol Evol 23(2):65–72

Green EM (2011) Fermentative production of Butanol: the industrial perspective. Curr Opin Biotechnol 22:1–7

Hill J, Nelson E, Tilman D, Polasky S, Tiffany D (2006) Environmental, economic and energy costs and benefits of biodiesel and ethanol fuels. PNAS 103(30):11206–11210

Huang Z, Zhnag Y, Zeng K, Liu B, Wang Q, Jiang D (2006) Measurements of laminar burning velocities for natural gas-hydrogen-air mixtures. Combust Flame 146:302–311

International Energy Statistics (2015). https://www.eia.gov/beta/international/data. Accessed on 13 Jan 2019

Isar J, Rangaswamy V (2012) Improved n-butanol production by solvent tolerant *C. beijerinckii*. Biomass Bioenergy 37:9–15

Jin C, Yao M, Liu H, Lee CF, Ji J (2011) Progress in the production and application of n butanol as a biofuel. Renew Sust Energ Rev 15(8):4080–4106

Kapasi ZA, Nair AR, Sonawane S, Sk S (2010) Biofuel- an alternate source of energy for present and future. J Adv Sci Technol 13(3):105–108

Kim S, Dale BE (2005) Life cycle assessment of various cropping systems utilized for producing biofuels: bioethanol and biodiesel. Biomass Bioenergy 29(6):426–439

Kraan S (2013) Mass cultivation of carbohydrate rich microalgae, a possible solution for sustainable biofuel production. Mitig Adapt Strateg Glob Chang 18:27–46

Lakaniemi AM, Tuovinen OH, Puhakka JA (2012) Production of electricity and butanol from microalgal biomass in microbial fuel cells. Bioeng Resour 5:481–491
Lee SY, Park JH, Jang SH, Neilson LK, Kim J, Jung KS (2008) Fermentative butanol production from *Clostridia*. Biotechnol Bioeng 101(2):209–228
Lin YL, Blaschek HP (1983) Butanol production from butanol tolerant strain by *C. acetobutylicum* in extruded corn broth. Appl Environ Microbiol 45(3):966–973
Lindsey R (2009) Climate and earth's energy budget. https://earthobservatory.nasa.gov/features/EnergyBalance. Accessed on 11 Jan 2019
Lombard LP, Ortiz J, Pout C (2008) A review on buildings energy consumption information. Energ Buildings 40:394–398
Longwell JP, Rubin ES, Wilson J (1995) Coal: energy for the future. Prog Energy Combust Sci 21(4):269–360
Maity JM, Bundschuh J, Chen CY, Bhattacharya P (2014) Microalgae for third generation biofuel production, mitigation of greenhouse gas emissions and wastewater treatment: present and future perspectives: a mini review. Energy 30:1–10
Makogon YF (2010) Natural gas hydrates – a promising source of energy. J Nat Gas Sci Eng 2(1):49–59
Miller BG (2005) Introduction to coal. In: Coal energy. Elsevier Academic Press, New York, pp 1–26
Monot F, Martin JR, Petitdemange H, Gay R (1982) Acetone and butanol production by *C. acetobutylicum* in a synthetic medium. Appl Environ Microbiol 44(6):1318–1324
Morse E, Turgeon A (2018) Petroleum. https://www.nationalgeographic.org/encyclopedia/petroleum/. Accessed on 12 Jan 2019
Neilson DR, Leonard E, Yoon SH, Tseng HC, Yuan C, Prather JKL (2009) Engineering alternative butanol production platforms in heterologous bacteria. Metab Eng 11(4–5):262–273
Pao HT, Fu HC (2013) Renewable energy, non-renewable energy and economic growth in Brazil. Renew Sust Energ Rev 25:381–394
Park JBK, Craggs RJ, Shilton AN (2011) Wastewater treatment high rate algal ponds for biofuel production. Bioresour Technol 102(1):35–42
Potts T, Du J, Paul M, May P, Beitle R, Hestekin J (2011) The production of butanol from Jamaica bay microalgae. Environ Prog Sustain Energy 31:1–3
Qureshi N (2009) Beneficial biofilms; waste water and other industrial applications. Biofilms in food and beverage industries, pp 474–498
Qureshi N, Saha BC, Cotta MA (2007) Butanol production from wheat straw hydrolysate using *C. beijerinckii*. Bioprocess Biosyst Eng 30:419–427
Qureshi N, Saha BC, Hector RE, Hughes SR, Cotta MA (2008) Butanol production from wheat straw by simultaneous saccharification and fermentation using Clostridium beijerinckii: part I—batch fermentation. Biomass Bioenergy 32:168–175
Qureshi N, Saha BC, Dein B, Hector RE, Cotta MA (2010) Production of butanol (a biofuel) from agricultural residues: part-I use of barley straw hydrolysate. Biomass Bioenergy 34:559–565
Radakovits R, Jinkerson RE, Darzins A, Posewitz MC (2010) Genetic engineering of algae for enhanced biofuel production. Eukaryot Cell 9(4):486–501
Ragauskas AJ, Williams CK, Davison BH, Britovsek G, Cairney J, Eckert CA, Frederick WJ, Hallett JP, Leak DJ, Liott CL, Mielenz JR, Murphy R, Templer R, Tschaplinski T (2008) The pathway forward for biofuel and biomaterial. Science 311:484–489
Raouf NA, Al-Homaidan AA, Ibraheem IBM (2012) Microalgae and waste water treatment. Saudi J Biol Sci 19(3):257–275
Rosen MA (2001) Energy and energy based comparison of coal fired and nuclear steam power plants. Exergy Int J 1(3):180–192
Selmani N, Mirghani MES, Alam MZ (2013) study the growth of microalgae in palm oil mill effluent waste water. 4th International conference on energy and environment. IOP publishing. https://doi.org/10.1088/1755-1315/16/1/012006
Shafiee S, Topal E (2009) When will fossil fuel reserves be diminished? Energy Policy 37(1):181–189

Smith VH, Sturm BSM, DeNoyelles FJ, Billings SA (2010) The ecology of algal biodiesel production. Trends Ecol Evol 25(5):301–309
Solan ED (2003) Fundamental principles and applications of natural gas hydrates. Nature 426:353–363
Subhadra B, George G (2010) Algal bio refinery-based industry: an approach to address fuel and food insecurity for a carbon-smart world. J Sci Food Agric 91(1):2–13
Szulczyk KR (2010) Which is a better transportational fuel: butanol or ethanol? Int J Energy Environ 1(3):501–512
Tao L, Aden A (2009) The economics of current and future biofuels. In Vitro Cell Dev Biol Plant 45:199–217
U.S. Energy information administration (2018) What is natural gas? https://www.eia.gov/energyexplained/index.php?page=natural_gas_home. Accessed 10 Jan 2019
Vasudevan P, Sharma S, Kumar A (2005) A liquid fuel from biomass: an overview. J Sci Ind Res 64:822–831
von Blottnitz H, Curran MA (2007) A review of assessments conducted on bio-ethanol as a transportation fuel from a net energy, greenhouse gas, and environmental life cycle perspective. J Clean Prod 15(7):607–619
Ward OP, Singh A (2002) Bioethanol technology: development and perspectives. Adv Appl Microbiol 51:53–80
Xue C, Zhao XQ, Liu CG, Chen LJ, Bai FW (2013) Prospective and development of butanol as an advanced biofuel. Biotechnol Adv 31(8):1575–1584

Suitability of the Lantana Weed as a Substrate for Biogas Production

3

Madan L. Verma, Raj Saini, Sneh Sharma, Varsha Rani, and Asim K. Jana

Abstract

Biogas technology is a source of renewable energy. It involves the biomethanation of biological waste to energy in the form of biogas which is a boon for sustainable agriculture to meet its partial needs in rural areas. There are so many substrates which can be used in biogas plants. In India, primarily cattle dung is the organic substrate used for the production of biogas. Rural and urban areas of developing countries produce various cellulosic biomass (agricultural residues, cattle dung, etc.), having a very good potential to meet the energy demand. Biogas technology is contributing towards socio-economic development as well as environmental protection. However, availability of this substrate in optimum quantity is one of the major impediments in the successful operation of biogas plants, and therefore, majority of the biogas digesters are underfed. Thus, there is a need to explore the biogas potential of other available organic substrates. The present chapter discusses the various substrates employed for biogas production.

M. L. Verma (✉)
Department of Biotechnology, Dr Y. S. Parmar University of Horticulture and Forestry, Hamirpur, Himachal Pradesh, India

Department of Biotechnology, School of Basic Sciences, Indian Institute of Information Technology, Una, Himachal Pradesh, India

R. Saini
Department of Basic Sciences, Dr Y. S. Parmar University of Horticulture and Forestry, Nauni, Himachal Pradesh, India

S. Sharma
Department of Biotechnology, Dr Y. S. Parmar University of Horticulture and Forestry, Hamirpur, Himachal Pradesh, India

V. Rani
Department of Biotechnology, Shoolini University, Solan, Himachal Pradesh, India

A. K. Jana
Department of Biotechnology, National Institute of Technology, Jalandhar, India

N. Srivastava et al. (eds.), *Substrate Analysis for Effective Biofuels Production*, Clean Energy Production Technologies, https://doi.org/10.1007/978-981-32-9607-7_3

Specific emphasis is given to the suitability of lantana plant for biogas production. The main challenges and potential opportunities in biogas production using an invasive lantana weed are also discussed.

Keywords
Plant · Weed · Substrate · Toxicity · Challenges · Bioenergy

3.1 Introduction

Biogas technology has become a popular and alternative method for clean and renewable energy production. Biogas technology can be employed for converting organic wastes which are obtained from livestock, agriculture and industries into manure and green energy (Alkanok et al. 2014; Angelidaki et al. 2018). Biogas obtained from organic waste and other waste residues can contribute a vital role in the energy future. Biogas plant converts fermentable organic matter such as farmyard manure, organic waste materials, energy crops and waste from slaughter houses into combustible gas and organic manure. It works by subjecting the organic matter to microbial decomposition in the absence of air, yielding carbon dioxide, methane and water. This process is known to occur naturally and is called as anaerobic decomposition. Anaerobic digestion has four biological and chemical stages, followed by the process of hydrolysis, acidogenesis, acetogenesis and methanogenesis of the waste materials for the production of biogas from biomass sources (Yadvika et al. 2004). Animal farms are producing a large quantity of manure in the recent years which is causing environmental issues. Therefore, it becomes very vital to find potential ways for treating massive manure (Fierro et al. 2014). Biogas technology is an optimum process for the organic waste treatment obtained from agriculture and animal husbandry which in turn is reducing the environmental pollution as well as the emission of greenhouse gases (Angelidaki et al. 2018; Chae et al. 2008; Khalid et al. 2011). Manure from animal farms is a beneficial source for the production of renewable fuel and is a rich source of organic fertilizers (Paolini et al. 2018; Kundwa et al. 2018).

Biogas technology is environmentally beneficial and its feedstock cost is also cost-effective (Mao et al. 2015). Animal manure is necessary for bacterial growth due to the presence of a variety of nutrients in it. It has been reported that animal manure alone is not appropriate for producing biogas because of its lower carbon and nitrogen ratio (Toma et al. 2016). The nitrogen content of livestock manure is higher, such as chicken manure (1%), fresh goat manure (1%), dairy manure (0.4%) and swine manure (0.2%) (Zhang et al. 2013).

The present chapter discusses the suitability of the lantana plant as one of the promising substrates for biogas production. The main challenges and potential opportunities in biogas production using an invasive lantana weed are also discussed.

3.2 Biogas as a Renewable Energy

Biogas obtained through this process is commonly referred to as gobar (cow dung) gas and also known by several names like fuel gas, sewer gas, marsh gas, swamp gas or wet gas. Biogas technology is a simple, low-cost and economically feasible process in rural areas, where organic waste is generated aplenty, which otherwise pollutes the environment and poses several health hazards (Khoiyangbam et al. 2011; Paolini et al. 2018; Sekoai et al. 2019). Anaerobic digestion is a process that involves a series of biotransformation reactions by using different microbial consortia. Crops and agro-industrial wastes, when aerobically digested, produce a large quantity of biogas, and biogas is obtained from animal waste and sludge to a lesser extent (Tricase and Lombardi 2009). As fuel crisis and environmental pollution associated with fossil fuels are increasing day by day, biogas has attracted wide attention. Biogas is an environment-friendly energy source because it acts as a global waste epidemic as well as helps in meeting the global energy demand. Biogas is utilizing nature's elegant tendency of recycling organic waste into productive resources. Biogas can be used directly for heating and lighting purposes as well as in an engine-driven generator to generate electricity (Kundwa et al. 2018). Effluent released from the biogas plant is an excellent fertilizer, which improves the physical properties of the soil like aeration, moisture holding and water infiltration capacities (Paolini et al. 2018). Biogas plants help improve the ecology and the environment via safe disposal of sewage and animal and human faeces in urban and rural areas. Biogas technology is providing solutions to counter the problem of global warming by minimizing fossil fuel consumption and curbing the greenhouse gas emission (Paolini et al. 2018; Nishio and Nakashimada 2007; Kundwa et al. 2018). Moreover, biogas is a viable and environment-friendly technology, which is not as popular as it should be (Khoiyangbam et al. 2011; Nishio and Nakashimada 2007).

Biofuels such as bioethanol, biodiesel and biogas are renewable energy sources to meet the world energy requirement (Verma 2017; Verma et al. 2011, 2012, 2013a, b, c,2016, 2008;Verma and Kanwar 2008, 2012; ; Kanwar and Verma 2010; Kanwar et al. 2006; 2007a, b, 2008a, b). Energy for biogas generation comes from the sun, through photosynthesis by plants (Angelidaki et al. 2018) (Fig. 3.1). Plant biomass is the storehouse of the solar energy which is being used either directly as feedstock or after partial digestion in animal guts to run the biogas plant. Biogas is obtained from animal waste after it is subjected to anaerobic digestion in digesters. Natural biopolymers, primarily present in the form of lipids, protein and carbohydrates, in the biomass are fermented anaerobically to produce biogas inside the biogas plant, which mainly consists of carbon dioxide and methane (Muñoz et al. 2015). Biogas is a very important alternative source of conventional energy throughout the globe, particularly in developing countries (Khoiyangbam et al. 2011).

Conventional energy sources are causing so much pollution, adversely affecting human health and the environment. Besides this, fossil fuels are expensive and finite, and therefore, power generation based on fossil fuels cannot be sustained in the long run. This highlights the urgency of substituting fossil fuels with renewable resources and using fuel-efficient devices (Kundwa et al. 2018). Advantages of

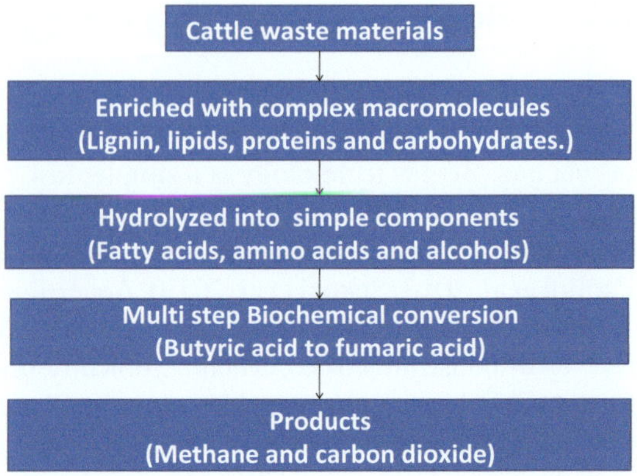

Fig. 3.1 Biogas production overview

using renewable energy are its perennial nature; use of locally available resources that do not need elaborate arrangements for transportation; modular nature, that is, small-scale units and systems can be almost as economical as large-scale ones; suitability for decentralized applications and use in remote areas; low gestation and less capital-intensive nature; environment-friendly nature; and effective usage both for augmenting the availability of power and as a tool for rural development and social justice (Angelidaki et al. 2018). Production of the greenhouse gas methane can be reduced by converting organic wastes into biogas (Khoiyangbam et al. 2011). Methane is found to be 21 times more effective in trapping heat inside the atmosphere as compared to carbon dioxide, so biogas combustion is reducing the greenhouse gas emissions. Additionally, biogas production also reduces insects, pathogens associated with traditional manure stockpile and the odours. Therefore, there has been a growing consensus worldwide which is favouring the use of renewable energy resources as clean and sustainable sources of energy (Lohri et al. 2010).

Biogas technology is providing smoke-free and ash-free kitchens to women and children so that they are not prone to respiratory infections and they are spared from the burden of gathering firewood (Katuwal and Bohara 2009). Biogas technology can reduce the amount of pollutants in the immediate environment and increases sanitation (Lohri et al. 2010). Biogas plant offers a hundred percent greener profile to the major food and industries by converting waste slurry of waste water plants into biogas production. It is not only a matter of supplying energy for the production of facilities or for citizens, but it is also reducing the industrial waste impact on the environment. Biogas technology is the recirculation of organic waste from industry and households in an environmentally friendly way. The slurry produced from biogas technology, when spread on the fields, is less odour producing as compared to the normal slurry (Holm-Nielsen et al. 2009). Plants absorb the fermented slurry better than that of the normal slurry which increases the yield. Slurry produced from biogas technology improves the nitrogen exploitation of soil which reduces leaching and thereby drinking water contamination.

3.3 Various Substrates for Biogas Production

In order to meet the energy requirement of the world today, renewable and eco-friendly sources of energy are needed to be continuously explored (Yadvika et al. 2004). Even though production of biogas is a traditional practice, innovation in these renewable sources further needs to be explored to fulfil the growing need of the increasing population at the global level. The main aspects of biogas production are shown in Fig. 3.2.

Efficient production of biogas needs a complex microbial process. Different microbial consortia are required for producing biogas (Table 3.1). These microorganisms have to work closely together like a team for efficient biogas production (Gopinath et al. 2014). Any disturbance in the teamwork of microorganisms results in reduced biogas production which can lead to the poor performance of the process. For regulating an efficient biogas process, an insight of functioning of microorganisms and the microbiology behind the biogas process should be known. Microorganisms need an appropriate culture medium, for their proper growth and

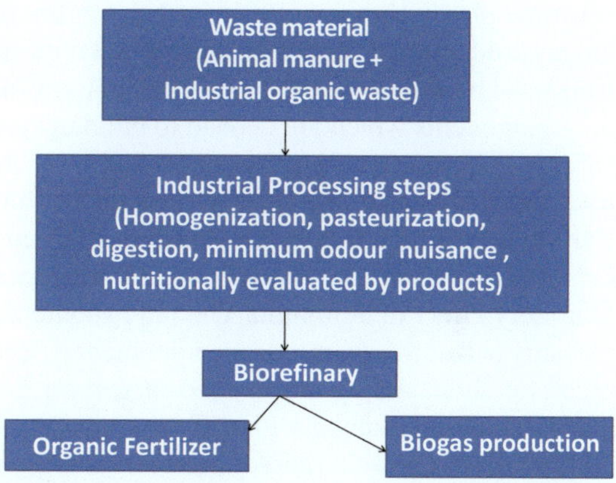

Fig. 3.2 Main aspects of biogas production

Table 3.1 Microorganism involved in biomethane production

Process	Bacteria involved	Product	References
Step 1: hydrolysis	Anaerobic hydrolysis bacteria	Monosaccharides, amino acids and fatty acids	Gopinath et al. (2014)
Step 2: acidity increased	Acid-forming bacteria	Organic acids, carbon dioxide	Bajpai (2017)
Step 3: acetic acid formation	Acetic acid-forming bacteria: *Clostridium* sp. *Lactobacillus*, *Actinomyces*, etc.	Acetic acid, carbon dioxide and hydrogen	Bajpai (2017); Wirth et al. (2012)
Step 4: methane formation	Methanogenic bacteria: *Methanobacillus*, *Methanococcus*, *Methanosarina*, etc.	Methane	Ali Shah et al. (2014); Gopinath et al. (2014); Nurlina et al. (2018)

functioning. A substrate acts as nutrient for the microorganisms which ideally contain an array of elements like energy source, electron acceptors, building blocks (i.e. monomer for new cell synthesis), vitamins and trace elements. Microorganisms can build up new cells as well as produce energy for their growth via accessing substrate. The organic waste used in biogas production acts as a broader substrate for various microorganisms. Organic materials with varied composition can provide more components for growth of microorganisms and thus are suitable for the growth of a diverse range of microorganisms. Composition of organic materials should not vary significantly, and thus a large range of microbial consortia are evolved to substrate-specified behaviour (Schnuera and Jarvis 2010).

Waste materials act as broader substrate for microorganisms for biogas production, and physical and chemical properties of the substrate regulate the process, stability and efficiency of the fermentation process. Substrate composition plays a very important role in the amount as well as the quality of the gas produced (Kushkevych et al. 2018). We can improve the bioprocess yield and maximize bioenergy production and biofertilizer quality based on the selection of right material (substrate) (Schnuera and Jarvis 2010). The physical and chemical properties of the substrate play a significant role in the stability of the bioprocess and subsequently biogas production. Carbon and nutritional requirements of the working microorganisms must be fulfilled by the substrate, in terms of energy source as well as the various components which are needed to build up new cells. The substrate must include various components which are necessary for the microbial enzyme system, like trace elements and vitamins (Osuna et al. 2003; Climehaga and Banks 2008). Carbon to nitrogen (C/N ratio) ratio is of great importance in biogas production. C/N ratio should not be too low, so that the process can easily slow down from the inhibitory effect of ammonia. The ratio should not be too high, so that the microorganisms will suffer from nitrogen deficiency (Yen and Brune 2007). It is difficult to ascertain about the optimum ratio of carbon to nitrogen as it varies with process reaction condition along with different substrates (Schnuera and Jarvis 2010).

Fruit and vegetable market produces a large number of wastes that spread nuisance because of their high biodegradability in municipal landfills (Viturtia et al. 1989). The most promising alternative to process these wastes is the anaerobic digestion. High moisture content and organic content of these wastes are very important for anaerobic digestion. The main advantage of anaerobic digestion is biogas production which can be employed for cooking as well as for the production of electricity. The effluent obtained from this processing (biogas technology) after minor treatment acts as a good source of soil enrichment. Biogas technology offers a cost-effective route for converting waste biomass to meet the global energy demand (Covarti et al. 1999; Semeonove 1999). Callaghan et al. (2002) reported that cattle slurry co-digestion with the waste materials of fruit and vegetable, along with chicken manure, provides the optimal combinations for a batch process in terms of biomethane production and volatile solid reduction. It is reported that cow dung produced more biogas compared to other substrates like tomato waste, food waste and spinach at pH between 6.5 and 7.1 and average temperature of 32.5 °C (Maile and Muzenda 2014).

The popularity of the biomethane technology is receiving more attention with time due to its benefits of converting waste into sustainable biofuel with the possibility of reducing harmful emissions and cogeneration. It is always challenging to operate a large-scale biogas plant, as the main substrate properties can vary with time as well as the new substrate availability demands reformulation of the mixture (Popscue and Jurcoane 2014). High output feedstock crops are needed for operating a biogas plant, due to the presence of micronutrients, as well as for obtaining predictable as well as a constant output. Crop selection for biogas is equally important in getting higher yield of biomethane production. A proportion of the food waste is being utilized by the anaerobic digestion units, which is beneficial for waste management as there is no need to spend money for the evacuation of waste matter. The effluent obtained by this process of anaerobic digestion is beneficial for the environment as it can be used as a valuable fertilizer in agricultural fields (Burgess and Witherford 2013).

Use of a single feedstock presents many challenges in the gas output stability so the main aim while operating a biogas plant should be to use a mixed feedstock. Using a mixed feedstock in anaerobic digestion process allows the process to occur effectively and also maximizes the output. Biomass with high content of sugar or starch is quicker to ferment as compared to the lignocellulosic substrates (Burgess and Witherford 2013). Stillage, which is a distillation waste product from ethanol production, can also be used for the production of biogas. Energy efficiency can be enhanced by linking the biogas and bioethanol production processes (Börjesson and Mattiasson 2007). Stillage as a substrate can work well for a biogas plant, but ammonia concentration becomes too high while using stillage as substrate. But it is very necessary to monitor the concentration of ammonia produced during the process if stillage is being used as a substrate in the anaerobic digestion process. Stillage can work efficiently as a substrate when provided a polysaccharide-enriched co-substrate (Schnuera and Jarvis 2010). Individual feedstocks such as energy crops, manure, fruit waste, vegetable waste and food waste have been reported to work less efficiently and lead to process instability. Co-digestion has been recommended by Sibiya et al. (2017) for enhancing the output of anaerobic digestion process and to deal with process instability.

Algae biomass obtained from waste water media is also reported to act as a substrate for the production of biogas via anaerobic digestion (Zhong et al. 2012). For example, eutrophicated lakes of China are explored for blue-green algae, and algae biomass has been harvested from this natural aquifer for the production of biogas (Zhong et al. 2012; Zeng et al. 2010). Intensive blooming of blue-green algae is a frequent recurrence throughout the world observed in many aquifers. Natural functioning of the aquifers is getting disturbed by the functioning of blue-green algae which is undermining the tourism development and recreation as well as diminishing their industrial applicability (Qin 2009). Guo (2007) described the intensive blue-green algae blooming phenomena in Lake Taihu which is an important lake in China, providing drinking water to the population of over two million people. So, the problem-causing blue green algae in Lake Taihu should be treated, and it is being utilized as a substrate for the production of biogas.

Biogas production from six different energy crops such as sorghum, maize, sunflower, amaranth and sugar beet with pig slurry was measured and sunflower as a substrate was found to produce the highest biomethane. The sorghum substrate and maize produced less biomethane as compared to the sunflower. However, sugar beet as a substrate produced the least biomethane (Mursec et al. 2009). A variety of substrates like chicken dungs, sewage sludge, goat dungs and palm oil mill effluent were analysed and it was found that anaerobic co-digestion of goat dungs produced the highest amount of biogas (Ali et al. 2015). So, it can be concluded that biogas technology is providing a better solution of managing organic waste by keeping the waste material out of landfills as well as reducing our dependence on fossil fuel consumption.

Biogas energy can be used for the improvement of the energy requirement of a farm itself, or the excess energy generated via anaerobic digestion can be sold out to an electricity network. Maize offers higher yields of about 30 tons of total solids per hectare (Landbeck and Schmidt 2005; Amon et al. 2003; Amon et al. 2007). Poor-quality maize grains can also be used for the production of biogas. Non-acidified maize grain silage when processed in an anaerobic chamber exhibited stability as compared to the acidified maize grain silage. Maximum specific biogas production is reported to be 0.72 m^3/kg, from non-acidified maize at 35 °C (Hutnan et al. 2009). Anaerobic digestion technology is not only solving the environmental issues but is also contributing towards green energy production as well as resolving social and economic issues. Cow and pig manure potential has been investigated via anaerobic digestion for biogas production with temperature of 35 ± 1.5 °C, and retention time was kept to be 20 days. Pig manure was found to produce higher biogas (about 2.5 m^3·batch-1) as compared to cow manure (about 2 m^3·batch-1) (Toma et al. 2016).

Straw and forest residues are reported as substrates for biogas production, but the pretreatment of these substrates with organic solvents leads to higher methane yields as compared to the untreated substrates. Pretreatments of lignocelluloses (straw and forest residues) with ethanol and acetic acid produced considerable biomethane (Kabir et al. 2013a). Wool textile wastes were also subjected to anaerobic fermentation for the production of biogas. As we know, wool is mainly composed of keratin which is an extremely strong and resistant structural protein. Thermal, enzymatic and combined treatments were performed to increase the methane yield. Combined thermal and enzymatic treatments showed significantly positive effects on wool degradation, which results in higher methane yields, i.e. 10–20-fold higher methane production, as compared to the yield obtained from the untreated samples (Kabir et al. 2013b). A biogas plant at Fiborgtangen, Norway, is utilizing several kinds of different substrates for biogas production, including animal manures, fish silage, sludge and straw. Characterization of these substrates exhibited that substrates with higher nitrogen contents should be used to the manure to avoid nitrogen deficiency. It has been recommended from the study that a higher proportion of the animal manure should be used to start the fermentation process and fibre sludge should be left out of the reactor feed because it is lower in nutrient value as well as in methane production (Svensson 2012). It is reported that carbohydrates of sugar beet silage which are easily degradable improved the anaerobic digestion process of the grass silage more profoundly. The co-digestion of sugar beet silage with that of the maize silage was insignificant without sugar beet silage (Ahmed et al. 2016). When straw and food wastes are collectively co-digested, higher volumes of methane have been reported as compared to the mono-digestion of the food wastes without causing any effect on the hydraulic retention time. Higher methane yield was reported from the food waste material co-digested with the straw as compared to the food waste alone, indicating that co-digestion of the food waste material with straw showed synergistic effects. Positive co-digestion effects were seen using the straw and food waste when compared with the straw briquettes at higher loading rates. Straw addition during the process of co-digestion with food waste showed lower effects on the microbes (Horvath et al. 2017).

Biogas technology is not just providing fuel, but it is also utilizing biomass from animal husbandry, fishery and forestry and evaluating the agricultural economy, environment protection, recycling of agriculture materials and improvement in the sanitary conditions of rural areas. Developing biogas technology is an alternative programme for providing green energy to rural areas. Concerted efforts are required for the development of biogas technology which will be useful for farming and domestic use as well as can be utilized in small-scale industrial applications (Omer 2017).

3.4 Lantana as a Substrate for Biogas Production

3.4.1 Biology of the Lantana Plant

Lantana plant, belonging to the family Verbenaceae, is a perennial weed of neotropical origin that was spread worldwide through the ornamentals' trade (Hiremath 2018; Goyal et al. 2018). The weed has subsequently occupied rangelands and natural areas in Africa, Asia and Australia due to its invasiveness (Day and Zalucki 2009; Taylor et al. 2019). Some of the major species of lantana weed are *Lantana camara* Linn., *L. indica* Roxb., *L. trifolia*, *L. crenulata*, *L. involucrata*, *L. sellowiana* and *L. lilacina*. *L. camara* var. aculeata, which bears red flowers and commonly known as lantana or wild sage, is the most prevalent (Shackleton et al. 2017; Sharma et al. 1991) (Fig. 3.3). The lantana plant can grow up to 2–5 m and its branches have curved prickles. The leaves may be ovate or oblong in shape with serrated margins and grow 5–9 cm in length. The upper surface of leaves become rough with maturity, cause skin irritation on touching and give off an unpleasant smell. The flowers are 2–6 cm in diameter and in large umbel round shape; the fruits are small drupes, about 3–5 mm in diameter, appearing greenish-blue initially and varying in colour from blue to black on ripening (Nanjappa et al. 2005). The fruits are small drupes, fleshy, about 3 mm in diameter, varying in colour from blue to black.

Lantana camara is a woody shrub that has invaded forests, pastures, orchards, tea gardens, grasslands and wasteland ecosystems in various parts of the major

Fig. 3.3 *Lantana camara* L. plant

climatic regions – tropical, subtropical and temperate – of the world (Angiras 2014). It is an ornamental shrub native of tropical America and was introduced to the different parts of the world from Mexico and Central America, the generic epicentre of lantana, via Europe (Parimoo et al. 2014). It is a fast-spreading noxious weed in the five continents – *Asia*, *Africa*, *North America*, *South America* and *Australia* – where it has posed serious ecological problems (Saha et al. 2018; Day et al. 2003). The lantana weed was introduced in India as an exotic ornamental plant in the nineteenth century (Angiras 2014; Kannan et al. 2013; Kohli et al. 2006), has spread almost all over the country and is a serious weed in the hilly terrains as well as in the plains (Raj et al. 2018; Angiras 2014).

Lantana can survive in very adverse soil and weather conditions and has become naturalized nearly all over the world due to its prolific growth and wide adaptability (Goyal et al. 2018; Mungi et al. 2018). It propagates with seeds as well as by cuttings. Natural propagation occurs through birds and other animals consuming lantana fruit and can spread its seeds across large distances (Diaz 2017; Ramaswami et al. 2016). Lantana plant has a number of adverse interactions with the biosphere. Lantana is considered to be one of the worst weeds because of their ease of invasiveness, possibility for spread ability and socio-economic ecological concerns (Wasantha Rathnayake 2019; Abebe 2018; Fandohan et al. 2015; Dobhal et al. 2010). Dense thickets of lantana overdominate the native bushland and other pastures that render productivity loss of invaded native areas by their competition for resources and allelopathic effects (Anwar et al. 2018; Mishra 2015), further aggravating the problems of farmers and leading to fodder scarcity.

Lantana is documented as one of the most noxious weeds in the world (Parimoo et al. 2014). Though not readily eaten, the plant may be fed upon by cattle in case of scarcity of grazing land (Kumar et al. 2018). The prevalent red-flower variety is the noxious, causing severe toxicity in grazing animals (Sharma et al. 2017; Baruti et al.

2018). The plant causes hepatotoxicity in animals and has adverse skin interactions with humans as well (Kumar et al. 2016). The ripe blue-black fruit of lantana are consumed by children and sometimes by adults without any overt harmful effect, but ingestion of the green berry has led to human fatalities (Diaz 2017). Recently, lantana stem secretions are found to adversely affect birds which also foraged in lantana (Taylor et al. 2019).

The toxic constituents present in the leaves of lantana have been identified as pentacyclic triterpenoids called lantadenes and are attributed to the noxious effects of lantana in grazing animals (Sharma et al. 2017). Lantadenes present in the lantana plant are in the form of lantadene A to D and reduced lantadene A to B which are esters of 22-hyroxyoleanonic acid (Fig. 3.4) which form the core structure of lantadenes that differ in their side chains at C-22 (Kumar et al. 2018). Lantadene A and lantadene B are the most abundant triterpenoids in a number of varieties of lantana plant and the two related putative toxins, viz. lantadene C and lantadene D, are also present in sizeable amounts while reduced lantadene A and B are the minor constituents of the leaves of *L. camara* L. (Sharma et al. 2000a, b).

Some researchers have advocated eradication of the lantana plant while others have laid emphasis on its utilization for various purposes. The alien invasive plants like lantana have been detected and mapped for their occurrence, spatial distribution and abundance through satellite remote sensing recently which is gaining

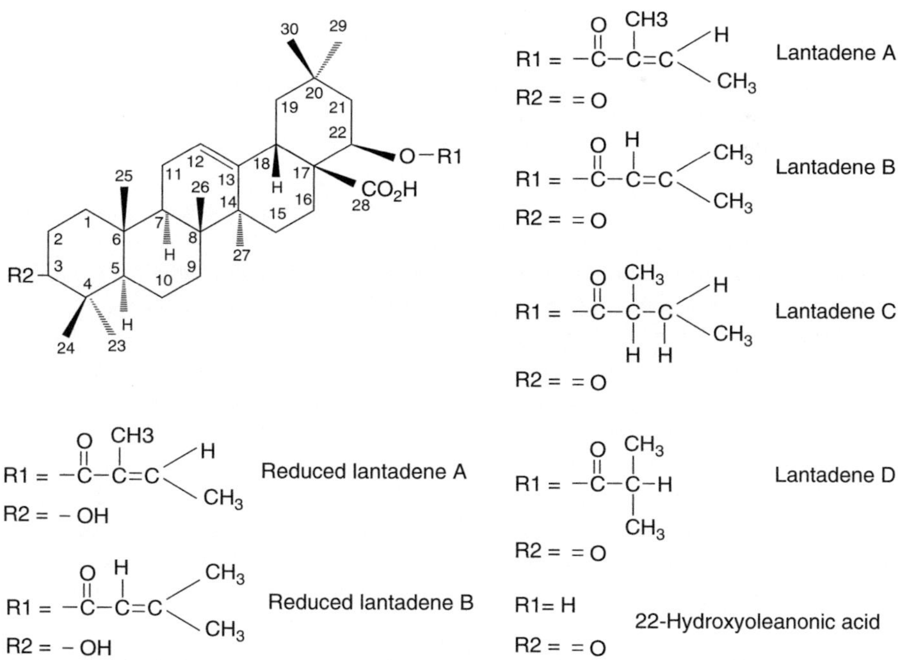

Fig. 3.4 Chemical structures of lantadenes

substantial attention globally (Royimani et al. 2018). The methods used for eradication and control of the plant include mechanical removal, chemical control with herbicides and biological control via lantana-eating insects (Raj et al. 2018; van Wilgen and Brian 2018; Schwarzländer et al. 2018). The continued use of chemicals can lead to selection of varieties resistant to herbicides and result in numerous toxicological and carcinogenic consequences on the environment (Rao et al. 2018); hence an integrated approach is the best management option for lantana. Singh et al. (2018) advocated eco-restoration of the invaded forest/plantation communities through artificial introduction of native species that will help in strengthening the management of forest invasive species like lantana in protected areas and plantation. The field experiments conducted showed that *Olax scandens*, *Petalidium barlerioides* and *Helicteres isora* have potential to outcompete lantana.

Taking into account its profuse growth, lantana weed has been utilized for various purposes. Lantana biomass has potential for generating organic fertilizer via composting and vermicomposting and its use as an organic manure was found effective in improving soil physical properties resulting in higher crop yields (Hussain et al. 2016; Rameshwar and Argaw 2016; Bhushan and Sharma 2002). The chemical compounds attributed to toxic and allelopathic activities of lantana weed are degraded during vermicomposting process converting into an organic fertilizer that has improved the growth and fruit production as compared to inorganic fertilizer (Hussain et al. 2015; Karthikeyan et al. 2014). *Lantana* leaves and twigs are frequently employed for mulching in India (Thankamani et al. 2016; Kumar et al. 2009).

Lantana exhibits antimicrobial activities also. The different extracts from the leaves of *L. camara* were found to show their effect against specific bacteria like *Staphylococcus aureus* and *Pseudomonas aeruginosa* and fungal strains, viz. *Aspergillus fumigatus* and *A. flavus*, with methanolic extracts giving best results (Naz and Bano 2013). Bio-agents based on lantana extracts were used to manage downy mildew of pea (Patil et al. 2017) and anthracnose caused in mango by *Colletotrichum gloeosporioides* (Deressa et al. 2015).

Lantana camara also holds a potential of wide range of actions against a number of plant pests and insects and can be used in the development of bioinsecticides. High doses of the impurified aqueous leaf extract of *L. camara* caused considerable stoppage of root galling and egg production by *Meloidogyne incognita* on a susceptible tomato cultivar, making this an environment-friendly approach suitable for root-knot disease management in tomato (Udo et al. 2014). Assessment of *L. camara* extracts as seed soaking treatment against reniform nematode infesting cowpea reduced the reproduction and populations of the nematode (Patil et al. 2016). Lantana leaf extracts specifically at 1:1 leaf to water ratio were found effective in controlling corn borer and corn earworm (Lucas et al. 2018). The *hexane extract* of *L. camara* caused mortality, slowed its development and growth and resulted in birth anomalies in *D. koenigii* (Kayesth and Gupta 2018). Oils and extracts serve as mosquito repellent. The combinations of *Ocimum gratissimum* and *L. camara* were tested as natural repellents against mosquitoes and can be an alternative to conventional repellents (Pavela and Benelli 2016; Keziah et al. 2015). Besides this the extracts also exhibited larvicidal activity against mosquito species (Hemalatha et al. 2015). Assessment of lantana plant extracts for its antihelminthic activity revealed promising effects against *Fasciola hepatica* (Alvarez-Mercado et al. 2015).

L. camara have been utilized in the activated carbon process and can be used for adsorptive removal of pollutants and heavy metals (Saini et al. 2017) and an effective adsorbent for removal of phenol from aqueous solutions (Girish and Murty 2015). High yields of fermentable sugars from lantana indicated its feasibility in the production of alcoholic biofuels (Borah et al. 2016). Further, it has been used in the fabrication of gold and silver nanoparticles which can be used for antimicrobial and catalytic activities (Shriniwas and Subash 2017; Kumar et al. 2016). The *Lantana camara* Linn. root extract-derived gold nanoparticles possessed significant in vitro antioxidant and cytotoxic properties and could be considered as potential alternate for the research and development of anticancer molecules (Ramkumar et al. 2017). Owing to their antioxidant, anti-inflammatory, antipyretic and analgesic attributes, the extracts from *L. camara* can be used for various medicinal purposes (de Sousa et al. 2018; Shamsee et al. 2018; Arbiastutie et al. 2017).

Lantana exhibits the potential for extraction of heavy metals and has been proved as an efficient phytoextractor for copper, zinc, chromium and manganese from fly ash-amended soil (Pandey et al. 2016; Pandey and Bhattacharya 2018). The Irula tribe in southern India are known for their crafting skills using this invasive shrub (Rangasamy et al. 2019), and its local use has also offered an opportunity for its control as reduced density of lantana was reported in harvested sites (Kannan et al. 2016).

3.4.2 Lantana as a Substrate for Biogas

Biogas production is an anaerobic digestion technology which involves a biochemical process that changes the waste biomaterials into a clean and renewable bioenergy (Hagos et al. 2017). Besides providing energy and manure, biomethanation of decomposable biological waste materials offers several social and environmental benefits (Mittal et al. 2018; Siegmeier et al. 2015). At present, the total biogas production in India is not up to the mark, and production per year is considerably low, i.e. 2.07 billion m^3, as compared to its estimated potential per year of 29–48 billion m^3 (Mittal et al. 2018). In India, mainly cattle dung is used as an organic substrate for the production of biogas. However, insufficient supply of this substrate is a primary issue in the successful operation of biogas plants and poor acceptance of the technology among the rural communities. In areas with shortage of cattle manure, anaerobic co-digestion can be the likely option where alternative feedstocks can be utilized for the anaerobic digestion process (Raha et al. 2014). Co-digestion benefits by balancing several parameters of reaction condition such as carbon-to-nitrogen ratio, pH, bacterial diversity, etc. that results in maximizing biogas generation if the ideal cheap substrate of organic waste mixture composition is identified (Muske and Rao 2019). Moreover, cattle dung is a poor substrate for biogas production as it is already digested in rumen; its supplementation with complex organic wastes has resulted in an augmentation in the quantity as well as quality of the biogas (Tufaner and Avsar 2016). So other available organic substrates have been explored for their biogas potential. The biomass of lantana is one such substrate evaluated for anaerobic digestion (Saha et al. 2018).

Lantana is a luxuriantly growing weed throughout the world and its biomass (leaves) is available almost throughout the year (Raghubanshi et al. 2005). Due to its high potential for growth and easy availability, research was done to work out the feasibility of producing methane from lantana biomass. Since lantana is a woody weed, leaves and soft twigs have been reported to be used for biogas production (Saha et al. 2018). The variation in the monthly chemical profile of lantana leaves was studied by Saini (2001), and it was reported that the total and volatile solids were in the range of 28.36–38.85% and 88.38–92.68%, respectively. There was a consistent rise in total solids with the maturity of leaves in every subsequent month. The monthly variation in volatile solids was found less compared to that in total solids.

The lignocellulosic structures in the fibrous materials of the various substrates used for biogas production may hinder their biodegradability (Yadav et al. 2019a; Yadav et al. 2019b). The substrates, therefore, require pretreatment(s) before the anaerobic digestion process which may facilitate their hydrolysis, improving the yield of biogas production (Venturin et al. 2019). Lantana biomass also needs to be predigested or pretreated before anaerobic digestion as reported by some workers (Saini et al. 2003; Kalia and Singh 1998; Kanwar and Kalia 1988; Dar and Tandon 1987). Pretreatment is a prerequisite step to obtain improved yield of biogas production. Pretreatment is done by converting the complex organic structure into a simpler structure that is accessible to microbial degradation (Yadvika et al. 2004). For predigestion, the chopped biomass was decomposed for 1 week in an open environment condition at room temperature (21–25 °C) by making heaps with its native microflora in some studies (Kanwar and Kalia 1988; Kalia and Singh 1998; Saini et al. 2003), whereas other researchers gave alkali treatment to the lantana biomass (Dar and Tandon 1987; Deshmukh 2013). Similar pretreatments have also been used in other substrates used for biogas production. Dahunsi et al. utilized *Chromolaena odorata* shoot for biogas generation and reported an increase in biogas from the substrate on thermo-alkaline pretreatment yielding over 49.20% biogas. The augmentation in methane production by 29% from sunflower stalks was also reported by Monlau et al. (2015) after alkaline pretreatment. The pretreatment enhances the solubilization in the substrate; the soluble sugars play a vital role during the hydrolysis and acidogenesis stages of anaerobic digestion leading to an increase in the population, diversity and activity of microbes which results in the production of intermediate acids for the acetogenesis and methanogenesis stages (Monlau et al. 2012).

Saini et al. (2003) used both fresh and predigested lantana for biogas production. The comparison of the chemical profile of fresh and predigested lantana in the study showed an increase in total solids and a reduction in volatile solids after predigestion of the lantana biomass. Total and volatile solids in fresh lantana were reported to be 35.18 and 90.25%, respectively, as compared to 37.87 and 88.47% in the case of predigested lantana. The carbon and nitrogen contents of fresh lantana were

50.14 and 1.4%, respectively, with a C:N ratio of 35.81, and that of partially digested lantana were 49.15 and 1.54%, respectively, with a C:N ratio of 31.90.

A reduction of volatile solids after the process of aerobic decomposition led to comparatively low carbon content in predigested lantana with respect to fresh lantana (Saini et al. 2003; Saini 2001). However, they found a slight increase in nitrogen content on predigestion. These changes resulted in lowering of the C/N ratio which probably helped in developing conducive conditions for the generation of biogas from this substrate. Partial aerobic decomposition aids in tissue disintegration making it available for microbial breakdown. Manifold increase in the microbial populations of bacteria, fungi and actinomycetes after predigestion might be responsible for the changes observed in the chemical profile of a substrate (Saini 2001). Similar changes were observed in the chemical profile of lantana after aerobic decomposition by other workers (Kanwar and Kalia 1988) also. The same trend was noticed in the case of *Ageratum* after predigestion of 4–5 days (Kalia and Kanwar 1990).

The predigested and fresh lantana were co-digested with cattle manure for generation of biogas by Saini et al. (2003) and reported that the cattle dung had total and volatile solids less than the fresh and predigested lantana (i.e. 20.18 and 84.90%, respectively); the carbon and nitrogen contents were slightly lower (47.17 and 0.98%, respectively) with a carbon to nitrogen ratio of 48.13. To optimize the best C/N ratio, dairy manure is mixed with substrate for higher biogas yield (Saha et al. 2018). The carbon to nitrogen ratio has great effect on biogas production as reported by Syafrudin et al. (2018) in their studies on biogas production for water hyacinth.

Partially decomposed lantana foliage in combination with cow dung has been utilized in biogas production in different studies (Kalia 1983; Kanwar and Kalia 1988). Dar and Tandon (1987) utilized wheat straw, lantana residue, apple and peach leaf litter pretreated with 1% solution of sodium hydroxide for 7 days prior to anaerobic fermentation with cattle dung at ambient temperature. High degradability of organic matter was achieved in the slurries with pretreated plant residues and lantana residues which resulted in twofold increase in biogas production. Lantana slurry showed the highest yields with 63–66% of biomethane and improved efficiency was 31–42% higher as compared to cattle manure alone. Deshmukh (2013) also used air-dried lantana plant samples pretreated with 1% sodium hydroxide solution for 8 days at 32–34 °C for anaerobic digestion. Pretreated lantana supplemented with cattle dung showed 13.40 L/100 g biogas production in 7 weeks with 58.3% methane and 11.72% carbon dioxide, while control (cattle dung alone) showed 10.60 L/100 g biogas. Maximum gas production was reported in the 5th week.

To study the suitability of lantana substrate for maximizing biogas production, laboratory-scale anaerobic batch digestion studies were performed with both fresh and predigested lantana leaves co-digested with cattle dung for a duration of 50 days at 35 ± 1 °C by Saini et al. (2003). They subjected different concentrations of fresh lantana (50%, 75%) and predigested lantana (25%, 50%, 75%) biomass along with cattle dung to anaerobic batch digestion keeping total solids constant in each digester and seeding with 10–15% (v/v) inoculum (obtained from a biogas plant continuously operating on pure cattle dung). Saini et al. (2003) reported gas production only in case of digesters fed with pure cattle dung and those supplemented with predigested lantana, but only up to a concentration of 50% lantana biomass (w/w, on dry weight basis), where the initial and final pH values were found within the desired range. The initial pH values in digesters supplemented with 25 and 50% predigested lantana were 7.88 ± 0.02 and 7.30 ± 0.02, respectively, which increased to 8.63 ± 0.03 and 8.55 ± 0.04, respectively, after anaerobic digestion of 50 days. While the fall in pH implies the synthesis and accumulation of volatile fatty acids, the increase indicates utilization of volatile fatty acids with the aid of methanogenesis process (Adanikin et al. 2017) leading to efficient methane production and operation of the digesters. The acidity and alkalinity levels were found within the optimum limits (Riggio et al. 2017) required for successful operation of digesters by Saini et al. (2003).

The total accumulated biogas from digesters fed with 25 and 50% predigested lantana was 8.37 and 60.70% more than the digester fed with only pure cattle dung, the total methane content being 13.18 and 21.09 l in contrast to 10.58 l from pure cattle dung (Saini et al. 2003). Hence, the supplementation of cattle dung with an organic substrate results in improvement in biogas production (Awasthi et al. 2018; Barua et al. 2019; Hagos et al. 2017). A comparison of the various organic wastes to derive an optimal ratio for biogas production revealed that a mixture of 50% poultry dropping and 50% weeds produce high biogas production (Okonkwo et al. 2018).

The weekly profile of the biogas production with lantana as substrate was also studied by Saini (2001). The major portion of biogas (about 70% of the total biogas) was produced during the early phase (first and fourth week) of digestion with pure cattle dung. However, a lag phase was seen in biogas production in digesters supplemented with predigested lantana at 25 and 50%. This lag phase was more conspicuous in the digesters fed with 50% lantana which was explained by the fact that the organisms in the inoculum (procured from biogas plant run on pure cattle dung) used were already acclimatized to the substrate, i.e. pure cattle dung, and initiated methanogenesis in a short duration, whereas in cases where lantana was supplemented with cattle dung, they probably needed more time for their adjustment to the new substrate lantana for its biomethanation. A similar lag phase was reported when cattle dung was supplemented with *Parthenium* (Gunaseelan 1987). The adaptation/acclimatization of microbial population during biomethanation of different organic wastes has been shown to accelerate the process of biogas production (Kalia et al. 1992; Kanwar and Guleria 1995).

The biogas potential of partially decomposed *Lantana* is not less than that of other weeds. The biogas produced per kg of total solids in case of lantana is 339.6 l as compared to 430.0, 241.0 and 94.0 l for *Parthenium* (Gunaseelan 1987), *Ageratum* (Kalia and Kanwar 1990) and *Eupatorium* (Jagadeesh et al. 1990), respectively. All the above studies were in batch digesters but with different levels of supplementation and treatments. To sum up, lantana is suitable as a substrate with cattle manure for the production of biogas, on the condition that it is pretreated before the anaerobic digestion process and added up to a concentration of 50% (w/w, on dry weight basis) (Saini et al. 2003). Use of lantana substrate for biogas may help in overcoming the acute problem of underfeeding of biogas digesters to a certain extent and also in the management of the weed.

For the bioconversion of any complex organic waste into methane, sufficient quantity of inoculum is required to complete the process (Saha et al. 2018). Since methanogenesis requires distinct groups of bacteria (Castellano-Hi nojosa et al. 2018; Dong et al. 2019; Yadvika et al. 2004), the presence of an adequate population of required bacteria to break down the primary as well as intermediate fermentation products must be taken care of. Biodegradable organic wastes of any kind can be assessed for its biogas production potential by fermenting the material in batch digesters under anaerobic conditions with an appropriate amount of inoculum. Córdoba et al. (2018) demonstrated that in cases where a low amount of inoculum is used, the adaptation of microorganisms resulted in much higher than presumed delay in methane production and extended the time required to attain satisfactory performance of the process.

Saini (2001) studied different inoculum/substrate (I/S) ratios to determine their effect on production of biogas from substrate lantana with its (predigested lantana leaf powder) concentrations ranging from 15.10 to 66.50 g VS/l and keeping inoculum (procured from biogas plant operating on pure cattle dung) volume constant, i.e. 500 ml. The experiment revealed that with the decrease in I/S ratio, biogas production increased as a higher volume of inoculum accelerates the digestion process and the rate of gas production (Abudi et al. 2018). However, the total methane produced showed a consistent pattern between I/S ratio of 50.10–79.11 but further increase in I/S ratio reflected a reduction in total accumulated methane content (Fig. 3.3). The substrate availability per unit of inoculum for biomethanation is decreased with the increase in I/S ratio. The quality of the biogas produced also improved with the increment in substrate load in the digesters. Rapid progress of anaerobic batch digestion for methane production depends primarily on a favourable pH range. Other studies have also indicated an increase in methane yield up to an optimum value of I/S and suggested I/S ratio as a vital component for the assessment of biodegradability (Bhui et al. 2018). Similarly, biochemical methane potential test for the terrestrial weed *Ageratum conyzoides* carried out to optimize ideal food to microorganism (F/M) ratio revealed that the F/M ratio of two acquired maximum biogas production from the anaerobic digestion of *A. conyzoides* biomass with cow dung as the source of microorganisms and further focused attention to the significance of having adequate population of necessary microbial flora, enzymes and growth factors to break down complex biomolecules (Saha et al. 2018).

As microorganisms act on lantana biomass for biomethanation, biotransformation of lantana toxins after the anaerobic digestion process has also been reported by Saini et al. (2003). An overall reduction in the amounts of lantadene A, lantadene B and reduced lantadene B was found in lantana biomass on HPLC analysis. However, an elevated level of reduced lantadene A was also observed, suggesting the conversion of lantadene A to reduced lantadene A during the anaerobic digestion process; formation of an unknown compound was also reported in digesters where biomethanation had occurred (Saini et al. 2003). Similar biotransformation of lantadene A to reduced lantadene A was detected by Sharma et al. (1999). Researchers used extracts of caecum and large intestine at the time of lantadene A administration to guinea pigs. In vitro study of lantadene A with caecal and intestine of guinea pig elicited conversion of lantadene A to reduced lantadene A under anaerobic conditions. However, it was not as marked as that found in in vivo conditions. The anaerobic digestion of lantana in batch digesters exhibited marked conversion, as reported earlier under in vivo conditions (Sharma et al. 1999; Sharma et al. 2000a, b). Degradation of lantana toxin, lantadene A, by aerobic Gram-negative bacteria, i.e. *Pseudomonas pickettii* (Sharma et al. 1997) and *Alcaligenes faecalis* (Singh et al. 1999; Singh et al. 2001a), and seven white-rot fungi (Singh et al. 2001b) have also been reported.

3.5 Challenges in Lantana for Potential Biogas Production

The collection of lantana biomass is not easy as only the soft twigs and leaves are suitable for biogas production (Saini et al. 2003). The hard stems of the plant have to be sorted out. Moreover, certain compounds in the plant may cause skin and allergy problems (Ghosal et al. 2016). The lantana biomass requires pretreatment before anaerobic digestion as mentioned earlier because the lignocellulosic structures may hinder its biodegradation and biogas formation (Yadav et al. 2019a, b).

The fresh biomass of lantana has not been found suitable for biogas generation by Saini (2001). The fresh lantana at 50 and 75% with cattle dung did not produce any biogas in laboratory-scale studies conducted by Saini et al. (2003). They noticed a drop in pH in digesters supplemented with fresh lantana biomass after anaerobic digestion for 50 days at 35 ± 1 °C. The initial pH (7.33 and 7.55) in these digesters lowered to 5.08 and 5.03, respectively, after 50 days of digestion. This sudden reduction in pH made the conditions unfavourable for the methane-producing microorganisms. The growth rate of methanogens reduces considerably at pH levels below 6.5 and thus affects methane production. The optimum pH during the anaerobic digestion process augments the required microbial activities leading to improvement in biogas yield (Riggio et al. 2017; Mao et al. 2015; Zonta et al. 2013). A pH in the range of 6.5–8.0 is necessary for efficient performance of the methanogens during the process (Zonta et al. 2013; Kothari et al. 2014). Saini et al. (2003) reported that though predigested lantana produced biogas up to 50% concentration along with cattle dung, a higher concentration of predigested lantana, i.e. 75%, showed no biogas production. Here also similar toxic conditions emerged as observed in case of digesters fed with fresh lantana; the pH dropped from 7.22 to 5.10 after 50 days of anaerobic digestion.

The lower values for pH in anaerobic digesters result from VFA accumulation (Castellano-Hinojosa et al. 2018). Such conditions were noticed by Saini et al. (2003) in digesters supplemented with predigested lantana at 75% and fresh lantana (at 50%, 75%) which failed to produce any gas. Here the volatile fatty acids (acidity) were very high (925 ± 64.4, 1410 ± 29.3 and 1130 ± 12.6 mg/l, respectively) after the anaerobic digestion, as compared to the digesters supplemented with predigested lantana at 50% and 75% where volatile fatty acids were 113 ± 19.5 and 120 ± 9.3, respectively, and produced biogas. The volatile fatty acids in the failed digesters cause reduction of methane-producing microorganisms as reported by Guo et al., hence no biogas production. Moreover, the buffering capacity of these digesters measured in terms of alkalinity (Saini et al. 2003) was also not enough to safeguard against the drop in pH due to accumulated volatile fatty acid; for optimum performance of anaerobic digesters, the ratio of volatile fatty acids to total alkalinity should be maintained <0.1–0.2 (Grasiuset al. 1997), but in the case of digesters which failed to produce any gas, it was more than 0.4. Similar results have been noticed in other studies when fresh *Ageratum* was subjected to anaerobic digestion (Kalia and Kanwar 1990). The higher concentration of volatile fatty acids during anaerobic digestion leads to a decrease in pH values and results in low production of methanogenesis (Franke-Whittle et al. 2014; Capson-Tojo et al. 2017; Bhui et al. 2018). The volatile fatty acids are the intermediate products in the anaerobic digestion process (Wagner et al. 2014) and are known to play a major role in methane production.

3.6 Conclusion

The lantana plant offers a promising substrate for biogas production, in particular to highly invaded areas of the weedy shrubs. However, the major source of lantana biomass for biogas production is currently restricted to certain parts of weedy plant, in particular to the soft twigs and leaves, in which sorting out the hard stems of the plant needs labour. Moreover, certain compounds present in the lantana plant may cause skin and allergy problems and, thus, need a special attention at the time of collection in order to avoid any potential hypersensitive skin reactions. The lantana biomass requires pretreatment prior to anaerobic digestion because the lignocellulosic structures hinder its biodegradation and biogas formation. Few studies have been done in the bioprospecting of biogas production. More consolidated efforts need to be done in understanding the correlation of lantana biomass pretreatment with respect to methane production in the near future.

References

Abebe FB (2018) Invasive *Lantana camara* L. shrub in Ethiopia: ecology, threat, and suggested management strategies. J Agric Sci 10(7):184–195

Abudi ZN, Hu Z, Abood AR, Liu D, Gao A (2018) Effects of alkali pre-treatment, total solid content, substrate to inoculum ratio, and pH on biogas production from anaerobic digestion of mango leaves. Waste Biomass Valoriz:1–11

Adanikin BA, Ogunwande GA, Adesanwo OO (2017) Evaluation and kinetics of biogas yield from morning glory (*Ipomoea aquatica*) co-digested with water hyacinth (*Eichhorniacrassipes*). Ecol Eng 98:98–104

Ahmed S, Einfalt D, Kazda M (2016) Co-digestion of sugar beet silage increases biogas yield from fibrous substrates. Biomed Res Int 2016:1–11

Ali Shah F, Mahmood Q, Maroof Shah M, Pervez A, Ahmad Asad S (2014) Microbial ecology of anaerobic digesters: the key players of anaerobiosis. Sci World J 2014:183752. https://doi.org/10.1155/2014/183752

Ali MYM, Hanafiah MM, Wen YH, Idris M, Aziz NIHA, Halim AA, Lee KE (2015) Biogas production from different substrates under anaerobic conditions. Third international conference on chemical, agricultural and medical sciences, Singapore, pp 54–56

Alkanok G, Demirel B, Onay TT (2014) Determination of biogas generation potential as a renewable energy source from supermarket wastes. Waste Manag 34:134–140

Alvarez-Mercado JM, Ibarra-Velarde F, Alonso-Díaz MÁ, Vera-Montenegro Y, Avila-Acevedo JG, García Bores AM (2015) In vitro antihelmintic effect of fifteen tropical plant extracts on excysted flukes of *Fasciola hepatica*. BMC Vet Res 11(1):45

Amon T, Kryvoruchko V, Amon B, Moitzi G, Buga S, Lyson DF, Hackl E, Jeremic D, Zollitsch W, Potsch E (2003) Biogas production from the energy crops maize and clover grass institute fur land- und UmveltundEnergietechnik. Universitat fur Bodenkultur, Wienna

Amon T, Amon B, Kryvoruchko V, Machmuller A, Hopfner-Sixt K, Bodiroza V, Hrbek R, Friedel J, Potsch E, Wagentristl H, Schreiner M, Zollitsch W (2007) Methane production through anaerobic digestion of various energy crops grown in sustainable crop rotations. Bioresour Technol 98:3204–3212

Angelidaki I, Treu L, Tsapekos P, Luo G, Campanaro S, Wenzel H, Kougias PG (2018) Biogas upgrading and utilization: current status and perspectives. Biotechnol Adv 36:452–466

Angiras NN (2014) Management of perennial weeds under non-cropland hill ecosystems. Indian J Weed Sci 46(1):52–60

Anwar T, Ilyas N, Qureshi R, Munazir M, Anwar P, Rahim BZ, Ansari K, Panni MK (2018) Allelopathic potential of *lantana camara* against selected weeds of wheat crop. Appl Ecol Environ Res 15:5405–5421

Arbiastutie Y, Marsono D, Hartati MS, Purwanto R (2017) The potential of understorey plants from Gunung Gede Pangrango National Park (West Java, Indonesia) as cervixs anticancer agents. Biodiversitas 18(1):109–115

Awasthi SK, Joshi R, Dhar H, Awasthi MK, Varjani S, Sarsaiya S, Zhang Z, Kumar S (2018) Improving methane yield and quality via co-digestion of cow dung mixed with food waste. Bioresour Technol 251:259–263

Bajpai P (2017) Anaerobic technology in pulp and paper industry, SpringerBriefs Applied Sciences and Technology. https://doi.org/10.1007/978-981-10-4130-3_2

Barua VB, Rathore V, Kalamdhad AS (2019) Anaerobic co-digestion of water hyacinth and banana peels with and without thermal pretreatment. Renew Energy 134:103–112

Baruti M, Singh B, Bhuyan M, Borthakur A, Bhuyan D, Chutia JP (2018) Management of *Lantana camara* poisoning in a bull. Int J Chem Stud 6(1):950–952

Bhui I, Mathew AK, Chaudhury S, Balachandran S (2018) Influence of volatile fatty acids in different inoculum to substrate ratio and enhancement of biogas production using water hyacinth and salvinia. Bioresour Technol 270:409–415

Bhushan L, Sharma PK (2002) Long-term effects of lantana (*Lantana* spp. L.) residue additions on soil physical properties under rice–wheat cropping. Soil Tillage Res 65(2):157–167

Borah AJ, Singh S, Goyal A, Moholkar VS (2016) An assessment of the potential of invasive weeds as multiple feedstocks for biofuel production. RSC Adv 6(52):47151–47163

Börjesson P, Mattiasson B (2007) Biogas as a resource-efficient vehicle fuel. Trends Biotechnol 26:7–13

Burgess J, Witherford S (2013) Biogas in practice, Hanover

Callaghan FJ, Wase DAJ, Thayanithy K, Forster CF (2002) Continuous co digestion of cattle slurry with fruit and vegetable wastes and chicken manure. Biomass Bioenergy 27:71–77

Capson-Tojo G, Ruiz D, Rouez M, Crest M, Steyer JP, Bernet N, Delgenès JP, Escudié R (2017) Accumulation of propionic acid during consecutive batch anaerobic digestion of commercial food waste. Bioresour Technol 245:724–733

Castellano-Hinojosa A, Armato C, Pozo C, González-Martínez A, González-López J (2018) New concepts in anaerobic digestion processes: recent advances and biological aspects. Appl Microbiol Biotechnol 102:5065–5076

Chae KJ, Jang A, Yim SK, Kim IS (2008) The effects of digestion temperature and temperature shock on the biogas yields from the mesophilic anaerobic digestion of swine manure. Bioresour Technol 99:1–6

Climehaga MA, Banks CJ (2008) Anaerobic digestion of catering wastes: effect of micronutrients and retention time. Water Sci Technol 57:698–692

Córdoba V, Fernández M, Santalla E (2018) The effect of substrate/inoculum ratio on the kinetics of methane production in swine wastewater anaerobic digestion. Environ Sci Pollut Res 25:21308–21317

Covarti A, Delborghi A, Zilli M, Arni S, Delborghi M (1999) Anaerobic digestion of the vegetable fraction of the municipal refuses: mesophilic verses thermophilic conditions. Bioprocess Eng 21:371–376

Dar GH, Tandon SM (1987) Biogas production from pretreated wheat straw, lantana residue, apple and peach leaf litter with cattle dung. Biol Wastes 21:75–83

Day MD, Zalucki MP (2009) *Lantana camara* Linn. (Verbenaceae). In: Muniappan R, GVP R, Raman A (eds) Biological control of tropical weeds using arthropods. Cambridge University Press, Cambridge, pp 211–246

Day MD, Wiley CJ, Playford J, Zalucki MP (2003) *Lantana*: current management, status and future prospects, vol 5. Australian Centre for International Agricultural Research, Canberra, pp 1–20

de Sousa EO, de Almeida SCX, Damasceno SS, Nobre CB, da Costa JGM (2018) *Lantana camara* L. and *Lantana montevidensis* (Spreng.) Briq. In: Albuquerque U, Patil U, Máthé Á (eds) Medicinal and aromatic plants of South America. Medicinal and aromatic plants of the world, vol 5. Springer, Dordrecht

Deressa T, Lemessa F, Wakjira M (2015) Antifungal activity of some invasive alien plant leaf extracts against mango (Mangifera indica) anthracnose caused by *Colletotrichum gloeosporioides*. Int J Pest Manag 61(2):99–105

Deshmukh HV (2013) Studies on the biomethanogenic potential of some ornamental plants. Bioscan 8(2):417–420

Diaz JH (2017) The ham and eggs plant, *Lantana camara*. Wilderness Environ Med 28(3):273–274

Dobhal PK, Kohli RK, Batish DR (2010) Evaluation of impact of *Lantana camara* L. invasion on four major woody shrubs along Nayar river of Pauri Garhwal in Himalaya. Int J Biodivers Conserv 2:166–172

Dong L, Cao G, Guo X, Liu T, Wu J, Ren N (2019) Efficient biogas production from cattle manure in a plug flow reactor: a large scale long term study. Bioresour Technol 278:450–455

Fandohan AB, Koko IKED, Avocevou-Ayisso C, Gouwakinnou GN, Savi MK, Assogbadjo AE, Kakai RG (2015) *Lantana camara* (Verbenaceae): a potential threat to the effectiveness of protected areas to conserve flora and fauna in Benin. Agron Afr 27(2):115–126

Fierro J, Gómez X, Murphy JD (2014) What is the resource of second-generation gaseous transport biofuels based on pig slurries in Spain. Appl Energy 114:783–789

Franke-Whittle IH, Walter A, Ebner C, Insam H (2014) Investigation into the effect of high concentrations of volatile fatty acids in anaerobic digestion on methanogenic communities. Waste Manag 34(11):2080–2089

Ghosal KJ, Saha B, Bhattacharya SG (2016) Clinical and immuno-proteomic approach on *Lantana camara* pollen allergy—a major health hazard. Allergy Asthma Clin Immunol 12:33

Girish CR, Murty VR (2015) Adsorption of phenol from aqueous solution using Lantana camara, forest waste: packed bed studies and prediction of breakthrough curves. Environ Process 2(4):773–796

Gopinath LR, Christy PM, Mahesh K, Bhuvneshwari R, Divya D (2014) Identification and evaluation of effective bacterial consortia for efficient biogas production. IOSR J Environ Sci Toxicol Food Technol 8:80–86

Goyal N, Esler KJ, Sharma GP (2018) What drives performance potential of *Lantana camara* L. (sensulato) in the invaded range? Trop Ecol 59(1):57–68

Gunaseelan VN (1987) *Parthenium* as an additive with cattle manure in biogas production. Biol Wastes 21:195–202

Guo L (2007) Doing battle with the green monster of Taihu Lake. Science 317(5842):1166

Hagos K, Zong J, Li D, Liu C, Lu X (2017) Anaerobic co-digestion process for biogas production: Progress, challenges and perspectives. Renew Sust Energ Rev 76:1485–1496

Hemalatha P, Elumalai D, Janaki A, Babu M, Velu K, Velayutham K, Kaleena PK (2015) Larvicidal activity of Lantana *camaraaculeata* against three important mosquito species. J Entomol Zool Stud 3(1):174–181

Hiremath AJ (2018) The case of exploding *lantana* and the lessons it can teach us. Resonance 23(3):325–335

Holm-Nielsen JB, Al Seadi T, Oleskowicz-Popiel P (2009) The future of anaerobic digestion and biogas utilization. Bioresour Technol 100:5478–5484

Horvath IS, Castillo MDP, Schnurer A, Agnihotri S, Ylitervo P, Edstrom M (2017) Utilization of straw pallets and briquettes as co-substrates at biogas plants. Energiforsk Rep 2017:1–62

Hussain N, Abbasi T, Abbasi SA (2015) Vermicomposting eliminates the toxicity of Lantana (*Lantana camara*) and turns it into a plant friendly organic fertilizer. J Hazard Mater 298:46–57

Hussain N, Abbasi T, Abbasi SA (2016) Transformation of toxic and allelopathic lantana into a benign organic fertilizer through vermicomposting. Spectrochim Acta A Mol Biomol Spectrosc 163:162–169

Hutnan M, Spalkova V, Bodik I, Lazor M (2009) Biogas production from maize grains and maize silage. J Environ Stud 19:323–329

Jagadeesh KS, Geeta GS, Reddy TKR (1990) Biogas production by anaerobic digestion of *Eupatorium odoratum*L. Biol Wastes 33(1):67–70

Kabir MM, del Pilar CM, Taherzadeh MJ, Sarvari Horvath I (2013a) Effect of the N-methylmorpholine-N-oxide (NMMO) pre-treatment on anaerobic digestion of forest residues. Bioresources 8:5409–5423

Kabir MM, Forgacs G, Sarvari Horvath I (2013b) Enhanced methane production from wool textile residues by thermal and enzymatic pre-treatment. Process Biochem 48:575–580

Kalia AK (1983) Bioconversion of *Lantana camara* to biogas. Extended abstracts: National Solar Energy Convention. Solar Energy Society of India, 15–18 December

Kalia AK, Kanwar SS (1990) Anaerobic fermentation of Ageratum for biogas production. Biol Wastes 32:155–158

Kalia AK, Singh SP (1998) Operation of 1m^3 plug flow plant on mixture of *Lantana camara* and cattle dung for biogas production. 74:13–16

Kalia VC, Kumar A, Jain SR, Joshi AP (1992) Biomethanation of plant materials. Bioresour Technol 41(3):209–212

Kannan R, Shackleton CM, Shaanker RU (2013) Reconstructing the history of introduction and spread of the invasive species, *Lantana*, at three spatial scales in India. Biol Invasions 15(6):1287–1302

Kannan R, Shackleton CM, Krishnan S, Uma Shaanker R (2016) Can local use assist in controlling invasive alien species in tropical forests? The case of *Lantana camara* in southern India. For Ecol Manag 376:166–173

Kanwar SS, Guleria RL (1995) Biogas production form mixtures of poultry litter and cattle dung with an acclimatized inoculum. Biogas Forum 60:21–22

Kanwar SS, Kalia AK (1988) *Lantana camara*- a potential source of biogas. In: Singh CM, Angiras NN (eds) Proceedings of seminar on control of *Lantana* and *Ageratum*. HPKV, Palampur, pp 55–57

Kanwar SS, Verma ML (2010) Engineering of microbial lipases. Encyclopedia of industrial biotechnology: bioprocess, bioseparation, and cell technology (EIB). Wiley Publishers, USA

Kanwar SS, Verma HK, Pathak S, Kaushal RK, Kumar Y, Verma ML, Chimni SS, Chauhan GS (2006) Enhancement of ethyl propionate synthesis by poly (AAc-co-HPMA-cl-MBAm)-immobilized *Pseudomonas aeruginosa* MTCC 4713 exposed to Hg^{2+} and NH_4^+ ions. Acta Microbiol Immunol Hung 53:195–207

Kanwar SS, Kaushal RK, Verma ML, Kumar Y, Azmi W, Gupta R, Chimni SS, Chauhan GS (2007a) Synthesis of ethyl oleate employing synthetic hydrogel-immobilized lipase of *Bacillus coagulans* MTCC-6375. Indian J Biotechnol 6:68–73

Kanwar SS, Verma ML, Maheshwari C, Chauhan S, Chimni SS, Chauhan GS (2007b) Properties of *poly* (AAc-*co*-HPMA-*cl*-EGDMA) hydrogel-bound lipase of *Pseudomonas aeruginosa* MTCC-4713 in synthesis of methyl acrylate. J Appl Polym Sci 104:183–191

Kanwar SS, Gehlot S, Verma ML, Gupta R, Kumar Y, Chauhan GS (2008a) Synthesis of geranyl butyrate employing *poly* (AAc-*co*-HPMA-*cl*-EGDMA) hydrogel-immobilized lipase of *Pseudomonas aeruginosa* MTCC-4713. J Appl Polym Sci 110:2681–2692

Kanwar SS, Sharma C, Verma ML, Chauhan S, Chimni SS, Chauhan GS (2008b) Short-chain ester synthesis by transesterification employing *poly* (MAc-*co*-DMA-*cl*-MBAm) hydrogel-bound lipase of *Bacillus coagulans* MTCC-6375. J Appl Polym Sci 109:1063–1071

Karthikeyan M, Hussain N, Gajalakshmi S, Abbasi SA (2014) Effect of vermicast generated from an allelopathic weed lantana (*Lantana camara*) on seed germination, plant growth, and yield of cluster bean *(Cyamopsistetragonoloba*). Environ Sci Pollut Res 21(21):12539–12548

Katuwal H, Bohara AK (2009) Biogas: a promising renewable technology and its impact on rural households in Nepal. Renew Sust Energ Rev 13:2668–2674

Kayesth S, Gupta KK (2018) Impact of Lantana camara hexane extract on survival, growth and development of *Dysdercuskoenigii* Fabricius (Heteroptera: Pyrrhocoriedae). Acta Ecol Sin 38(3):187–192

Keziah EA, Nukenine EN, Danga SPY, Younoussa L, Esimone CO (2015) Creams formulated with *Ocimum gratissimum* L. and *Lantana camara* L. crude extracts and fractions as mosquito repellents against *Aedes aegypti* L. (Diptera: Culicidae). J Insect Sci 15(1):1–5

Khalid A, Arshad M, Anjum M, Mahmood T, Dawson L (2011) The anaerobic digestion of solid organic waste. Waste Manag 31:1737–1744

Khoiyangbam RS, Gupta N, Kumar S (2011) Biogas technology: towards sustainable development. The Energy and Resources Institute Publisher, New Delhi, pp 1–18

Kohli RK, Batish DR, Singh HP, Dogra KS (2006) Status, invasiveness and environmental threats of three tropical American invasive weeds (*Parthenium hysterophorus* L., *Ageratum conyzoides* L., *Lantana camara* L.) in India. Biol Invasions 8:1501–1510

Kothari R, Pandey AK, Kumar S, Tyagi VV, Tyagi SK (2014) Different aspects of dry anaerobic digestion for bio-energy: an overview. Renew Sust Energ Rev 39:174–195

Kumar P, Pant M, Negi GCS (2009) Soil physico-chemical properties and crop yield improvement following *Lantana* mulching and reduced tillage in rainfed croplands in the Indian Himalayan Mountains. J Sustain Agric 33:636–657

Kumar R, Katiyar R, Kumar S, Kumar T, Singh V (2016) *Lantana camara*: an alien weed, its impact on animal health and strategies to control. J Exp Biol Agric Sci 4(3S):321–337

Kumar R, Sharma R, Patil RD, Mal G, Kumar A, Patial V, Kumar P, Singh B (2018) Sub-chronic toxicopathological study of lantadenes of *Lantana camara* weed in Guinea pigs. BMC Vet Res 14:129

Kundwa MJ, Mahame C, Bigirimana T, Sanga P (2018) Contribution of modern biogas plant to energy source and environment protection in Rwanda. J Civil Environ Eng 8:4–7

Kushkevych I, Vitezova M, Vitez T, Kovac J, Kauka P, Jesionek W, Bartos M, Barton L (2018) A new combination of substrates: biogas production and diversity of the methanogenic microorganisms. Life Sci 13:119–128

Landbeck M, Schmidt W (2005) Energy maize-goals, strategies and first breeding successes. CD-ROM computer file. In: Proceedings of the first international energy farming congress, Papenburg, Germany

Lohri C, Vogeli Y, Oppliger A, Giusti A, Zurbrugg C (2010) Evaluation of biogas systems in Nepalese prisons. IWA-DEWATS conference, decentralized wastewater treatment solutions in developing countries conference and exhibition, Surabaya, Indonesia

Lucas MB, Cherry E, Salacup CE, Simon SR (2018) Efficacy of rose bangle (*Lantana Camara* L.) leaf extract in controlling corn borer (*Ostriniafurnacalis*) and corn earworm (*Helicoverpazea*) of sweet corn (*Zea mays* var. Saccharata). Asia Pac J Multidiscip Res 6(2):18–22

Maile II, Muzenda E (2014) Production of biogas from various substrates under anaerobic conditions. International conference on innovative engineering technologies, Thialand

Mao C, Feng Y, Wang X, Ren G (2015) Review on research achievements of biogas from anaerobic digestion. Renew Sust Energ Rev 45:540–555

Mishra A (2015) Allelopathic properties of *Lantana camara*. Int Res J Basic Clin Stud 3(1):13–28

Mittal S, Ahlgren EO, Shukla PR (2018) Barriers to biogas dissemination in India: a review. Energy Policy 112:361–370
Monlau F, Barakat A, Steyer JP, Carrere H (2012) Comparison of seven types of thermo-chemical pretreatment on the structural features and anaerobic digestion of sunflower stalks. Bioresour Technol 120:241–247
Monlau F, Kaparaju P, Trably E, Steyer JP, Carrere H (2015) Alkaline pretreatment to enhance one-stage CH_4 and two-stage H_2/CH_4 production from sunflower stalks: mass, energy and economical balances. Chem Eng J 260:377–385
Mungi NA, Coops NC, Ramesh K, Rawat GS (2018) How global climate change and regional disturbance can expand the invasion risk? Case study of *Lantana camara*invasion in the Himalaya. Biol Invasions 20(7):1849–1863
Muñoz R, Meier L, Diaz I, Jeison D (2015) A review on the state-of-the-art of physical/chemical and biological technologies for biogas upgrading. Rev Environ Sci Biotechnol 14:727–759
Mursec B, Vindis P, Janzekovic M, Brus M, Cus F (2009) Analysis of different substrates for processing into biogas. J Achiev Mater Manuf Eng 37:652–657
Muske AN, Rao PV (2019) Evaluation of biogas production potential by anaerobic co-digestion with substrate mixture of fruit waste, lawn grass, and manures. In: Rathinasamy M, Chandramouli S, Phanindra K, Mahesh U (eds) Water resources and environmental engineering II. Springer, Singapore
Nanjappa HV, Saravanane P, Ramachandrappa BK (2005) Biology and management of *Lantana camara* L.–a review. Agric Rev 26(4):272–280
Naz R, Bano A (2013) Phytochemical screening, antioxidants and antimicrobial potential of *Lantana camara* in different solvents. Asian Pac J Trop Dis 3:480–486
Nishio N, Nakashimada Y (2007) Recent development of anaerobic digestion processes for energy recovery from wastes. J Biosci Bioeng 103:105–112
Nurlina E, Sambasri S, Hartati E, Safitri R, Hodijat A (2018) Utilization of waste as biogas substrate by dominan microbes identified. J Phys Conf Ser 1013:012211
Okonkwo UC, Onokpite E, Onokwai AO (2018) Comparative study of the optimal ratio of biogas production from various organic wastes and weeds for digester/restarted digester. J King Saud Univ Eng Sci 30:123–129
Omer A (2017) Biogas technology for sustainable energy generation: development and perspectives. MOJ Appl Bionics Biomech 1:137–148
Osuna MB, Zandvoort MH, Iza JM, Lettinga G, Lens PNL (2003) Effects of trace element addition on volatile fatty acid conversion in anaerobic granular sludge reactors. Environ Technol 24:573–587
Pandey SK, Bhattacharya T (2018) Effect of two biodegradable chelates on metals uptake, translocation and biochemical changes of *Lantana camara* growing in fly ash amended soil. Int J Phytoremediation 20(3):214–224
Pandey SK, Bhattacharya T, Chakraborty S (2016) Metal phytoremediation potential of naturally growing plants on fly ash dumpsite of Patratu thermal power station, Jharkhand, India. Int J Phytoremediation 18(1):87–93
Paolini V, Petracchini F, Segreto M, Tomassetti L, Naja N, Cecinato A (2018) Environmental impact of biogas: a short review of current knowledge. J Environ Sci Health 53:899–906
Parimoo HA, Sharma R, Patil RD, Sharma OP, Kumar P, Kumar N (2014) Hepatoprotective effect of Ginkgo biloba leaf extract on lantadenes-induced hepatotoxicity in Guinea pigs. Toxicon 81:1–12
Patil J, Sharma MK, Yadav S (2016) Management of reniform nematode, *Rotylenchulusreniformison* cowpea by using botanicals. Indian J Ecol 43(2):613–614
Patil NB, Zacharia S, Kumari M (2017) Eco-friendly management of powdery mildew of garden pea (*Pisum sativum* L.). Int J Curr Microbiol App Sci 6(2):684–689
Pavela R, Benelli G (2016) Ethnobotanical knowledge on botanical repellents employed in the African region against mosquito vectors – a review. Exp Parasitol 167:103–108

Popscue C, Jurcoane S (2014) Evaluation of biogas potential of some organic substrates from agriculture and food industry and co-digestion in large scale biogas plant. Rom Biotechnol Lett 20:10648–10655

Qin B (2009) Lake eutrophication: control counter measures and recycling exploitation. Ecol Eng 35:1569–1573

Raghubanshi AS, Rai LC, Gaur JP, Singh JS (2005) Invasive alien species and biodiversity in India. Curr Sci 88(4):539–540

Raha D, Mahanta P, Clarke ML (2014) The implementation of decentralised biogas plants in Assam, NE India: the impact and effectiveness of the National Biogas and Manure Management Programme. Energy Policy 68:80–91

Raj R, Das TK, Kaur R, Singh R, Shekhawat K (2018) Invasive noxious weed management research in India with special reference to *Cyperusrotundus*, *Eichhorniacrassipes* and *Lantana camara*. Indian J Agric Sci 88(2):181–196

Ramaswami G, Kaushik M, Prasad S, Sukumar R, Westcott D (2016) Dispersal by generalist frugivores affects management of an invasive plant. Biotropica 48(5:638–644

Rameshwar HY, Argaw A (2016) Biodegradation of *Lantana camara* using different animal manures and assessing its manurial value for organic farming. South Indian J Biol 2(1):52–60

Ramkumar R, Balasubramani G, Raja RK, Raja M, Govindan R, Girija EK, Perumal P (2017) *Lantana camara* Linn root extract-mediated gold nanoparticles and their *in vitro* antioxidant and cytotoxic potentials. Artif Cells Nanomed Biotechnol 45(4):748–757

Rangasamy C, Mahajan M, Radhakrishnan A (2019) From weed to furniture. In: Pullanikkatil D, Shackleton C (eds) Poverty reduction through non-timber forest products, Sustainable development goals series. Springer, Cham

Rao AN, Singh RG, Mahajan G, Wani SP (2018) Weed research issues, challenges, and opportunities in India. Crop Prot. https://doi.org/10.1016/j.cropro.2018.02.003

Riggio S, Torrijos M, Debord R, Esposito G, van Hullebusch ED, Steyer JP, Escudié R (2017) Mesophilic anaerobic digestion of several types of spent livestock bedding in a batch leach-bed reactor: substrate characterization and process performance. Waste Manag 59:129–139

Royimani L, Mutanga O, Odindi J, Dube T, Matongera TN (2018) Advancements in satellite remote sensing for mapping and monitoring of alien invasive plant species (AIPs). Phys Chem Earth A/B/C. https://doi.org/10.1016/j.pce.2018.12.004

Saha B, Sathyan A, Mazumder P, Choudhury SP, Mishra U (2018) Biochemical methane potential (BMP) test for *Ageratum conyzoides* to optimize ideal food to microorganism (F/M) ratio. J Environ Chem Eng 6(4):5135–5140

Saini R (2001) Anaerobic digestion of lantana weed for generation of biogas. M.Sc. thesis, CSKHPKV, Palampur, Himachal Pradesh, India, p 67

Saini R, Kanwar SS, Sharma OP, Gupta MK (2003) Biomethanation of lantana weed and biotransformation of its toxins. World J Microbiol Biotechnol 19:209–213

Saini VP, Suthar S, Chaudhari K, Kumar K (2017) Valorization of toxic weed Lantana camara L. biomass for adsorptive removal of lead. J Chem 2017:1–12

Schnuera A, Jarvis A (2010) Microbiological handbook for biogas plants. Swedish gas center report, pp 1–207

Schwarzländer M, Hinz HL, Winston RL, Day MD (2018) Biological control of weeds: an analysis of introductions, rates of establishment and estimates of success, worldwide. BioControl 63(3):319–331

Sekoai PT, MoroOuma CN, Preez SP, Modisha P, Engelbrecht N, Bessaraboy DG, Ghimire A (2019) Applications of nanoparticles in biofuels an overview. Fuel 237:380–397

Semeonove IS (1999) Mathematical modelling and parameters estimation of fermentation process. Bioprocess Eng 21:377–381

Shackleton RT, Witt ABR, Aool W, Pratt CF (2017) Distribution of the invasive alien weed, *Lantana camara*, and its ecological and livelihood impacts in eastern Africa. Afr J Range Forage Sci 34(1):1–11

Shamsee ZR, Al-Saffar AZ, Al-Shanon AF, Al-Obaidi JR (2018) Cytotoxic and cell cycle arrest induction of pentacyclic triterpenoides separated from *Lantana camara* leaves against MCF-7 cell line in vitro. Mol Biol Rep:1–10

Sharma OP, Vaid J, Sharma PD (1991) Comparison of lantadenes content and toxicity of different taxa of the lantana plant. J Chem Ecol 17:2283–2291

Sharma OP, Dawra RK, Datta AK, Kanwar SS (1997) Biodegradation of lantadene a, the pentacyclic triterpenoid hepatotoxin by *Pseudomonas pickettii*. Lett Appl Microbiol 24:229–232

Sharma S, Sharma OP, Dawra RK, Bhat TK (1999) Disposition of lantadene a, the pentacyclic triterpenoid hepatotoxin, orally administered to Guinea pigs. Toxicol Lett 105:59–66

Sharma OP, Singh A, Sharma S (2000a) Levels of lantadenes, bioactive pentacyclic triterpenoids, in young and mature leaves of *Lantana camara*var. Aculeata Fitoterapia 71:487–491

Sharma S, Sharma OP, Singh B, Bhat TK (2000b) Biotransformation of lantadenes, the pentacyclic triterpenoid hepatotoxins of lantana plant, in Guinea pig. Toxicon 38(9):1191–1202

Sharma S, Sharma OP, Singh B, Bhat TK (2000c) Biotransformation of lantadenes, the pentacyclic triterpenoid hepatotoxins of lantana plant, in Guinea pig. Toxicon 38(9):1191–1202

Sharma R, Kumar R, Mal G, Patil RD, Kumar P, Singh B, Bhar R (2017) Association of lantadenes contents and *Lantana camara* weed toxicity in ruminants of Himachal Pradesh, India. Ruminant Sci 2(6):289–292

Shriniwas PP, Subash KT (2017) Antioxidant antibacterial and cytotoxic potential of silver nanoparticles synthesized using terpenes rich extract of *Lantana camara* L. leaves. Biochem Biophys Rep 10:76–81

Sibiya NT, Muzenda E, Mbohba C (2017) Evaluation of potential substrates for biogas production via anaerobic digestion: a review. Proc World Cong Eng Computer Sci San Francisco, USA

Siegmeier T, Blumenstein B, Möller D (2015) Farm biogas production in organic agriculture: system implications. Agric Syst 139:196–209

Singh A, Sharma OP, Dawra RK, Kanwar SS, Mahato SB (1999) Biotransformation of lantadene A (22 p – angeloyloxy-3-oxo Olean-12-en-28-oic acid), the pentacyclic triterpenoid, by *Alcaligenes faecalis*. Biodegradation 10:373–381

Singh A, Sharma OP, Kurade NP, Ojha S (2001a) Detoxification of lantana hepatotoxin, lantadene A, using *Alcaligenes faecali*. J Appl Toxicol 21(3):225–228

Singh A, Sharma OP, Bhat TK, Vats SK, Ojha S (2001b) Fungal degradation of lantadene A, the pentacyclic triterpenoid hepatotoxin of lantana plant. Int Biodeterior Biodegradation 47(4):239–242

Singh S, Khatri PK, Kumar D, Naik VR (2018) Biological-control of *Lantana camara* through crop competition using native species. Indian J Trop Biodivers 26(1):81–86

Svensson K (2012) Biogas production from multi substrate. Civil and Environmental Engineering. Norwegian University of Science and Technology

Syafrudin, Nugraha WD, Munkar G, HHA M, Budiyono (2018) The influence of C/N ratio to biogas production from water hyacinth as an aquatic weed plant during liquid anaerobic digestion (L-AD). Adv Sci Lett 24:9832–9834

Taylor PB, van Heerden FR, Thompson LJ (2019) A novel threat from invasive Lantana: stem exudate clogs the feet of small birds. Emu Austral Ornithol 119(1):90–94

Thankamani CK, Kandiannan K, Hamza S, Saji KV (2016) Effect of mulches on weed suppression and yield of ginger (*Zingiberofficinale* roscoe). Sci Hortic 207:125–130

Toma L, Voicu G, Ferdes M, Dinca M (2016) Animal manure as substrate for biogas production. 15th international scientific conference engineering for rural development Jelgava Latvia, pp 629–634

Tricase C, Lombardi M (2009) State of the art and prospects of Italian biogas production from animal sewage: technical economic consideration. Renew Energy 34:477–485

Tufaner F, Avsar Y (2016) Effects of co-substrate on biogas production from cattle manure: a review. Int J Environ Sci Technol 13:2303–2312

Udo IA, Osai EO, Ukeh DA (2014) Management of Root-Knot Disease on tomato with Bioformulated*Paecilomyceslilacinus* and leaf extract of *Lantana camara*. Braz Arch Biol Technol 57(4):486–492 van Wilgen, Brian W (2018) The management of invasive alien plants in South Africa: strategy. Progress and Challenges Outlooks on Pest Manage 29(1):13–17

Venturin B, Bonatto C, Damaceno FM, Mulinari J, Fongaro G, Treichel H (2019) Physical, chemical, and biological substrate pretreatments to enhance biogas yield. In: Treichel H, Fongaro G (eds) Improving biogas production. Biofuel and biorefinery technologies, vol 9. Springer, Cham, pp 25–44

Verma ML (2017) Nanobiotechnology advances in enzymatic biosensors for the Agri-food industry. Environ Chem Lett 15(4):555–560

Verma ML, Kanwar SS (2008) Properties and application of *Poly* (MAc-*co*-DMA-*cl*-MBAm) hydrogel immobilized *Bacillus cereus* MTCC 8372 lipase for synthesis of geranyl acetate. J Appl Polym Sci 110:837–846

Verma ML, Kanwar SS (2012) Harnessing the potential of thermophiles: the variants of extremophiles. Dyn Biochem Process Biotechnol Mol Biol 6(1):28–39

Verma ML, Azmi W, Kanwar SS (2008) Microbial lipases: at the interface of aqueous and non-aqueous media-a review. Acta Microbiol Immunol Hung 55:265–293

Verma ML, Azmi W, Kanwar SS (2011) Enzymatic synthesis of isopropyl acetate catalysed by immobilized *Bacillus cereus* lipase in organic medium. Enzym Res 2011, Article ID 919386, 7 pages

Verma ML, Barrow CJ, Kennedy JF, Puri M (2012) Immobilization of β-galactosidase from *Kluyveromyces lactis* on functionalized silicon dioxide nanoparticles: characterization and lactose hydrolysis. Int J Biol Macromol 50:432–437

Verma ML, Rajkhowa R, Barrow CJ, Wang X, Puri M (2013a) Exploring novel ultrafine Eri silk bioscaffold for enzyme stabilisation in cellobiose hydrolysis. Bioresour Technol 145:302–306

Verma ML, Naebe M, Barrow CJ, Puri M (2013b) Enzyme immobilisation on aminofunctionalised multi-walled carbon nanotubes: structural and biocatalytic characterisation. PLoS One 8(9):e73642

Verma ML, Chaudhary R, Tsuzuki T, Barrow CJ, Puri M (2013c) Immobilization of β-glucosidase on a magnetic nanoparticle improves thermostability: application in cellobiose hydrolysis. Bioresour Technol 135:2–6

Verma ML, Puri M, Barrow CJ (2016) Recent trends in nanomaterials immobilised enzymes for biofuel production. Crit Rev Biotechnol 36(1):108–119

Viturtia A, Mata-Alvarez J, Cecchi F, Fazzini G (1989) Two phase anaerobic digestion of a mixture of fruits and vegetable wastes. Biol Wastes 29:189–199

Wagner AO, Reitschuler C, Illmer P (2014) Effect of different acetate: propionate ratios on the methanogenic community during thermophilic anaerobic digestion in batch experiments. Biochem Eng J 90:154–161

Wasantha Rathnayake RM (2019) Threats to sandy shore habitats in Sri Lanka from invasive vegetation. In: Makowski C, Finkl C (eds) Impacts of invasive species on coastal environments, Coastal research library, vol 29. Springer, Cham

Wirth R, Kovacs E, Maroti G, Bagi Z, Rakhely G, Kovacs KL (2012) Characterization of a biogas-producing microbial community by short-read next generation DNA sequencing. Biotechnol Biofuels 5:41

Yadav M, Paritosh K, Pareek N, Vivekanand V (2019a) Coupled treatment of lignocellulosic agricultural residues for augmented biomethanation. J Clean Prod 213:75–88

Yadav M, Singh A, Balan V, Pareek N, Vivekanand V (2019b) Biological treatment of lignocellulosic biomass by *Chaetomium globosporum*: process derivation and improved biogas production. Int J Biol Macromol 128:176–183

Yadvika S, Sreekrishman TR, Kohli S, Rana V (2004) Enhancement of biogas production from solid substrates using different techniques-a review. Bioresour Technol 95:1–10

Yen HW, Brune D (2007) Anaerobic co-digestion of algal sludge and waste paper to produce methane. Bioresour Technol 98:130–134

Zeng SJ, Yuan XZ, Shi XS, Qui YL (2010) Effect of inoculums/ substrate ration on methane yield and orthophosphate release from anaerobic digestion of *Microcystis spp*. J Hazard Mater 178:89–93

Zhang T, Liu L, Song Z, Ren G, Feng Y, Han X, Yang G (2013) Biogas production by co-digestion of goat manure with three crop residues. PLoS One 8:e66845

Zhong W, Zhongzzhi Z, Yijing L, Wei Q, Meng X, Min Z (2012) Biogas productivity by co digesting Taihu blue algae with corn straw as an external carbon source. Bioresour Technol 114:281–286

Zonta Z, Alves MM, Flotats X, Palatsi J (2013) Modeling inhibitory effects of long chain fatty acids in the anaerobic digestion process. Water Res 47:1369–1380

Recent Progress in Emerging Microalgae Technology for Biofuel Production

4

John Jeslin, Antwin Koshy, Munusamy Chamundeeswari, and Madan Lal Verma

Abstract

Microalgae are potentially remarked as the efficient candidate for the economically valuable biofuel production. The microalgal biomass is capable of storing solar energy into a biochemical energy that forms a promising feedstock for biofuel production. The greatest threat in the production of biofuels from microalgae lies in the economic transition of microalgal biomass into biofuels. The global efforts have been initiated for the modifications of microalgal strains that would reasonably result in increased biomass yield thereby increasing the total lipid productivity. Furthermore, the commercialization of the microalgal-based biofuels enables the disposal of unsustainable and nonrenewable fossil fuels. In this chapter, we review the recent exploration of oleaginous microalgae for the increased production of biofuels by the enhancement of triacylglycerol (TAG)/lipid accumulation within microalgal cells. In addition to this, various aspects of genetic manipulation of microalgal species to induce enhanced TAG/lipid biosynthesis and accumulation have also been critically discussed.

Keywords

Biochemical energy · Fossil fuels · Oleaginous · Triacylglycerol

J. Jeslin · M. Chamundeeswari
Department of Biotechnology, St. Joseph's College of Engineering, Chennai, India

A. Koshy
Department of Mechanical Engineering, St. Xavier's Catholic College of Engineering, Nagercoil, India

M. L. Verma (✉)
Centre for Chemistry and Biotechnology, Deakin University, Burwood, VIC, Australia

Department of Biotechnology, School of Basic Sciences, Indian Institute of Information Technology, Una, Himachal Pradesh, India

N. Srivastava et al. (eds.), *Substrate Analysis for Effective Biofuels Production*, Clean Energy Production Technologies, https://doi.org/10.1007/978-981-32-9607-7_4

4.1 Introduction

Fossil fuel, the demanded energy source, is extracted and exploited globally due to rapid industrialization and urbanization. The major drawback of using fossil fuels is its negative environmental issues due to the excess emission of pollutant gases such as carbon monoxide, carbon dioxide, sulphur dioxide, nitrogen dioxide, etc. and high economic costs (Valipour et al. 2012). To overcome these pollutants, renewable energy sources such as wind/solar power/biomass are used.

Global fuel demand is on the rise with increase in population and their living standards. Economically, using biomass is an ideal alternative to fossil fuels (Duraiarasan et al. 2016). Limited resource for biofuel generation with consecutive price fluctuations seems to be challenging in biofuel production technology. Contribution of biofuel at global level is very rare due to highly implemented policy measures (Neuling and Kaltschmitt 2017). The need for biofuel is non-futile and futuristic since development in biofuel production can minimize the existing greenhouse gas emission and climate change issues (Buchspies and Kaltschmitt 2016; Wulf et al. 2017). Also, it generates green jobs for a sustainable society and creates a sustainable energy provision (Linares and Perez-Arriaga 2013; Awudu and Zhang 2012).

Microalgae, the species with 80% oil content, unicellularly/multicellularly grows widely in freshwater or saltwater, and it can be applied in biofuel generation. To be specific, biofuel production using microalgae technology comes under the third generation of biofuels that overcomes the challenges faced by the first-generation edible crop usage and second-generation lignocellulosic energy crop usage of biofuels (Saifullah et al. 2014; Htet et al. 2013; Rasool and Hemalatha 2016). Microalgae can be grown in open pond/closed photobioreactor (Fu et al. 2010; Surendhiran and Vijay 2014). To enhance the mass production commercially, various factors such as nutrient availability, light source, temperature, axenic culture maintenance, and scale-up cost should be considered.

Blue-green algae, one of the most important photosynthetic primary producers, synthesizes valuable biomolecules like proteins, lipids, and carbohydrates that are utilized for the production of biofuels (Duraiarasan et al. 2016). The biomass of microalgae is used for generating biofuel of different types such as bio-oil, biodiesel, biogas, and bioethanol. Currently, different technologies such as liquefaction, fermentation, pyrolysis, heterogeneous catalytic process (Yavuz et al. 2006), transesterification, and anaerobic digestion is used for the mass production of biofuels from algae (Takeda and Nishijima 2006; Gomes et al. 2015). Algae biomass can be produced naturally or artificially. Algal bloom is one of the drawbacks of natural cultivation, and it can be rectified by using artificial photobioreactors (Adeniyi et al. 2018). But this artificial method is not economically feasible due to high capital cost (Brennan and Owende 2010).

The biomass yield is greater in algae, which is estimated to be about 200 barrels of oil/hectare of land when compared to the high energy crops (Surendhiran et al. 2014). Hence, microalgae is one of the most promising renewable energy source with its eternal capability to fix the carbon dioxide by the phenomenon of photosynthesis (Li et al. 2011). For a sustainable biofuel generation from algae, energy-efficient methods are required for harvesting, algae farming, and biomass processing (Bligh and Dyer 1959; Mascolo et al. 2013). The challenges to be faced in microalgae technology includes the generation of toxic compounds/carcinogens/allergens (Shao et al. 2011; Hu et al. 2008). Efficient technology which is economically feasible to be commercialized must be developed and upgraded.

4.2 Microalgae for Biofuel Production

Biofuels are reasonable alternative sources of fossil fuels. These biofuels range from first generation to the fourth generation in which microalgae occupies the third generation. The valuable energy compounds from the intra- as well as extracellular regions of the microalgae are highly capable of producing economically important biofuels (Harun et al. 2010; Singh et al. 2017). Microalgae expose an enormous metabolic plasticity and biologically diverse nature that makes them effectively grow on a vast organic or inorganic carbon source (Cho et al. 2017; Chew et al. 2017). The major consideration that makes the microalgae as a major source for biofuel production is their ability to accumulate lipids either in the intra- or extracellular regions. This lipid accumulation can also be enhanced by the external environmental stress (Pereira et al. 2016; Savchenko et al. 2017). The bioproducts from microalgae include biohydrogen, biodiesel, bioethanol, bioelectricity, biobutanol, and other organic volatile compounds (Fig. 4.1).

However, the use of microalgae for biofuel production is limited due to the increased cultivation and harvesting cost (10–20-fold) (Laurens et al. 2017; Santos et al. 2016a, b; Zhu et al. 2017). Therefore, scale-up of biomass production along with the economically valuable coproducts can overcome the constraints in the microalgae-based biofuel production. Recent researches are carried out to minimize the drawbacks such as a larger cultivation area requirement, higher nutrient, and power and water consumption to enable large-scale biomass production that can serve the upcoming biofuel conversions.

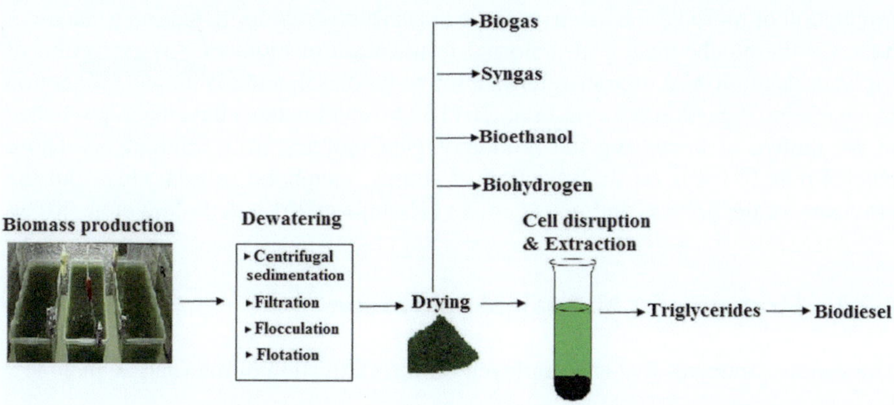

Fig. 4.1 Overview of different biofuel production from microalgae

4.2.1 Criteria for Cultivation System Selection

Different criteria for the appropriate selection of the microalgal cultivation system include (Chang et al. 2017):

- Tolerability of the microalgal strain
- Type of the end product and its quality
- Construction and maintenance cost
- Scalability and performance

The adaptability or tolerability of the microalgal strain is highly preferred for the scale-up process of biomass production. In the photobioreactor, the microalgae should withstand the outdoor environment, oxygen accumulation, and the shear forces developed within the system by aeration or pumping. In case of the open pond system, the microalgal cells should have the capacity to overcome the cross-contamination, environmental fluctuations, and photoperiods (Brennan and Owende 2010; Chang et al. 2017).

The quality or type of end product depends on the lipid or the carbohydrate accumulation in the biomass, which are cheaper than the fossil fuel production. The integrated bioreactor system that reduces the cost of production and lesser land utilization with higher productivity can enhance the biofuel economical value. The usage of the waste water source for microalgal cultivation can have a dual benefit in the production of biofuel commercially (Chang et al. 2017; Li et al. 2014a, b).

The overall cost of biofuel production lies in the kinetic performance during the scalability process. The relative cost of the open as well as closed cultivation systems determines the final value of the produced biofuel. Compared with the closed system, open system can be easily scaled up but possesses a drawback of reduced biomass production. Therefore, selection of the appropriate cultivation system and the cost of production travel in the same direction, affecting the final product quality (Chang et al. 2017).

The microalgal biomass should undergo any one of the processes for the efficient production of biofuels. The steps include (1) carbohydrate fermentation to bioalcohols, (2) thermochemical conversion or gasification of biomass, (3) extraction of triglycerides followed by transesterification to biodiesel, and (4) biogas production by anaerobic digestion (Craggs et al. 2011). Different patents have been published in the motive of increasing the productivity of biofuels from microalgae. Those patents mainly focus on the selection of strains, metabolic cellular flux, and the efficiency of the process and management (Thompson 2013; de la Jara et al. 2016).

4.2.2 Strategies for Microalgal Cultivation

The various strategies that are employed for the cultivation of microalgae include:

- Integrated wastewater treatment and mitigation of CO_2

- Feeding modes for cultivation
- Continuous two-stage cultivation
- Stress cultivation
- Co-culture
- Cultivation with water recycling

4.2.2.1 Integrated Wastewater Treatment and Mitigation of CO_2

The phosphorus- and nitrogen-rich waste water is highly desired for microalgal cultivation. The inorganic nitrogen is consumed by the nitrification process in the form of nitrates and the phosphorus is reduced by precipitation, absorption, sedimentation, or ion exchange (Zhu et al. 2013a; Ruiz-Marin et al. 2010). The microalgal growth can also be affected by the availability of the carbon source. CO_2 is the major carbon source that influences microalgal growth. The limited atmospheric CO_2 creates a necessity of externally supplying CO_2 for microalgal growth enabling mitigation of CO_2. About 1.8 tonnes of CO_2 are required for the production of one ton biomass (Razzak et al. 2013; Ho et al. 2012).

4.2.2.2 Feeding Modes for Cultivation

Different modes of inlet feed influencing the microalgal growth include batch, fed-batch, continuous, or semi-continuous modes. In batch mode, microalgal cells are cultivated for a limited time period and there will be exhaustion of the nutrient after a certain period of time. This mode of cultivation will lead to a low productivity and steady state cannot be maintained (Zhu et al. 2013a). In semi-continuous mode, the cultivation begins in batch mode following the intermittent biomass harvesting and replenishing the media into the system (Zhu et al. 2013b). While in continuous mode, the biomass harvest and nutrient supplement will simultaneously occur within the system (Zhu et al. 2013b). Both semi- and continuous modes are capable of maintaining a steady state throughout the process (twofold more than the batch cultivation) (McGinn et al. 2014).

4.2.2.3 Continuous Two-Stage Cultivation

The two-stage cultivation involves the first stage with either mixotrophic or heterotrophic production condition and the second stage with the autotrophic production condition. The mixotrophic production utilizes an inorganic and organic carbon source with both light and dark photoperiods. In heterotrophic and autotrophic production, the microalgae utilize organic carbon and illumination, respectively (Koller et al. 2012). Compared with the autotrophic production, heterotrophic production reproduces a faster growth rate and increased productivity. But the expensive organic carbon limits its cultivation range (Chi et al. 2011). Therefore, in the foremost mixotrophic or heterotrophic production, the microalgal cells are provided with the essential nutrients for cellular growth, and the second autotrophic stage is provided with CO_2, for lipid or other end product production (Zheng et al. 2012; Ra et al. 2015).

4.2.2.4 Stress Cultivation

External stress like nutrient depletion, light intensity, salinity, metal ion stress, and temperature can influence the cellular composition and the successive yield by the microalgae. The changes in the nutrient composition can alter the lipid accumulation in the production unit, which would further influence the overall biofuel conversion of the system (Devi and Mohan 2012). In this strategy of cultivation, the microalgae are supplemented with sufficient quantity of the nutrient and further will be limited or allowed to undergo starvation stress, which would sequentially alter the lipid accumulation in the production system (Ito et al. 2013; Yang et al. 2014).

4.2.2.5 Co-culture

The mixed culture or the co-culture of microalgae with other bacteria, fungi, or other species of microalgae can influence the microalgal growth performance and the waste water nutrient removal (Hena et al. 2015). The co-culture of microalgae with the bacterial system reduces the oxygen accumulation in the production unit and also enables the use of CO_2 released by the bacterial system during its subsequent growth. Meanwhile, the extracellular compounds released by the microalgae benefit the bacterial system and make the microalgae assimilate the nutrients during production. However, microalgal-bacterial growth can also have an inhibitory action increasing the complexity of their growth. The growth will mainly depend on the external culturing environment where they are co-cultured (Zhang et al. 2012).

The co-culture of microalgae with the fungal system causes symbiotic association, assisting the bio-flocculation during the microalgal harvest. The grown microalgal cell will aggregate with the filamentous fungi forming lichen thereby immobilizing in the pellets. This co-culture is also found to influence the lipid, biomass, and bioproduct yield (Mackay et al. 2015; Zhou et al. 2013; Xie et al. 2013).

The mixed or co-culture of microalgae with other microalgal species enables them to grow as a colony or consortium and there will be either competitive or mutualistic relationship. This mode of cultivation is found to remove >98% of the nutrient contents from dairy waste water (Sathish and Sims 2012; Hena et al. 2015).

4.2.2.6 Cultivation with Water Recycling

When waste water is used for the growth of microalgae, it is essential to prevent the microalgal outflow into the environment after the treatment process. The harvest water recycling will prevent the release of any untreated microalgal cells into the environment (Resurreccion et al. 2012; Park et al. 2013).

4.2.3 Different Types of Lipids Produced by Microalgae

The microalgal lipids can be categorized into two, namely:

- Storage lipids (non-polar/neutral lipids)
- Structural lipids (polar lipids)

The storage lipids include TAG as the predominant fatty acid that majorly consists of saturated fatty acids along with unsaturated fatty acids to some extent. Structural lipids consist of larger amount of polyunsaturated fatty acids. TAG forms the basic energy repository of microalgae and structural lipids forms the essential dietary supplement of humans and aquatic organisms (Sharma et al. 2012).

The structural lipids in microalgae include sterols and polar lipids like phospholipids. These structural components accommodate a major role in the permeable selective membrane barrier of the organelles and the microalgal cells. These membrane lipids carry out specific functions of being a matrix for various metabolic processes and actively participate in the membrane fusion. Apart from this, polar lipids such as sphingolipid, inositol lipid, and oxidative products form an intermediate for the cell signaling and are involved in the adverse environmental response reactions (Sharma et al. 2012).

The storage lipids, TAG being a storage product, provide metabolic energy through catabolism (Gurr et al. 2002). TAG molecules are synthesized in the presence of light that are stored in the cytosolic lipid bodies and are reused for the biosynthesis of polar lipids in the absence of light. Microalgae generally constitute TAG with saturated fatty acids along with some monounsaturated fatty acids. Certain high oil yielding species are found to possess PUFA (polyunsaturated fatty acids) as accumulated TAG molecules (Bigogno et al. 2002). This PUFA-abundant TAG is a metabolically active component and is found to be a specific fatty acid reservoir. The de novo pathway becomes declined during the adverse environmental conditions and the PUFA-abundant TAG helps in the reorganization of the membrane by donating distinct acyl groups to polar lipids and monogalactosyl diacylglycerol (Khozin-Goldberg and Cohen 2006).

4.3 Microalgal Bioprocessing

The production of biofuel from microalgae commence from the large-scale biomass yield to the effective extraction of an intermediate compound (triglycerides) to get converted into valuable bioproducts. Different stages of bioprocessing of microalgae for biofuel production are briefly discussed below:

4.3.1 Biomass Production

The cultivation of microalgae artificially with optimum growth conditions can enhance the biomass yield to the maximum extent than in the natural conditions. However, utilizing the natural source for microalgal growth can reduce the cost of processing and commercial production of biofuels (Janssen et al. 2003; Chew et al. 2017). The growth-limiting factor and light source can be made available by using the sunlight, but this may affect the biomass production at a continuous rate. The assimilation of carbon dioxide is the key characteristic of phototropic microalgae. It can make use of higher CO_2 concentration (about 150,000 ppmv) (Bilanovic et al.

2009; Chiu et al. 2009; Hu and Sato 2017). The soluble carbonates like $NaHCO_3$ and Na_2CO_3 as well as any external carbon dioxide source from power plant are used in the production unit (Hsueh et al. 2007; Doucha et al. 2005; Abu Hajar et al. 2017). It also requires other inorganic compounds such as silicon, phosphorus, and nitrogen (Suh and Lee 2003, Halim et al. 2011). There are three major mechanisms of biomass production based on its nutrient requirement such as:

- Photoautotrophic biomass production
- Heterotrophic biomass production
- Mixotrophic biomass production

4.3.1.1 Photoautotrophic Biomass Production

Photoautotrophic biomass production is the photosynthetic mode of growth, which utilizes inorganic salts and compounds in the presence of a light source. This mode of production is the economically viable process which can be carried out through two conventional methods as follows:

4.3.1.1.1 Open Pond Production Unit

The open pond cultivation of microalgae is the cheaper and convenient mode of production system that varies with their form and shape. They can be either a natural water system like lagoon or lake or an artificial one (Jiménez et al. 2003; Nayak et al. 2016). Different types of open pond systems used widely in the large-scale microalgae production include shallow big pond, closed pond, raceway pond, and circular pond tank (Harun et al. 2010; Apel et al. 2017). The highly preferred artificial pond system is raceway pond (Jiménez et al. 2003; Fernández et al. 2013; De Andrade et al. 2016). It is an earth-lined compact concrete system with an oval-shaped, closed-loop recirculation channel (0.2–0.5 m depth). A paddle wheel will be introduced to provide a continuous cycle of production for the stabilization of growth and its productivity (Fig. 4.2). The submerged aerator and surface air in turn will provide the sufficient CO_2 for the microalgal growth. The output in the biomass yield highly depends on the location of the pond system (Masojidek and Torzillo 2008; Chisti 2016). The growth-limiting factors in the open pond system include pH, DO concentration, culture contamination, oxygen accumulation temperature, and intensity of the light source. The contamination in the open pond system is the other major issue in the mass production of microalgae. Certain microalgae can be only cultivated in a specific condition, such as *Chlorella* in a high nutritious condition, *Spirulina* in a high alkaline condition, *Dunaliella* in a high saline condition to prevent the cross-contamination from the open environment (Tafreshi and Shariati 2006; Chiaramonti et al. 2013). Recently, state estimators (software sensor) are used for the accurate measurement of the biomass concentration in the outdoor cultivation of microalgae in raceway photobioreactors (García-Mañas et al. 2019). The pond cultivation system is highly scalable and possesses less construction cost, operating cost, and maintenance cost (Apel and Botz 2015). The reactor configurations such as depth of the pond, temperature, mixing, light availability, CO_2 delivery, and gas transfer rate limit the productivity rate (Sutherland et al. 2014; Zeng et al. 2016; Santos et al. 2016a, b).

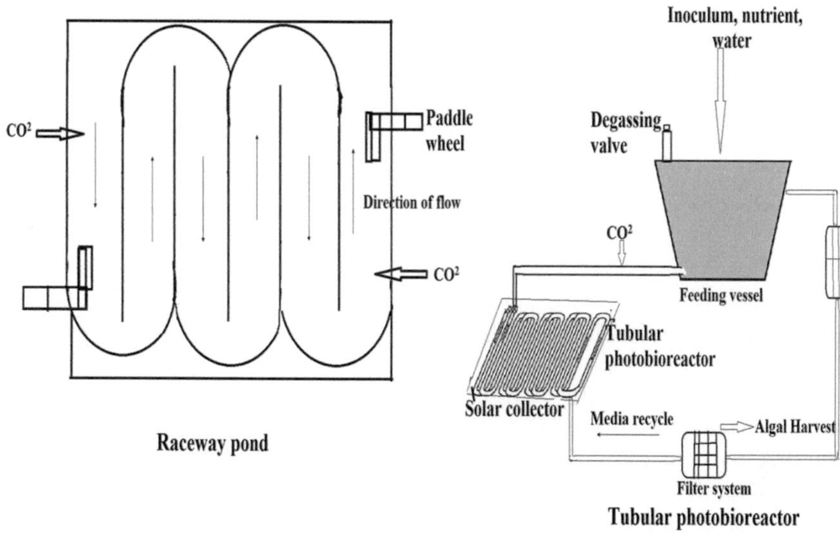

Fig. 4.2 Schematic representation of raceway pond and photobioreactor

4.3.1.1.2 Closed Photobioreactor Production Unit

The closed photobioreactor can overcome the hurdles in the open system such as cross-contamination by the open environment and maintaining the specific condition for the growth of microalgae with a systemized setup. It can maintain the single algal species under specified condition for an extended period at a continuous rate. However, the cost of production is high compared to the open system (Ugwu et al. 2008; Chisti 2007). In industries, closed photobioreactors can be column photobioreactor, flat-plate photobioreactor, and tubular photobioreactor (Fig. 4.2). In general, a photobioreactor consists of an array of parallel tubes made of plastic or glass (Ugwu et al. 2008). These tubes $\leq$0.1 m are arranged vertically, horizontally, helically, or inclined to absorb the maximum sunlight (Ugwu et al. 2002; Jorquera et al. 2010). They are also provided with continuous mixing and agitation for maximum gas exchange and productivity. The algal culture is again recirculated for oxygen and carbon dioxide exchange in the liquid media (Eriksen 2008a, b). These closed photobioreactors are highly preferred in the pilot-scale production of biomass due to its controllable environment.

Column photobioreactor is a closed system with high volumetric mass transfer rate, which can be easily maintained with low cost (Eriksen 2008a, b). The column photobioreactor provides an efficient aeration and mixing. It also consists of transparent walls that help in the light transport. The recirculation of the culture causes the gas exchange between the liquid medium (Sa´nchez Miro´n et al. 2002; Singh et al. 2015).

Flat-plate photobioreactor is the ancient type of closed system that provides illumination over a larger surface area with a thin culture layer that possesses less accumulation of dissolved oxygen (Ugwu et al. 2008). It allows high photoautotrophic microalgal growth (>80 g^{-1}) with higher efficiency of photosynthesis compared with the tubular one (Richmond et al. 2003; Khademi et al. 2015).

A tubular photobioreactor is a closed system owing to the largest surface area for sunlight illumination in the outdoor culture (Eriksen 2008a, b; Liao et al. 2016). This system possesses a drawback of its tubular dimensional length limitation that would result in the depletion of CO_2, pH variation, and accumulation of oxygen. Hence, multiple units are required in order to maintain large-scale microalgae production.

There are also other configurations like biofilm and thin-layer reactor configuration which need further development for the economically significant volumetric production (Apel et al. 2017; Blanken et al. 2017; Pruvost et al. 2017).

4.3.1.1.3 Hybrid Two-Stage Production Unit

The hybrid system utilizes both open and closed systems (more than two configurations), thereby reducing the disadvantages of both systems like reduced energy consumption and enhanced productivity and workload (Depra et al. 2019; Jacob-Lopes et al. 2016). The initial stage uses the photobioreactor for the controlled environment for the sustained growth that will overcome the open environmental contamination. The latter stage utilizes the open system for enhanced lipid accumulation by subjecting the cells to nutrient stress (Huntley and Redalje 2007; Rodolfi et al. 2008).

4.3.1.2 Heterotrophic Biomass Production

Heterotrophic production utilizes organic compounds like glucose for its energy and nutrient. This mode of biomass production is independent of the light source. Therefore, larger surface area is not necessarily required for growth and mass production (Miao and Wu 2006; Gomaa et al. 2016). The heterotrophic microalgae are generally cultured in the fermentor or in a stirred tank reactor (Eriksen 2008a, b; Francisco et al. 2015). This will produce high cell density with controlled growth at a reduced construction and harvesting cost. However, the energy utilization is high due to organic carbon production by photosynthesis (Chisti 2007; Maroneze et al. 2014). This mode of biomass production will produce larger biomass from heterotrophic microalgae with high lipid accumulation. For example, *Chlorella protothecoides* possess 55% lipid accumulation than the autotrophic microalgae (Miao and Wu 2006). The factors affecting the heterotrophic cultivation include oxygen supply, mixing, and viscosity (Santos et al. 2015)

4.3.1.3 Mixotrophic Biomass Production

In mixotrophic biomass production, the microalgae possess both the characteristics of autotrophic and heterotrophic biomass production, i.e., they undergo photosynthesis as well as organic carbon assimilation. Neither the light intensity nor the organic carbon source plays as a growth-limiting factor in mixotrophic biomass production. The major examples of mixotrophic mode of biomass production include green algae (*Spirulina platensis*) and the cyanobacterium (*Chlamydomonas reinhardtii*) (Andrade and Costa 2007; Chen et al. 2019). These organisms use light for photosynthesis and organic carbon for aerobic respiration thereby reducing the biomass loss in the dark phase as growth is regulated by the glucose utilization during diurnal phases (Andrade and Costa 2007). The growth yield will be higher for the mixotrophic when compared to the photoautotrophic but will be lesser in comparison with the heterotrophic biomass production. For example, *Spirulina* species shows higher growth yield in mixotrophic than others due to less photoinhibition and biomass loss during dark respiration period (Chojnacka and Noworyta 2004; Chen et al. 2013). Through mixotrophic method of cultivation, *Chlorella sorokiniana* MB-1-M12 is successfully grown for the production of lutein. With 75% media replacement ratio, the outdoor semicontinuous cultivation results in higher product productivity (Chen et al. 2019).

The biomass productivity of different microalgae and the characteristics of different production units are provided in Tables 4.1 and 4.2.

The different factors that impact the growth of microalgae and the biofuel production include:

- Productivity by photosynthetic efficiency
- Productivity by strain selection
- Lipid accumulation
- Production and harvesting cost

4.3.2 Biomass Recovery/Harvesting Microalgal Biomass

The biomass recovery occupies about 20–30% of the total production cost. The solid-liquid separation of microalgal biomass is a crucial step in biofuel production (Wang et al. 2008). This determines the economic value of the produced biomass. The harvesting technique depends on the size and strain selected for the biomass production and the cell density. Different solid-liquid separation techniques include:

- Centrifugal sedimentation
- Filtration
- Flocculation
- Flotation

The general harvesting techniques follow two major steps such as:

1. Bulk harvesting

The biomass separated from the bulk medium produces 2–7% solid biomass by a concentration factor of 100–800×. It depends on the harvesting method and the biomass concentration that are present initially. This aggregates the microalgal cells together, thereby enhancing the further processes.

Table 4.1 Microalgae and their productivity rate in different modes of cultivation

Microalgae	Production unit	Biomass productivity (g/L/d)	References
Nannochloropsis salina	Open raceway pond	0.013	Crowe et al. (2012)
Nannochloropsis gaditana	Tubular photobioreactor	0.59	San Pedro et al. (2014)
Scenedesmus sp.	Open raceway pond	0.170	de Godos et al. (2014)
Chlorella pyrenoidosa	Stirred bioreactor	3.45	Pleissner et al. (2013)
Chlorella minutissima	Fermenter	0.44	Katiyar et al. (2017)
Chlorella ellipsoidea	Bubble column	0.03–0.04	Wang et al. (2014)
Chlorella sorokiniana	Short light path flat plate	2.9–14.8	Tuantet et al. (2014)
Chlorella protothecoides	Fermenter	7.66	Lu et al. (2010)
Thermosynechococcus elongatus	Flat-plate airlift photobioreactor	2.9	Bergmann and Trösch (2016)
Chlorella vulgaris	Membrane photobioreactor	0.06	Marbella et al. (2014)
Phormidium autumnale	Bubble column	6.68	Francisco et al. (2015)
Scenedesmus obliquus	Bubble column	0.21	Maroneze et al. (2016)
Chlorella vulgaris	Flat-plate airlift	0.16–0.95	Münkel et al. (2013)
Chlorella vulgaris	Hybrid photobioreactor	0.05	Heidari et al. (2016)
Chlorella sp.	Airlift photobioreactor	0.25	Lal and Das (2016)
Chlorella vulgaris	Biofilm photobioreactor	7.07	Pruvost et al. (2017)
Scenedesmus acutus	Open raceway pond	0.066	Eustance et al. (2015)

2. Thickening

This step is the energy harnessing step that utilizes filtration, ultrasonic aggregation, and centrifugation techniques. This step helps in concentrating the slurry to form a thickened biomass.

4.3.2.1 Centrifugal Sedimentation

Centrifugal sedimentation is based on Stoke's law which states that settling characteristics of the solid suspension depends on Stoke's radius and density of the cells and the velocity of sedimentation (Schenk et al. 2008). This method is applicable for harvesting large-sized microalgae like *Spirulina*, which is approximately about >70 μm (Muñoz and Guieysse 2006). Centrifugal sedimentation is an

Table 4.2 Advantages and disadvantages of different production units

Production unit	Advantages	Disadvantages
Raceway pond (Chisti 2016; Chang et al. 2017; Singh and Sharma 2012)	Less power consumption	Large area requirement
	Economically easier in cleaning and maintenance	Limited surface illumination
	Less investment	Limited culture growth conditions
		Prone to contamination
Vertical photobioreactors (Koller 2015; Chang et al. 2017; Huang et al. 2017)	Scalability and easier to sterilize	Limited surface illumination
	Lesser photoinhibition	High construction and maintenance cost
	High mixing capacity and mass transfer rate	Complexity construction materials
Horizontal tubular photobioreactors (Chang et al. 2017; Torzillo et al. 2015)	Higher productivity	Large area requirement
	Larger surface illumination	Photoinhibition and decreased mass transfer
	Preferred for outdoor cultivation	
	Cheaper	
Heterotrophic bioreactor (Li et al. 2014a, b; Chang et al. 2017)	Low cost with high productivity	Limited microbial cultivation
	Scalability	Prone to contamination
Flat-plate photobioreactor (Faried et al. 2017; Li et al. 2014a, b; Bergmann and Trösch 2016)	Higher productivity and surface illumination	Difficulty in scalability and temperature control
	Less power consumption	Shear damage
	Preferred for outdoor cultivation	Photoinhibition and fouling
	Uniform distribution of light	
Hybrid photobioreactor (Heidari et al. 2016; Jacob-Lopes et al. 2016)	Higher stability and lesser operating cost	Limited microbial cultivation
	Lesser area requirement	
Biofilm photobioreactor (Gross et al. 2015; Hoh et al. 2015)	Lesser harvesting cost	Gradient formation and limitations in scalability
	Low water consumption and light limitations	
	Higher CO_2 mass transfer	
Membrane photobioreactor (Billad et al. 2015; Luo et al. 2016)	Easy operation	High cost
	Higher productivity	Limited studies
	Quality effluent treatment	

energy-dependent technique and the recovery process depends on the algal cell settling time, the residence time, and the settling depth of the centrifuge. The efficiency of the biomass recovery will be greater than 95%, and the cell viability is about 88–100% at 13,000×g (Brennan and Owende 2010).

4.3.2.2 Filtration

Filtration is the physical recovery process of microalgal cells that includes tangential flow filtration, dead end filtration, microfiltration, pressure filtration, ultrafiltration, and vacuum filtration. This method is applicable for harvesting larger microalgae of >70 μm like *Spirulina* and *Coelastrum.* It is also used for the recovery of smaller microalgae <30 μm through the membrane filtration method under applied pressure. For example, ultrafiltration and microfiltration are the widely used membrane filtration techniques. The conventional filtration technique for larger cell separation employs the use of any diatomaceous filter aid or other materials like cellulose to increase the efficiency of the biomass recovery (Molina-Grima et al. 2003).

4.3.2.3 Flocculation

Flocculation is the technique which, as exercised before any other harvesting techniques, can increase the efficiency of the biomass recovery (Chen et al. 2008; Chatsungnoen and Chisti 2016). Flocculation is the technique which induces aggregation of the algal cells by the addition of any cationic molecules such as multivalent cations, chitosan, or polymers. Organic flocculants can prevent the variations in the pH and has the advantage of using low concentrations of flocculant preventing environmental hazards (Sharma and Gupta 2006; Vandamme et al. 2013). The negative charge of the algal cells gets neutralized or decreased by the addition of cations thereby inducing the aggregation of the biomass. The cells may also aggregate through bridging process. Some examples for commonly used flocculants are metal salts such as aluminum sulfate ($Al_2(SO_4)_3$), ferric chloride ($FeCl_3$), and ferric sulfate ($Fe_2(SO_4)_3$) (Divakaran and Pillai 2002; Salim et al. 2011). Further subjecting the aggregated mass to other harvesting techniques can enhance the efficiency of the biomass recovery. The microalgal cells are also capable of undergoing the autoflocculation process. The autoflocculation process occurs by the precipitation of microalgal cells through the addition of phosphate, magnesium, carbonate, and calcium ions, thereby interrupting the CO_2 supply which in turn aggregates the algal cells by bridging process (Harun et al. 2010).

4.3.2.4 Flotation

Flotation is the physical technique that makes the algal cells float over the liquid medium naturally due to the accumulation of lipid content in cells. Flotation can also be acquired by the micro-sized air bubble dispersion on the liquid medium that enables the entrapment of algal cells to the surface. This method takes an advantage of devoid of chemical usage. However, the technical feasibility is less studied (Bruton et al. 2009).

4.3.3 Dewatering Process and Biomass Extraction

The wet biomass has to be further processed to extract the lipid from the inner cell. The dehydration process or drying process is widely employed for the conversion of wet biomass into a dried one for the easy extraction of biofuels. Subsequently, biofuel and coproducts (metabolites) are extracted and purified from the dried biomass.

Dehydration process helps in the enhancement of the viability of the desired final product. Different types of dehydration process include low-pressure shelf drying, drum drying, sun drying, fluidized bed drying, Refractance Window™ technology drying, spray drying, and freeze-drying (Wan et al. 2015; Wang et al. 2015a, b). Sun drying is the time-consuming, but ancient method used for dehydration process, while freeze- and spray drying are costly, but widely employed methods for drying large-scale high-value products. The optimum drying temperature has to be maintained in order to prevent the variation in the lipid yield and composition. The higher the temperature, the higher will be the changes in the lipid composition and yield. However, lipid TAG content remains unchanged on drying the biomass at 60 °C but alters the lipid yield. The lipid extraction from the biomass includes (Harris et al. 2018):

- Use of organic solvents such as acetone, diethyl ether, ethanol, benzene, and methanol
- Supercritical CO_2 extraction
- Gas-expanded liquid extraction
- Liquid CO_2 extraction
- Ionic liquid as the extraction solvent
- Switchable solvents for extraction

The extraction of metabolites and intracellular product involves the disruption of microalgal cells for the efficient release of the desired product. Physical, chemical, and mechanical cell disruption methods can be used for the release of intracellular content (Harris et al. 2018). The extraction process can also be carried out along with the cell disruption process simultaneously that includes techniques such as ultrasound-assisted extraction, microwave-assisted extraction, and surfactant-assisted extraction process.

4.3.4 Biofuel Conversion

The conversion of biomass to biofuel involves technically feasible methods involving biochemical, thermochemical, and chemical conversion and direct combustion process. Different feedstocks of microalgae and its conversion techniques along with its end product are given in Table 4.3.

The factors that affect conversion process are:

- Economic value of the product
- Quantity and type of the feedstock
- Energy form and the type of the end product

4.3.4.1 Biochemical Process

The conversion of algal biomass into a biofuel by a biochemical process involves the following methods:

Table 4.3 Different forms of microalgal feedstock and their conversion techniques for different end products (Pierobon et al. 2018)

Feedstock form	Conversion technique	End product
Microalgae in culture (<10% total suspended solids)	Anaerobic digestion	Methane
	Direct secretion	Hydrogen, alcohols, or alkanes
	Fermentation	Bioethanol
Microalgae slurry (>15% total suspended solids)	Supercritical water gasification	Syngas
	Hydrothermal liquefaction	Hydrothermal liquefaction
Dry powder	Direct combustion	Heat
	Pyrolysis	Bio-oil

Fig. 4.3 Anaerobic digestion

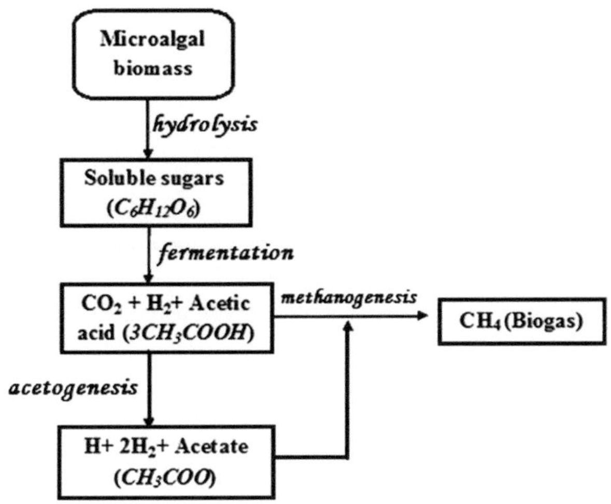

4.3.4.1.1 Anaerobic Digestion

This process involves anaerobic breaking down of organic biomass to release economically valuable biogas such as methane, carbon dioxide, and other trace gases, including hydrogen sulfide with an energy level 20–40% lesser than the feedstock heating value (Pittman et al. 2011). This process is applicable for the processing of wet biomass thereby overcoming the dehydration process. This process sequentially involves the following steps (Fig. 4.3):

(a) Hydrolysis: complex compounds (algal organic particulate substrate) → sugars (soluble)
(b) Fermentation: soluble sugars (in the presence of acidogenic bacteria) → $CO_2 + H_2$ + by-products (acetic acid, alcohols, volatile fatty acids)
(c) Acetogenesis: fermented products → acetate substrate

(d) Methanogenesis: $CO_2 + 4H_2$ (directly from step 2; in the presence of methanogens) → CH_4 (biogas) + H_2O (or) acetate → CH_4 (biogas)

These processes are carried out in an anaerobic digester in which the higher protein content can affect the digester performance (Sialve et al. 2009; Trivedi et al. 2015) and inhibits the anaerobic growth by ammonium production.

4.3.4.1.2 Alcoholic Fermentation

Alcoholic fermentation is the biological conversion of sugar containing biomass into carbon dioxide and bioethanol with the evolution of energy. Further purification of bioethanol is carried out by distillation in which the concentrated alcohol is condensed to produce an ethanolic solution. For example, high starch containing *C. vulgaris* (approx. 37% dry weight) is found to produce 65% efficiency of ethanolic conversion (Naik et al. 2010).

4.3.4.1.3 Photobiological Hydrogen Production

The eukaryotic microalgae have the potency to produce hydrogen gas naturally during light and dark reaction or during CO_2 fixation as an electron donor at anaerobic condition. The microalgae produce hydrogen anaerobically by conversion of the water molecules into hydrogen ions during photosynthesis and subsequently convert it to hydrogen by the hydrogenase enzyme. The photosynthetic hydrogen gas production initially involves the growth of microalgae by typical natural condition, and the latter stage involves depletion of sulfur, which in turn stimulates anaerobic condition and produces hydrogen gas constantly (Brennan and Owende 2010).

4.3.4.2 Thermochemical Process

The thermochemical process involves the thermal decomposition of organic matters in the microalgal biomass (Chen et al. 2015). Different types of thermochemical conversion process involve:

4.3.4.2.1 Gasification

In gasification process, the biomass undergo partial oxidation at a temperature of 800–1000 °C to produce syngas (combustible gas containing CO, CH_4, H_2, N, and CO_2) (Sanchez-Silva et al. 2013; Ramzan et al. 2011). The theoretical yield of methanol from 1 g of *Spirulina* is found to be 0.64 g by a gasification process at the temperature of 850–1000 °C (Goyal et al. 2008).

4.3.4.2.2 Liquefaction

Thermochemical liquefaction employs a low temperature of about 300–350 °C at a high pressure of 5–20 MPa in the presence of a catalyst and hydrogen to convert the biomass to produce biofuel (Shuping et al. 2010). This method is highly applicable for the conversion of wet microalgal biomass to yield liquid biofuel. For example, conversion of *B. braunii* biomass by thermochemical liquefaction at 300 °C is found to yield 64% dry weight basis of oil (Brennan and Owende 2010).

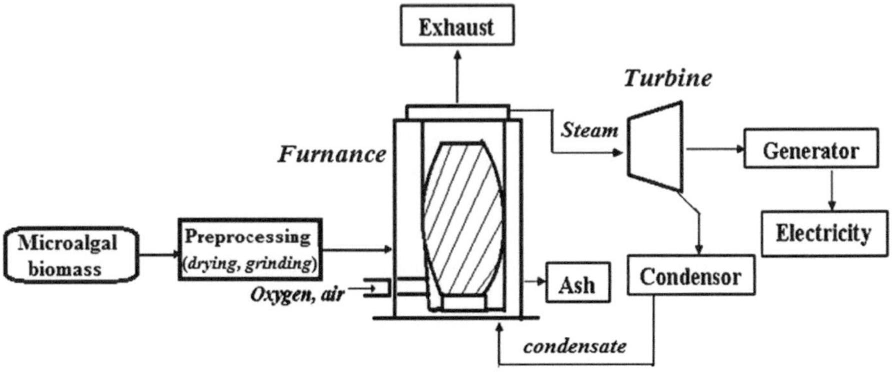

Fig. 4.4 Direct combustion process

4.3.4.2.3 Pyrolysis

Pyrolysis is the anaerobic process of conversion of the microalgal biomass into syngas, bio-oil, and charcoal at the temperature range of 350–700 °C. Pyrolysis process can be flash pyrolysis, slow pyrolysis, or fast pyrolysis process. The flash pyrolysis process is carried out at 500 °C with a vapor residence time of about one second, and the fast pyrolysis process is carried out at the same temperature with a vapor residence time of 10–20 s. Slow pyrolysis is carried out at 400 °C with a very long residence time (Bridgwater 2007; Sirajunnisa and Surendhiran 2016). The drawbacks in the pyrolysis of oil are its instability, acidity, and viscosity and require further upgrading techniques (Raheem et al. 2015; Casoni et al. 2016; Zhang et al. 2016).

4.3.4.3 Direct Combustion

The direct combustion is the process of burning biomass in the presence of oxygen or air. This helps in the conversion of chemical energy into hot gases usually carried out at a temperature of 800–1000 °C in a steam turbine, furnace, or boiler (Goyal et al. 2008; Suganya et al. 2016). The combustion of biomass should contain a moisture content of about <50% of its dry weight. Direct combustion would generate a power (100–300 MW) for domestic to industrial use. The pretreatment process is necessary for power generation like drying and grinding (McKendry 2002; Trivedi et al. 2015). The direct combustion of biomass is desired to produce both the power and heat simultaneously that will enhance plant efficiency. The schematic representation is shown in Fig. 4.4.

4.3.4.4 Chemical Reaction

Transesterification is the widely used conversion process for the biodiesel production from biomass. Transesterification is the chemical process of converting triglycerides into fatty acid methyl ester (FAME) and glycerol by reacting with methanol in the presence of an enzyme, acid, or base catalyst.

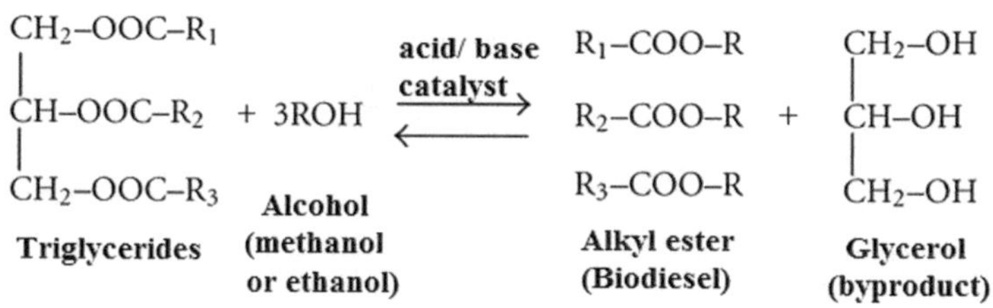

The biodiesel selectively separates from the mixture by the two-layer separation process. The glycerol layer is withdrawn to separate FAME of low molecular weight from the mixture (Cercado et al. 2018). This glycerol can be used in cosmetic or pharmaceutical industries.

The biodiesel produced from microalgae is highly viscous than petroleum diesel. This transesterification process can reduce the viscosity and increase the fluidity of the algal diesel (Al-lwayzy and Yusaf 2013). This can be blended with the petroleum diesel and can be used in engines without any further modifications. Direct and conventional transesterification are the two types of transesterification processes used for biodiesel formation.

4.3.4.4.1 Direct/In Situ Transesterification

The direct transesterification uses single-stage chemical conversion process without any pretreatments (Lee and Lee 2016; Ehimen et al. 2010). This chemical process involves simultaneous lipid extraction and transesterification process. In this direct transesterification process, the untreated wet biomass are fed directly into the reactor system for conversion process (Jazzar et al. 2015). The flow diagram of the direct transesterification process is shown in Fig. 4.5.

4.3.4.4.2 Conventional Transesterification

The conventional methods for the transesterification process involve microwave-assisted transesterification, ultrasound-assisted transesterification, or combination of microwave- and ultrasound-assisted transesterification. The combination of both methods is highly efficient in biodiesel production (Hsiao and Lin 2013). It is a two-stage process, such that it involves energy-intensive mechanical lipid extraction prior to the transesterification process (Al-lwayzy et al. 2014; Lee and Lee 2016;

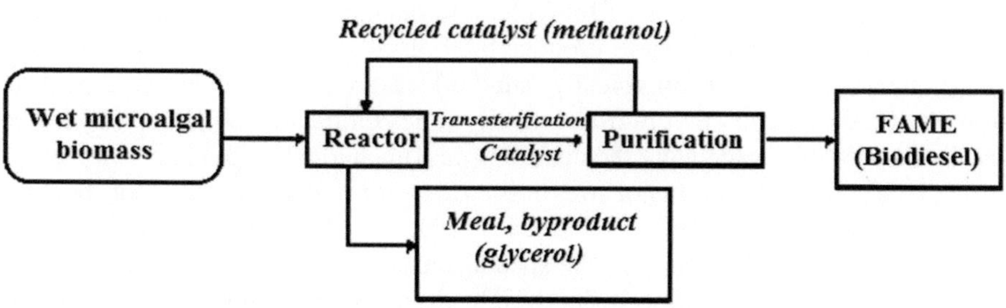

Fig. 4.5 Direct transesterification process

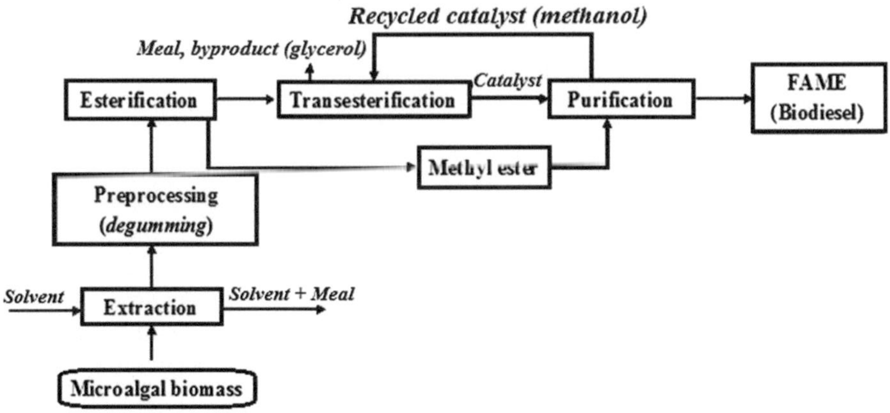

Fig. 4.6 Conventional two-stage transesterification process

Hossain et al. 2018). The conventional method can yield a refined biodiesel that can be applicable for a high-speed engine (Salam et al. 2016). However, the production cost of biodiesel is higher compared to the direct transesterification process due to the high energy and time-consuming process (Jazzar et al. 2015; Al-lwayzy et al. 2014). The flow diagram (Fig. 4.6) emphasizes the biodiesel production through conventional transesterification process.

4.4 Genetic Manipulation of Algae for Enhanced Lipid/TAG Production

There are different economical obstacles that prevent deployment of microalgal biofuels in a large scale to serve the global energy necessity. It starts from the upstream to the downstream production of biofuels for sustainable implementation into the environment. To overcome the difficulties in the large-scale production of microalgal biofuels, different approaches have been utilized. The highly preferred approach is the enhancement in the lipid/TAG accumulation. This can be achieved by the intensive knowledge in the induction of lipogenesis (lipid formation) with its biosynthetic pathway and the reactions related to it.

For the efficient production of biofuels, microalgae harness carbon dioxide and sunlight from the environment to produce biomolecules such as carbohydrates, lipids, proteins, and other economically value-added products. Biochemical conversion into biodiesel and other biofuels mainly relies on the degree of carbohydrates and lipids produced, especially the neutral lipids (uncharged lipids) like triacylglycerols (TAG). TAG plays a major role in the transesterification process for the successful formation of FAME (Lohman et al. 2015). TAG is the ester of fatty acid chains with a glycerol backbone. The specialized microalgal lipid bodies can accumulate about 20–60% of its cell dry weight during stress conditions (Klok et al. 2014). This TAG accumulation enables the microalgae to withstand the stress conditions such as chemical and physical conditions. The chemical stress conditions rely on the alterations in the salinity, CO_2 assimilation, exposure to heavy metals, and nutrient supplements. The physical stress conditions include temperature and light intensity shifts (Singh et al. 2016; Arora et al. 2017a, b). TAG maintains the energy cycle, cellular functions, and lipid homeostasis in the intracellular regions (Lenka et al. 2016). But the extended stress tolerance results in the photosynthesis breakdown. This in turn can cause chlorosis, restricted cell division, and reduced TAG productivity (Vonlanthen et al. 2015). In addition to this there will be a reduced lipogenesis process with the limited active growth of the microalgae. Thus, the biofuel production from microalgal cells highly depends on the optimum growth rate and its TAG accumulation (Davis et al. 2011).

4.4.1 General Strategies in Conventional Genetic Engineering Techniques

For the increased accumulation of lipid in microalgae, different genetic engineering techniques have been utilized for the formation of transgenic microalgae with modified lipogenetic pathways and other related pathways. Different genetic engineering tools such as electroporation and gene gun method for RNA silencing, DNA incorporation via homologous recombination, and integrative random selection marker are used for the successful transformation of microalgae for enhanced lipid production (Ghosh et al. 2016). The modifications are made within the chloroplast, mitochondria, or nuclear genome based on the genes deleted or inserted to form beneficial transformed microalgal cells (Gimpel et al. 2015). For instance, the engineered or modified enzymes in the plastid or nuclear genome responsible for secondary metabolism enable simpler and increased flexibility in the post-translational modifications. This creates an advantage of using untranslated codons and heterologous promoters. This would end in the reduced gene expressions due to gene silencing, reduced copy number, and positional effects (Gimpel et al. 2015). But it supports multiple gene expressions with the advantage of higher efficient gene knockout and homologous recombination.

Only a limited number of microalgae have been successfully transformed for the increased lipid accumulation through these genetic engineering techniques. Some of the successfully transformed microalgal cells are *Chlamydomonas reinhardtii* CC-849, *Cyclotella cryptic, Nannochloropsis gaditana, N. salina, Chlorella ellipsoidea, Dunaliella tertiolecta, Chlorella minutissimsa,* and *Phaeodactylum tricornutum* (Fei et al. 2017; Banerjee et al. 2016; Wei et al. 2017; Ho et al. 2014; Kang et al. 2017).

Lipid accumulation can be increased by any of the four major routes such as:

- Decrement in the TAG catabolism
- Improvement in carbon and energy metabolism
- Enhancement in the flux or fatty acid metabolic rate
- Reduction in the competing carbon synthesis pathway

Among those methods, flux enhancement is the most preferred recombinant technique used to increase lipid accumulation. Flux enhancement is carried out by the gene (genes responsible for lipid biosynthesis) overexpression. For example, gene acetyl-CoA carboxylase is the first gene overexpressed in *Cyclotella cryptic* and *N. saprophila* microalgae. But it doesn't result in enhanced lipid accumulation (Schuhmann et al. 2016). Furthermore, transformation of microalgae by the overexpression of lipid genes such as ME, fatty acid synthase, and isoforms of DGAT has also been carried out. The gene silencing of DGAT2-4 results in the increased lipid accumulation of approximately 30–50% (Scranton et al. 2015; Schuhmann et al. 2016). Along with these genes, thioesterase gene is also successfully overexpressed in the microalgae *Phaeodactylum tricornutum* nuclear genome that resulted in enhanced total fatty acid content (C12–C14) (Gimpel et al. 2015). Fatty acid-acyl carrier protein thioesterase gene when overexpressed in *Chlamydomonas reinhardtii* produced an increased total fatty acid content of about 14–15% (Wei et al. 2017). The gene expression of GDPI and LPAAT in *Chlamydomonas reinhardtii* leads to the increased total lipid content of about 23.6% and 17.4%, respectively (Wang et al. 2017), and the gene knockout of lipase gene results in the increased lipid accumulation of three times higher than the normal non-transformed cells of *Thalassiosira pseudonana* (Gimpel et al. 2015).

In case of the inhibition of competing carbon synthesis pathway (starch), the deletion of isoamaylase or AGPase gene results in the increased lipid/TAG accumulation in the mutant strain than the wild strain of *Chlamydomonas reinhardtii* (Schuhmann et al. 2016). However, the starch synthesis inhibition can result in decreased biomass production (Trentacoste et al. 2013).

4.5 TAG/Fatty Acid Biosynthesis

The modification of microalgae for the enhanced TAG accumulation creates a necessity to understand the TAG biosynthesis pathway that is summarized in Fig.4.7.

Fatty acids are amphipathic compounds that consist of a –COOH (carboxylic) group at one end and a variable hydrocarbon moiety at the other end (Li-Beisson et al. 2013). The inert lipid droplet, TAG, is a glycerolipid made up of three groups of fatty acyl molecules that are esterified with a backbone made of glycerol. The lipid metabolism of *Arabidopsis* is used for the basic understanding of fatty acid synthesis in the algae. But the biosynthesis of fatty acids in microalgae is found to vary in two different senses (Liu and Benning, 2012):

- The microalgae produce/synthesize TAG with unique acyl groups that are not present in plants (e.g., *Arabidopsis*).
- Prokaryotic plastidic TAG biosynthesis pathways are not found in plants, but are found in the microalgal lipid synthesis pathway.

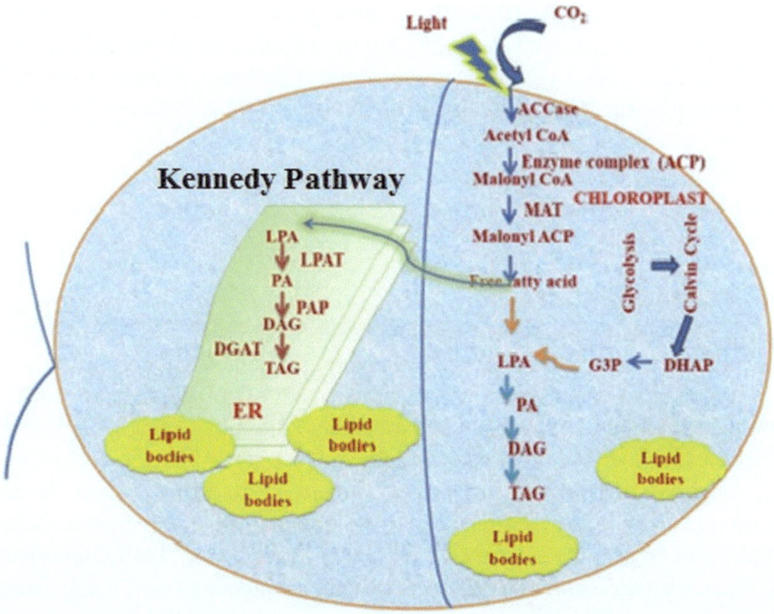

Fig. 4.7 TAG synthesis and accumulation cycle (*DAG* diacylglycerol, *DGAT* diacylglycerol acyltransferase, *G-3-P* glycerol-3-phosphate, *ACCase* acetyl-CoA carboxylase, *ACP* acyl carrier protein, *FFA* free fatty acid, *DHAP* dihydroxyacetone phosphate, *MAT* malonyl-CoA/ACPtransacylase, *PAT* lysophosphatidic acid acyltransferase, *LPA* lysophosphatidic acid, *PA* phosphatidic acid, *PAP* phosphatidic acid phosphatase, *TAG* triacylglycerol). (Adopted from Banerjee et al. 2016)

Acetyl-coenzyme A is the precursor molecule in lipid biosynthesis that occurs in the chloroplast through de novo pathway. Photosynthetic efficiency is the flux controlling trait that takes place in photoautotrophic conditions with acetyl-CoA as the carbon precursor, NADPH and NADH as the reducing agent, and ATP as the energy source.

Fatty acid biosynthesis initiates with the acetyl-CoA carboxylation to give malonyl-CoA that is catalyzed by the enzyme ACCase (acetyl-CoA carboxylase) (Cronan and Waldrop, 2002). This enzyme complex is the rate-limiting step in the biosynthesis of fatty acid in plants but not in algae (Huerlimann and Heimann 2013).

The malonyl-CoA enters the plastid of microalgae and gets transferred to ACP molecules (acyl carrier proteins). Further, there occurs a sequence of acyl chain elongation that yields C_{16} or C_{18} fatty acid chains. This sequence of reactions is catalyzed by fatty acid synthase enzyme (FAS) with multiple subunits.

The chain elongation of fatty acids produced terminates by the catalytic reaction of two enzymes: chloroplast acyltransferase and acyl-ACP thioesterase. The acyltransferase enzyme catalyzes the acyl group removal from the ACP molecule, immediately after which the nascent fatty acid gets transferred from the ACP molecule to form glycerol-3-phosphate. The succeeding enzyme, thioesterase, is involve in the hydrolysis of acyl-ACP, thereby releasing the free fatty acid molecule. This free fatty acid is thus taken away to the chloroplast outer envelope and gets renewed to acyl-CoA. This further yields the glycerolipids (Bellou et al. 2014; Li et al. 2013).

TAG biosynthesis majorly occurs through two different pathways, namely:

1. Kennedy pathway
2. Acyl-CoA independent pathway

4.5.1 Kennedy Pathway

The foremost primary pathway for the biosynthesis of TAG/glycerolipids in microalgae is Kennedy's biosynthesis pathway. It is an acyl-CoA-dependent pathway that initiates the acylation of glycerol-3-phosphate and the subsequent acylation of lysophosphatidic acid (LPA). This step results in phosphatidic acid formation. The formed phosphatidic acid will further get converted into diacylglycerol by dephosphorylation process in the presence of DGAT enzyme (diacylglycerol acyltransferase). The final step is the conversion of diacylglycerol into TAG (Banerjee et al. 2016). Thus, the formation of TAG occurs by the sequence of acyl group transfer at varying positions of G-3-P from the acyl-CoA molecule (Li-Beisson et al. 2013). It is necessary to focus on the genetic regulations in Kennedy's biosynthesis pathway for the genetic modifications in microalgae for enhanced TAG accumulation.

4.5.1.1 Acyltransferase

Acyltransferase is the key enzyme that defines the end product and the composition of acyl in the produced TAG during Kennedy's pathway. Acyltransferase enzymes found in Kennedy's pathway include acyl-CoA/glycerol-3-phosphate acyltransferase (GPAT), acyl-CoA/diacylglycerol acyltransferase (DGAT), and acyl-CoA/lysophosphatidic acid acyltransferase (LPAAT).

The initial step of acylation is catalyzed by the GPAT enzyme, in which the fatty acids are transferred from the acyl-ACP in the plastids to the glycerol-3-phosphate (position: *sn*-1) to form lysophosphatidic acid. The algae are found to possess only homologs of plastid GPAT, while plants are found to possess both the plastid and ER-GPAT enzyme (Vieler et al. 2012). For example, algal sources such as *Chlamydomonas reinhardtii, Phaeodactylum tricornutum,* and *Ostreococcus tauri* are found to contain only plastid GPAT enzyme. There is about 50% increment in the TAG accumulation in *Chlamydomonas* when GPAT gene from the green oleaginous microalgae, *Lobosphaera incises*, is overexpressed beyond any growth compromise (Iskandarov et al. 2015).

LPAAT enzyme is responsible for the second acyl group shift to lysophosphatidic acid (position: *sn*-2). This converts lysophosphatidic acid to phosphatidic acid. This phosphatidic acid gets dephosphorylated into diacylglycerol (*sn*-1, 2) in the presence of enzyme phosphatidic acid phosphatase. These phosphatidic acid and diacylglycerol are used for the subsequent formation of TAG or the membrane lipid molecules. LPAAT enzyme plays a key role in the production of the neutral lipid,

TAG; the gene knockdown of this LPAAT leads to approximately 20% decrement in the TAG production in *Chlamydomonas* species by means of artificially created microRNA (Lv et al. 2013).

The third acyltransferase enzyme, DGAT, is responsible for the catalysis reaction to assemble TAG by the addition of acyl group to diacylglycerol (position: *sn*-3). This step is the rate-limiting step in algal and plant TAG biosynthesis. There are different subtypes of DGAT enzymes (Vieler et al. 2012; Radakovits et al. 2012). But the regulation and role of these subtypes are not clearly understood since there is a difference in the presence of subtypes in different organisms. For example, one DGAT type 2 is present in the genome of *Arabidopsis*, while there are five subtypes of DGAT type 2 enzyme found in the genome of *C. reinhardtii*. Recent research is made to develop an in vitro radiolabel-free assay to understand the functions of CrDGTT (*C. reinhardtii* DGTT) (Liu et al. 2016). There are also *in vivo* studies made to understand the functions of DGAT through gene overexpression and knockdown (Iwai et al. 2014; Deng et al. 2012). DGAT type 2 is usually found to be residing in the endoplasmic region for the assembly of TAG molecules (Chapman and Ohlrogge 2012). The role of DGAT type 2 is also studied in the microalgae *Nannochloropsis* (heterokont oleaginous). DGAT type 2 genes are found to be upregulated in the microalgae *Nannochloropsis oceanica* IMET1, and TAG accumulation is found to be enhanced under the stress condition of nitrogen deprivation (Li et al. 2014a, b).

4.5.2 Acyl-CoA Independent Pathway

The acyl-CoA independent pathway is the alternative pathway for the synthesis of TAG molecules. It uses phospholipid that acts as an acyl donor molecule and diacylglycerol acts as the acyl acceptor molecule. This independent pathway is characterized by the acyl transfer that is catalyzed in the presence of PDAT [phospholipids: DGAT (diacylglycerol acyltransferase)] (Banerjee et al. 2016). The gene encoding this enzyme from *C. reinhardtii* (Cr-PDAT) is functionally characterized by cloning through reverse genetic methods. The multifunctional PDAT enzyme is responsible for the lipid turnover in the membrane and the biosynthesis of TAG in *C. reinhardtii*. It is found that under nitrogen depletion stress condition, the 65% gene knockdown of this Cr-PDAT gene reduced the TAG accumulation of about 58% (Yoon et al. 2012).

The in vitro synthesis of TAG uses membrane lipids in the chloroplast as the substrates such as sulfoquinovosyl diacylglycerol, phosphatidylglycerol, and monogalactosyl diacylglycerol (Yoon et al. 2012).

4.6 Fatty Acid and TAG Enhancement by Genetic Manipulations

In yeast and plants, the quality and quantity of TAG and fatty acids are improved by five general genetic manipulation strategies (Hoffmann et al. 2008):

1. β-Oxidation inhibition of fatty acids in peroxisome or reduction of the lipase enzyme hydrolysis
2. Thioesterase regulations for the optimization of the fatty acid chain length
3. Fatty acid desaturase enzyme regulation for the control of saturation profile of the fatty acid
4. Increment in the fatty acid precursors and glycerol-3-phosphate
5. Enzyme overexpressions for the effective biosynthesis of fatty acids and TAG

The above strategies are also found to be efficient in the genetic manipulation of microalgae for the TAG and fatty acid biosynthesis with varying results. The gene overexpression in Kennedy's pathway and the transcriptional regulators results in the enhanced TAG production and its composition.

4.6.1 Genes in Kennedy Pathway and in the Biosynthesis of Fatty Acid

For the enhancement of fatty acid biosynthesis, genes such as Kas II (β-ketoacyl-ACP synthase II) and ACC1 (acetyl-CoA carboxylase 1) are independently modified to increase the yield. It is carried out by providing an increased amount of precursor molecules to acyl- ACP. The ACCase enzyme is found to be the control point for lipid biosynthesis in potatoes and *E. coli* (Klaus et al. 2004; James and Cronan, 2004). However, the ACCase enzyme overexpression is not found to improve the lipid yield in *Chlorella vulgaris, Navicula saprophila, Chlorella sorokiniana,* and *Cyclotella cryptic* (Leyva et al. 2015; Radakovits et al. 2010; Wan et al. 2011). But Kas II overexpression in *C. reinhardtii* is found to induce C18 compositional modification (Blatti et al. 2013).

Another enzymatic complex that is responsible for the enhancement of TAG accumulation is a multifunctional interdependent FAS (fatty acid synthase) enzyme complex with multiple subunits. For example, the microalgae *C. reinhardtii* is found to accumulate more TAG by the transcription enhancement of FAS enzyme complex genes (along with the acyl ACP thioesterase 1 and ACP genes) that is grown in nitrogen depleted environment (Miller et al. 2010).

Modifications of genes in Kennedy's pathway include DGAT gene overexpression. It is found to have an enhanced neutral lipid (TAG) production of about 20–44% by the native gene DGAT2 overexpression in *C. reinhardtii* as well as in *P. tricornutum* (Niu et al. 2013; Deng et al. 2012). Other gene cassettes from the yeast such as GPAT, G3PDH, PAP, LPAT, and DGAT are successfully used to modify Kennedy's pathway that in turn resulted in the twofold enhancement in the TAG accumulation in *Chlorella minutissima* (Hsieh et al. 2012). Similarly, overexpression of *Lobosphaera incisa* GPAT leads to 50% increase in the TAG production in *Chlamydomonas reinhardtii* (Iskandarov et al. 2015). Thus, studies confirm that the modifications or regulation of genes in Kennedy's pathway can exert a great impact in the accumulation or production of TAG.

4.6.2 Desaturase Enzymatic Gene and Substrate Specificity

The fatty acid desaturase enzyme plays a key role in fatty acid desaturation that catalyzes the bond conversion from C–C to C=C in fatty acyl chain at certain positions (Li-Beisson et al. 2015). The detailed functional validation of genes that has the efficiency for fatty acid desaturation can enable the rational engineering technique to unsaturate the fatty acids.

The fatty alkane end product depends upon the specificity in the substrate for acyl-ACP thioesterase. The substrate decides the property and the length of the produced carbon chain (Radakovits et al. 2011).

4.6.3 Transcription and Regulatory Genes

Biochemical and molecular interpretations enable the genetic modification of microalgae for enhanced biofuel production. The computational and sequencing techniques enable the researchers to understand the genomic and transcriptomic level of microalgae that furthermore gives a clear exploration on the transcriptional as well as regulatory proteins involved in the fatty acid and TAG synthesis process. The genomic and transcriptomic analysis has enabled the reconstruction of the central metabolism of carbon and the fatty acid biosynthesis pathway in *Chlorella pyrenoidosa.* It is proposed that the enzyme NAD(H) kinase 3 controls the NADH supply for the production of fatty acid or lipid molecules and by overexpressing this enzyme leads to higher lipid production in cells of about 110.4% (Fan et al. 2015). On overexpressing the transcription factor, DOF type in *C. reinhardtii* leads to enhanced lipid or fatty acid production twice more than that of the wild type (Ibanez-Salazar et al. 2014).

4.6.4 Other Regulatory Genes

Excluding the biosynthetic genes for fatty acids or TAG, there are also other genes that are responsible for the genetic manipulation of microalgae for biofuel production. It includes the genes bound toward TAG accumulation, photosynthesis (PEPC1), lipid droplet formation, and TCA cycle, structural genes that could alter the TAG synthesis carbon flux. For instance, overexpressing the PtPNPLA3 gene (lipid body integrated gene) resulted in enhanced lipid formation of about 70% in the microalgae *Phaeodactylum tricornutum* (Wang et al. 2015a, b). Similarly, other genes such as CIS or PEPC1 when downregulated in *C. reinhardtii* also resulted in enhanced TAG formation of about 169.5% and 20% than the wild type (Deng et al. 2013, 2014). The CIS inhibition redirected the carbon flux in *Chlorella protothecoides* into the biosynthesis of lipid, thereby enhancing the TAG biosynthesis (Wu et al. 2015).

4.7 Lipid Induction Through Environmental Stress

The growth of microalgae at optimum condition can induce higher biomass productivity, but leads to low lipid synthesis. However, higher lipid-producing microalgal species is generally slow in its growth. Therefore, microalgae with high lipid induction in a short span of time are highly opted for biofuel production.

TAG molecules are naturally an energy storage component of the cell that allows microalgae to withstand the stress environment. The physiological biosynthesis pathway has to be reprogrammed for higher lipid induction. Different adverse conditions such as pH stress, temperature stress, nutrient stress, chemical stress, heavy metal stress, osmotic stress, and radiation stress are implemented for high TAG induction in microalgae as summarized in Table 4.4. When the growth of microalgae starts to decline, the components for the new membrane formation do not occur; rather the microalgal cells synthesize and accumulate TAG from the deposited fatty acids (Sharma et al. 2012). TAG molecules in turn act as the protective agent during adverse stress conditions.

Therefore, in adverse environmental conditions, microalgae are capable of modifying the lipid synthesis pathway that favors 20–50% dry cell weight of TAG production and accumulation is shown in Fig. 4.8.

4.8 Factors Affecting the Biomass Production

Microalgae depend on biotic/abiotic factors (such as toxic chemicals, salinity, temperature) and their complex interactions for its effective growth and nutrient consumption (Markou and Georgakakis, 2011; Abou-Shanab et al. 2016). Apart from these factors, high density of microalgal growth also leads to auto-inhibitor accumulation, reduced efficiency of photosynthesis, and self-shading (Xu et al. 2015).

4.8.1 Nutrient Supplements

Carbon: The vital element essential for effective microalgal growth is carbon. Inorganic carbon molecules are utilized by the concentrating carbon dioxide mechanism from the extracellular environment (Markou and Georgakakis 2011; Hwnag et al. 2011). HCO_3 and CO_2 are utilized by the microalgae as an inorganic carbon source. HCO_3 gets converted into carbon dioxide in the presence of carbonic anhydrase enzyme (Markou and Georgakakis 2011).

Nitrogen: Nitrogen forms the macroelement that is highly utilized for microalgal growth. The range of nitrogen content in the biomass of microalgae is from 1.0% to 10%. The range mainly depends on the nitrogen availability and the type of available nitrogen. It can be in the form of N_2, NO_2^-, NH_4^+, or NO_3^-, and this nitrogen deficiency enables the microalgae to improve the lipid accumulation within the cells (Xu et al. 2015). The deficiency in microalgae causes the following changes:

Table 4.4 Examples of enhanced lipid accumulation by environmental adverse condition (Sharma et al. 2012)

Microalgae	Stress condition	Lipid profile change
Phaeodactylum tricornutum	Limitation in nitrogen	69–75% increase in the TAG accumulation
Phaeodactylum tricornutum	UV irradiation	Increased PUFA and EPA content
Chlorella vulgaris	Limitation in nitrogen	16.41% increase in total lipid productivity
Chlorella sp.	Limitation in urea	0.124 g/L/d total lipid productivity
Chlorella vulgaris	Fe^{3+}	Increased total lipid productivity to 56.6%
Euglena gracilis	Cadmium, zinc, copper	Increased total lipid productivity
Chaetoceros sp.	Temperature limitation of 25 °C	16.8% increase in total lipid productivity
Isochrysis galbana	Temperature increase from 15 °C to 30 °C	Increased neutral lipid accumulation
Coelastrella sp.	Limitation in phosphorous and nitrogen – combined effect	Increased TAG accumulation
Tichocarpus crinitus	Low light intensity	Increased TAG accumulation
Chlamydomonas reinhardtii	Sulphur limitation	Increased TAG accumulation
Ochromonas danica	Temperature increase from 15 °C to 30 °C	Increased total lipid productivity
Thalassiosira pseudonana	Continuous or dark/light cycle at stationary phase (strong light)	Increased TAG accumulation
C. simplex	High UV-B irradiation	Increased total lipid productivity
Nannochloropsis oculata	Temperature increase from 20 °C to 25 °C	14.92 % increase in total lipid productivity
Dunaliella tertiolecta	29 g/L to 58 g/L NaCl (salinity stress)	Increased lipid and TAG content
Chlorella sp.	Alkaline pH	Increased TAG content
Cyclotella cryptica	Silicon starvation	27.6% to 54.1% increase in total lipid productivity

- Decrement in the thylakoid cellular content
- Acyl hydrolase enzyme activation
- phospholipid hydrolysis stimulation

These changes cause the increase in intercellular fatty acid content and the nitrogen deficiency produces TAG molecules from acyl-CoA and the activation of diacylglycerol acyltransferase. Thus, this results in the enhanced lipid/TAG accumulation in microalgal cells (Xu et al. 2015).

Phosphorous: Phosphorous is also considered as one of the vital macronutrients for microalgal growth. It is one of the vital limiting elements for its growth apart from its lesser concentration requirement (less than 1%). It is found to play an

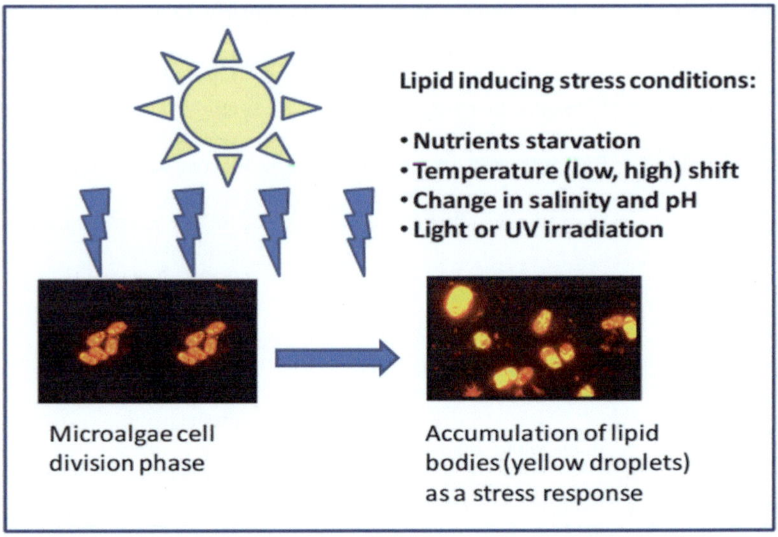

Fig. 4.8 Enhanced lipid accumulation through external stress conditions (Sharma et al. 2012)

essential role in photosynthesis, signal transduction, respiration, and transfer of energy (Singh and Saxena 2015). Furthermore, it results in the enhancement of TAG and lipid accumulation during the phosphorous depleted growth of certain microalgae such as *Pavlova lutheri, Phaeodactylum tricornutum, Isochrysis galbana, Chlorella* sp., and *Chaetoceros* sp. (Sharma et al. 2012; Liang et al. 2015)

Trace elements: The growth of microalgae also depends on the concentration of trace elements available in the nutrient media such as zinc, nickel, boron, molybdenum, cobalt, boron, copper, manganese, and chloride (Yang et al. 2016). The enzymatic reactions in microalgae depend on the iron content for effective photosynthetic process in PS I and II (Cao et al. 2014). The increase in the concentration of iron content in the nutrient media resulted in enhanced lipid as well as TAG accumulation in microalgae such as *Chlorella vulgaris* and *Scenedesmus obliquus* (Liu et al. 2008; Abd El Baky et al. 2012). However, the major challenge lies in the large-scale biomass production either in open system or closed system with the continuous supply of nutrients along with the trace elements and water. But the large-scale biomass production without any economic and environmental profit like waste water treatment is undesirable (Pittman et al. 2011; Lundquist et al. 2010).

4.8.2 Light Source and Intensity

Photosynthesis and microalgal growth are directly affected by light, i.e. the intensity, duration, and type of light source (Vadiveloo et al. 2015). The photosynthetic efficiency of microalgae depends on the light or dark cyclic phase (Shen 2014).

Optimum light illumination is essential for the higher biomass yield and lipid productivity, and the light illumination beyond the optimum requirement does not influence microalgal growth and can result in reduced cell viability (Harun et al. 2014; Simionato et al. 2013). The green microalgae consist of chlorophyll a and b as a light harnessing element that could readily absorb red and blue light sources (Schulze et al. 2014). Shading is the major issue in microalgal growth for higher biomass production. After the fourth day of microalgal cultivation, there will be decrease in the light penetration and light distribution. Different solutions for the prevention of shading limitation have been encountered such as (Bosma et al. 2014; Zhu et al. 2016; Ooms et al. 2016):

- Variations in the reactor design
- Use of different sources of light like halogen tungsten and LEDs
- Changes in the illumination of light incidence either on one side of the photobioreactor or on the double sides of the culturing photobioreactor
- Increasing the light path by changing the light and the direction of light fall on the reactor system
- Scattering microalgal culture by producing gas bubbles

The venture of isolating and identifying microalgae that can adapt to the environment with shorter light wavelength can overcome the light scarcity for the efficient growth and development of microalgae. Also, the use of low concentration of phytohormone chemicals can induce the microalgae to push the microalgal cells toward the light to enhance cell division, thereby, reducing the shelf shading due to high cell density (El-Salama et al. 2014).

4.8.3 Temperature Effect

Temperature is the other major factor that influences microalgal growth and lipid induction. It also influences the size of the cell, the nutrient consumption, and the biochemical composition of microalgal cells (Singh and Singh, 2015). The range of temperature for effective microalgal growth is from 15 °C to 40 °C based on the type of microalgal species and the place where they are isolated (Juneja et al. 2013). The major role of temperature includes photoinhibition that would in turn affect microalgal growth and biomass productivity. The photoinhibition mechanism of microalgae includes:

- Reduction in the carbon dioxide fixation as reduced temperature influences the electron transport in the specified flux.
- Reduced temperature can protect photosystem II by inhibiting active oxygen species formation and photoinhibition.
- Reduced temperature also impacts the assemblage of PS II active complex by interfering the production of D1 protein molecules.

The changes in the temperature influence the process of lipid deposition in microalgal cells (Singh and Singh 2015).

4.9 Conclusion and Future Perspectives

Microalgae are determined as the wide-spectrum biofuel producing source and are evaluated by many pilot-scale studies. Different NGOs as well as different global role players are actively participating in the conversion of microalgal biofuels as an alternative sustainable energy source for the efficient transport system. Various associations including ABO (Algae Biomass Organization), EABA (European Algae Biomass Association), and NAA (National Algae Association) are actively participating in the upgrade of microalgae-based biofuels. The microalgal biomass attributes higher productivity than that of higher plants. The viability of economically viable biofuel production is not yet thoroughly formulated, and there lies a necessity of optimizing the transition step of biomass into valuable biofuel. The reduction in the capital investment for the reactor design and microalgal cultivation forms the crucial part in the upscaling of pilot studies for biofuel production. The overestimation of productivity and underrating other factors such as the geographical location, biological interactions, and competitions also form the major issues in the upscalability of biofuel production.

Biofuel production can be enhanced by microalgal strain development, bioprospecting, as well as various breeding techniques. The bioprospecting of high oil yielding microalgae with short time span growth and with reduced shelf-shading forms the impact to overcome the difficulties in the high productive biofuel conversion. Bioprospecting can be carried out through various genetic and phenotypic characterizations of wild-type microalgal species. Some of the essential factors that influence microalga-based biofuel production are shear stress resilience, high cellular non-polar lipid content, and higher microalgal growth rate. Further, genetic modification of microalgae with the desired characteristics can create a mutagenic source for biofuel production. Staining of microalgal cells (florescent dyes: BODIPY 505/515 Nile red) can enable to differentiate the lipid yielding cells from others through FACs or flow cytometry methods. Raman micro-spectroscopy is another vital technique that would give the iodine number along with efficient differentiation high lipid yielding microalgal cells. Nuclear magnetic resonance techniques, ^{13}C- NMR, and ^{1}H- NMR spectroscopy are also found to be efficient in determining lipid content in microalgal cells.

The production of biofuel alone from the microalgal biomass would be economically inefficient without any lucrative possibilities such as bioremediation in terms of waste water treatment. The synergistic effect of biofuel production along with the waste water utilization for microalgal growth and biomass production increases the economic significance of microalgae in global industries.

References

Abd El Baky HH, El-Baroty GS, Bouaid A, Martinez M, Aracil J (2012) Enhancement of lipid accumulation in *Scenedesmus obliquus* by optimizing CO2 and Fe3+ levels for biodiesel production. Bioresour Technol 119:429–432

Abou-Shanab RAI, Singh M, Rivera-Cruz A, Power G, Bagby-Moon T, Das K (2016) Effect of Brachionus rubens on the growth characteristics of various species of microalgae. Electron J Biotechnol 22:68–74

Abu Hajar HA, Riefler RG, Stuart BJ (2017) Cultivation of the microalga Neochloris oleoabundans for biofuels production and other industrial applications (a review). Appl Biochem Microbiol 53(6):640–653

Adeniyi OM, Azimov U, Burluka A (2018) Algae biofuel: current status and future applications. J Renew Sustain Energy 90:316–335

Al-lwayzy S, Yusaf T (2013) Chlorella protothecoides microalgae as an alternative fuel for tractor diesel engines. Energies 6:766

Al-lwayzy SH, Yusaf T, Al-Juboori RA (2014) Biofuels from the fresh water microalgae Chlorella vulgaris (FWM-CV) for diesel engines. Energies 7:1829–1851

Andrade MR, Costa JAV (2007) Mixotrophic cultivation of microalga Spirulina platensis using molasses as organic substrate. Aquaculture 264(1–4):130–134

Apel AC, Botz DW (2015) Engineering solutions for open microalgae mass cultivation and realistic indoor simulation of outdoor environments. Bioprocess Biosyst Eng 38:995–1008

Apel AC, Pfaffingera CE, Basedahla N, Mittwollena N, Göbela J, Sautera J, Brückb T, Weuster-Botz D (2017) Open thin-layer cascade reactors for saline microalgae production evaluated in a physically simulated Mediterranean summer climate. Algal Res 25:381–390

Arora N, Gulati K, Patel A, Pruthi PA, Poluri KM, Pruthi V (2017a) A hybrid approach integrating arsenic detoxification with biodiesel production using oleaginous microalgae. Algal Res 24:29–39

Arora N, Patel A, Sharma M, Mehtani J, Pruthi PA, Pruthi Fei X, Li X, Li P, Deng X (2017b) Involvement of Chlamydomonas DNA damage tolerance gene UBC2 in lipid accumulation. Algal Res 22:148–159

Awudu I, Zhang J (2012) Uncertainties and sustainability concepts in biofuel supply chain management: a review. Renew Sust Energ Rev 2:1359–1368

Banerjee C, Dubey KK, Shukla P (2016) Metabolic engineering of microalgal based biofuel production: prospects and challenges. Front Microbiol 7:432

Bellou S, Baeshen MN, Elazzazy AM, Aggeli D, Sayegh F, Aggelis G (2014) Microalgal lipids biochemistry and biotechnological perspectives. Biotechnol Adv 32:1476–1493

Bergmann P, Trösch W (2016) Repeated fed-batch cultivation of Thermosynechococcus elongatus BP-1 in flat-panel airlift photobioreactors with static mixers for improved light utilization: influence of nitrate, carbon supply and photobioreactor design. Algal Res 17:79–86

Bigogno C, Khozin-Goldberg I, Cohen Z (2002) Accumulation of arachidonic acid-rich triacylglycerols in the microalga *Parietochloris incisa* (trebuxiophyceae, chlorophyta). Phytochemistry 60:135–143

Bilanovic D, Andargatchew A, Kroeger T, Shelef G (2009) Freshwater and marine microalgae sequestering of CO_2 at different C and N concentrations—response surface methodology analysis. Energy Convers Manag 50(2):262–267

Billad MR, Arafat HA, Vankelecom IFJ (2015) Membrane technology in microalgae cultivation and harvesting: a review. Biotechnol Adv 32:1283–1300

Blanken W, Schaap S, Theobald S, Rinzema A, Wijffels RH, Janssen M (2017) Optimizing carbon dioxide utilization for microalgae biofilm cultivation. Biotechnol Bioeng 114(4):769–776

Blatti JL, Michaud J, Burkart MD (2013) Engineering fatty acid biosynthesis in microalgae for sustainable biodiesel. Curr Opin Chem Biol 17:496–505

Bligh EG, Dyer WJ (1959) A rapid method of total lipid extraction and purification. Can J Biochem Physiol 37:911–917

Bosma R, de Vree JH, Slegers PM, Janssen M, Wijffels RH, Barbosa MJ (2014) Design and construction of the microalgal pilot facility AlgaePARC. Algal Res Biomass Biofuels Bioprod 6:160–169
Brennan I, Owende P (2010) Biofuels from microalgae. A review of technologies for production, processing and extractions of biofuels and co-products. Renew Sust Energ Rev 14:557–577
Bridgwater AV (2007) IEA bioenergy 27th update: biomass pyrolysis. Biomass Bioenergy 31:7–18
Bruton T, Lyons H, Lerat Y, Stanley M, Rasmussen MB (2009) A review of the potential of marine algae as a source of biofuel in Ireland. Sustainable Energy Ireland, Dublin, p 88
Buchspies B, Kaltschmitt M (2016) Life cycle assessment of bioethanol from wheat and sugar beet discussing environmental impacts of multiple concepts of co-product processing in the context of the European Renewable Energy Directive. Biofuels:1–24
Cao J, Yuan H, Li B, Yang J (2014) Significance evaluation of the effects of environmental factors on the lipid accumulation of *Chlorella minutissima* UTEX 2341 under low-nutrition heterotrophic condition. Bioresour Technol 152:177–184
Casoni AI, Zunino J, Piccolo MC, Volpe MA (2016) Valorization of Rhizoclonium sp. algae via pyrolysis and catalytic pyrolysis. Bioresour Technol 216:302–307
Cercado PA, Ballesteros FC, Capareda SC (2018) Biodiesel from three microalgae transesterification processes using different homogenous catalysts. Int J Technol 4:645–651
Chang JS, Show PL, Ling TC, Chen CY, Ho SH, Tan CH et al (2017) Photobioreactors. In: Larroche C, Sanroman M, Du G, Pandey A (eds) Current developments in biotechnology and bioengineering: bioprocesses, bioreactors and controls. Elsevier, Atlanta, pp 313–352
Chapman KD, Ohlrogge JB (2012) Compartmentation of triacylglycerol accumulation in plants. J Biol Chem 287:2288–2294
Chatsungnoen T, Chisti Y (2016) Harvesting microalgae by flocculation–sedimentation. Algal Res 13:271–283
Chen L, Li P, Liu Z, Jiao Q (2008) The released polysaccharide of the cyanobacterium Aphanothece halophytica inhibits flocculation of the alga with ferric chloride. J Appl Phycol 21(3):327–331
Chen CY et al (2013) Microalgae-based carbohydrates for biofuel production. Biochem Eng J 78:1–10
Chen W, Lin B, Huang M, Chang J (2015) Thermochemical conversion of microalgal biomass into biofuels: a review. Bioresour Technol 184:314–327
Chen JH, Chen CY, Hasunuma T, Kondo A, Chang CH, Ng I, Chang JS (2019) Enhancing lutein production with mixotrophic cultivation of *Chlorella sorokiniana* MB-1-M12 using different bioprocess operation strategies. Bioresour Technol 278:17–25
Chew KW, Yap JY, Show PL, Suan NH, Juan JC, Ling TC, Lee DJ, Chang J (2017) Microalgae biorefinery: high value products perspectives. Bioresour Technol 229:53–62
Chi Z, Zheng Y, Jiang A, Chen S (2011) Lipid production by culturing oleaginous yeast and algae with food waste and municipal wastewater in an integrated process. Appl Biochem Biotechnol 165:442–453
Chiaramonti D, Prussi M, Casini D, Tredici MR, Rodolfi L, Bassi N, Zittelli GC, Bondioli P (2013) Review of energy balance in raceway ponds for microalgae cultivation: rethinking a traditional system is possible. Appl Energy 102:101–111
Chisti Y (2007) Biodiesel from microalgae. Biotechnol Adv 25(3):294–306
Chisti Y (2016) Large-scale production of algal biomass: raceway ponds. In: Bux F, Chisti Y (eds) Algae biotechnology: products and processes. Springer, New York, pp 21–40
Chiu SY, Kao CY, Tsai MT, Ong SC, Chen CH, Lin CS (2009) Lipid accumulation and CO_2 utilization of Nanochloropsis oculata in response to CO_2 aeration. Bioresour Technol 100(2):833–838
Cho DH, Choi JW, Kang Z, Kim BH, Oh HM, Kim HS et al (2017) Microalgal diversity fosters stable biomass productivity in open ponds treating wastewater. Sci Rep 7:1–11
Chojnacka K, Noworyta A (2004) Evaluation of Spirulina sp. growth in photoautotrophic, heterotrophic and mixotrophic cultures. Enzym Microb Technol 34(5):461–465
Craggs R, Heubeck S, Lundquist T, Benemann J (2011) Algal biofuels from wastewater treatment high rate algal ponds. Water Sci Technol 63(4):660–665

Cronan JE, Waldrop GL (2002) Multi-subunit acetyl-CoA carboxylases. Prog Lipid Res 41:407–435

Crowe B, Attalah S, Agrawa S, Waller P, Rya R, Van Wagenen J et al (2012) A comparison of Nannochloropsis salina growth performance in two outdoor pond designs: conventional raceways versus the arid pond with superior temperature management. Int J Chem Eng Appl 2012:9–21

Davis R, Aden A, Pienkos PT (2011) Techno-economic analysis of autotrophic microalgae for fuel production. Appl Energy 88:3524–3531

De Andrade GA, Berenguel M, Guzmán JL, Pagano DJ, Acién FG (2016) Optimization of biomass production in outdoor tubular photobioreactors. J Process Control 37:58–69

de Godos I, Mendoza JL, Acién FG, Molina E, Banks CJ, Heaven S et al (2014) Evaluation of carbon dioxide mass transfer in raceway reactors for microalgae culture using flue gases. Bioresour Technol 153:307–314

de la Jara A, Assunção P, Portillo E, Freijanes K, Mendoza H (2016) Evolution of microalgal biotechnology: a survey of the European Patent Office database. J Appl Phycol 28(5):2727–2740

Deng XD, Gu B, Li YJ, Hu XW, Guo JC, Fei XW (2012) The roles of acyl-CoA: diacylglycerol acyltransferase 2 genes in the biosynthesis of triacylglycerols by the green algae Chlamydomonas reinhardtii. Mol Plant 5:945–947

Deng X, Cai J, Fei X (2013) Effect of the expression and knockdown of citrate synthase gene on carbon flux during triacylglycerol biosynthesis by green algae Chlamydomonas reinhardtii. BMC Biochem 14:38

Deng X, Cai J, Li Y, Fei X (2014) Expression and knockdown of the PEPC1 gene affect carbon flux in the biosynthesis of triacylglycerols by the green alga Chlamydomonas reinhardtii. Biotechnol Lett 36:2199–2208

Depra MC, LGR M, de Menezes CR, Zepka LQ, Jacob-Lopes E (2019) A new hybrid photobioreactor design for microalgae culture. Chem Eng Res Des 144:1–10

Devi MP, Mohan SV (2012) CO2 supplementation to domestic wastewater enhances microalgae lipid accumulation under mixotrophic microenvironment: effect of sparging period and interval. Bioresour Technol 112:116–123

Divakaran R, Pillai VNS (2002) Flocculation of algae using chitosan. J Appl Phycol 14(5):419–422

Doucha J, Straka F, Lívanský K (2005) Utilization of flue gas for cultivation of microalgae (Chlorella sp.) in an outdoor open thin-layer photobioreactor. J Appl Phycol 17(5):403–412

Duraiarasan S, Razack SA, Manickam A et al (2016) Direct conversion of lipids from marine microalga C. salina to biodiesel with immobilised enzymes using magnetic nanoparticle. J Environ Chem Eng 4:1393–1398

Ehimen EA, Sun ZF, Carrington CG (2010) Variables affecting the in situ transesterification of microalgae lipids. Fuel 89:677–684

El-Salama S, Kabra AN, Ji MK, Kim JR, Min B, Jeon BH (2014) Enhancement of microalgae growth and fatty acid content under the influence of phytohormones. Bioresour Technol 172:97–103

Eriksen N (2008a) Production of phycocyanin – a pigment with applications in biology, biotechnology, foods and medicine. Appl Microbiol Biotechnol 80(1):1–14

Eriksen N (2008b) The technology of microalgal culturing. Biotechnol Lett 30(9):1525–1536

Eustance E, Badvipour S, Wray JT, Sommerfeld MR (2015) Biomass productivity of two Scenedesmus strains cultivated semi-continuously in outdoor raceway ponds and flat-panel photobioreactors. J Appl Phycol 28:1471–1483

Fan J, Ning K, Zeng X, Luo Y, Wang D, Hu J et al (2015) Genomic foundation of starch-to-lipid switch in oleaginous Chlorella spp. Plant Physiol 169:2444–2461

Faried M, Samer M, Abdelsalam E, Yousef RS, Attia YA, Ali AS (2017) Biodiesel production from microalgae: processes, technologies and recent advancements. Renew Sust Energ Rev 79:893–913

Fei X, Li X, Li P, Deng X (2017) Involvement of Chlamydomonas DNA damage tolerance gene UBC2 in lipid accumulation. Algal Res 22:148–159

Fernández FGA, Sevilla JMF, Grima EM (2013) Photobioreactors for the production of microalgae. Rev Environ Sci Biotechnol 12:131–151

Francisco EC, Franco TT, Zepka LQ, Jacob-Lopes E (2015) From waste-to-energy: the process integration and intensification for bulk oil and biodiesel production by microalgae. J Environ Chem Eng 3:482–487

Fu CC, Hung TC, Chen JY et al (2010) Hydrolysis of microalgae cell walls for production of reducing sugar and lipid extraction. Bioresour Technol 101(22):8750–8754

García-Mañas F, Guzmána JL, Berenguela M, Acién FG (2019) Biomass estimation of an industrial raceway photobioreactor using an extended Kalman filter and a dynamic model for microalgae production. Algal Res 37:103–114

Ghosh A, Khanra S, Mondal M, Halder G, Tiwari ON, Saini S, Kumar T, Gayen K (2016) Progress toward isolation of strains and genetically engineered strains of microalgae for production of biofuel and other value added chemicals: a review. Energy Convers Manag 113:104–118

Gimpel JA, Henríquez V, Mayfield SP (2015) In metabolic engineering of eukaryotic microalgae: potential and challenges come with great diversity. Front Microbiol 6:1–14

Gomaa MA, Al-Haj L, Abed RMM (2016) Metabolic engineering of Cyanobacteria and microalgae for enhanced production of biofuels and high-value products. J Appl Microbiol 121(4):919–931

Gomes MCS, Arroyo PA, Pereira NC (2015) Influence of oil quality on biodiesel purification by ultrafiltration. J Membr Sci 496:242–249

Goyal HB, Seal D, Saxena RC (2008) Bio-fuels from thermochemical conversion of renewable resources: a review. Renew Sust Energ Rev 12(2):504–501

Gross M, Jarboe D, Wen Z (2015) Biofilm-based algal cultivation systems. Appl Microbiol Biotechnol 99:5781–5789

Gurr MI, Harwood JL, Frayn KN (2002) Lipid biochemistry: an introduction, 5th edn. Blackwell, Oxford, p 320

Halim R, Gladman B, Danquah MK, Webley PA (2011) Oil extraction from microalgae for biodiesel production. Bioresour Technol 102(1):178–185

Harris J, Viner K, Champagne P, Jessop PG (2018) Advances in microalgal lipid extraction for biofuel production: a review. Biofuels Bioprod Bioref:1–18

Harun R, Singh M, Forde GM, Danquah MK (2010) Bioprocess engineering of microalgae to produce a variety of consumer products. Renew Sust Energ Rev 14:1037–1047

Harun I, Yahya L, Chik MN, Kadir NNA, Pang MA (2014) Effects of natural light dilution on microalgae growth. Int J Chem Eng Appl 5(2):112–116

Heidari M, Kariminia HR, Shayegan J (2016) Effect of culture age and initial inoculums size on lipid accumulation and productivity in a hybrid cultivation system of Chlorella vulgaris. Process Saf Environ 104:111–122

Hena S, Fatimah S, Tabassum S (2015) Cultivation of algae consortium in a dairy farm wastewater for biodiesel production. Water Resourc Ind 10:1–14

Ho SH, Chen CY, Chang JS (2012) Effect of light intensity and nitrogen starvation on CO2 fixation and lipid/carbohydrate production of an indigenous microalga Scenedesmusobliquus CNW-N. Bioresour Technol 113:244–252

Ho S, Ye X, Hasunuma T, Chang J, Kondo A (2014) Perspectives on engineering strategies for improving biofuel production from microalgae – a critical review. Biotechnol Adv 32:1448–1459

Hoffmann M, Wagner M, Abbadi A, Fulda M, Feussner I (2008) Metabolic engineering of omega3-very long chain polyunsaturated fatty acid production by an exclusively acyl-CoA-dependent pathway. J Biol Chem 283:22352–22362

Hoh D, Watson S, Kan E (2015) Algal biofilm reactors for integrated wastewater treatment and biofuel production: a review. Chem Eng J 287:466–473

Hossain FM, Rainey TJ, Ristovski Z, Brown RJ (2018) Performance and exhaust emissions of diesel engines using microalgae FAME and the prospects for microalgae HTL biocrude. Renew Sust Energ Rev 82:4269–4278

Hsiao M, Lin C (2013) Optimization of the production of biodiesel assisted by ultrasonic and microwave. Int J Electr Power 2(2):54–59

Hsieh HJ, Su CH, Chien LJ (2012) Accumulation of lipid production in Chlorella minutissima by triacylglycerol biosynthesis-related genes cloned from Saccharomyces cerevisiae and Yarrowia lipolytica. J Microbiol 50:526–534

Hsueh HT, Chu H, Yu ST (2007) A batch study on the bio-fixation of carbon dioxide in the absorbed solution from a chemical wet scrubber by hot spring and marine algae. Chemosphere 66(5):878–886

Htet MZ, Ling LY, Yun SH et al (2013) Biofuel from microalgae: a review on the current status and future trends. Int J Adv Biotechnol Res 4(3):329–341

Hu JY, Sato T (2017) A photobioreactor for microalgae cultivation with internal illumination considering flashing light effect and optimized light-source arrangement. Energy Convers Manag 133:558–565

Hu G, Pan JL, Hu YL et al (2008) Preparation and evaluation of solid phase microextraction fiber based on molecularly imprinted polymers for trace analysis of tetracyclines in complicated samples. J Chromatogr A 1188:97–107

Huang Q, Jiang F, Wang L, Yang C (2017) Design of photobioreactors for mass cultivation of photosynthetic organisms. Engineering 3:318–329

Huerlimann R, Heimann K (2013) Comprehensive guide to acetyl-carboxylases in algae. Crit Rev Biotechnol 33:49–65

Huntley M, Redalje D (2007) CO_2 mitigation and renewable oil from photosynthetic microbes: a new appraisal. Mitig Adapt Strat Glob Chang 12(4):573–608

Hwnag JH, Choi JA, Abou-Shanab RA, Raghavulu VS, Salama ESA, Dempsey BA, Jeon BH (2011) Influences of CO2 and light wavelength on the acceleration of microalgal biomass as raw materials for biodiesel production. Abstracts of papers of the American Chemical Society. American Chemical Society 1155, Washington, DC, USA 20036

Ibanez-Salazar A, Rosales-Mendoza S, Rocha-Uribe A, Ramirez-Alonso JI, Lara-Hernandez I, HernandezTorres A et al (2014) Over-expression of Dof-type transcription factor increases lipid production in Chlamydomonas reinhardtii. J Biotechnol 184:27–38

Iskandarov U, Sitnik S, Shtaida N, Didi-Cohen S, Leu S, Khozin-Goldberg I et al (2015) Cloning and characterization of a GPAT-like gene from the microalga Lobosphaera incisa (Trebouxiophyceae): overexpression in Chlamydomonas reinhardtii enhances TAG production. J Appl Phycol 28:907–919

Ito T, Tanaka M, Shinkawa H, Nakada T, Ano Y, Kurano N et al (2013) Metabolic and morphological changes of an oil accumulating trebouxiophycean alga in nitrogen-deficient conditions. Metabolomics 9:178–187

Iwai M, Ikeda K, Shimojima M, Ohta H (2014) Enhancement of extraplastidic oil synthesis in Chlamydomonas reinhardtii using a type-2 diacylglycerol acyltransferase with a phosphorus starvation-inducible promoter. Plant Biotechnol J 12:808–819

Jacob-Lopes E, Zepka LQ, Merida LGR, Maroneze MM, Neves C (2016) Fotobiorreator híbrido industrial e bioprocesso realizado por meio do dito fotobiorreator. Patent PT2016041028

James ES, Cronan JE (2004) Expression of two Escherichia coli acetyl-CoA carboxylase subunits is autoregulated. J Biol Chem 279:2520–2527

Janssen M, Tramper J, Mur LR, Wijffels RH (2003) Enclosed outdoor photobioreactors: light regime, photosynthetic efficiency, scale-up, and future prospects. Biotechnol Bioeng 81(2):193–210

Jazzar S, Quesada-Medina J, Olivares-Carrillo P, Marzouki MN, Acién-Fernández FG, Fernández-Sevilla JM et al (2015) A whole biodiesel conversion process combining isolation, cultivation and in situ supercritical methanol transesterification of native microalgae. Bioresour Technol 190:281–288

Jiménez C, Cossío BR, Labella D, Xavier Niell F (2003) The feasibility of industrial production of Spirulina (Arthrospira) in southern Spain. Aquaculture 217:179–190

Jorquera O, Kiperstok A, Sales EA, Embiruçu M, Ghirardi ML (2010) Comparative energy life-cycle analyses of microalgal biomass production in open ponds and photobioreactors. Bioresour Technol 101:1406–1413

Juneja A, Ceballos RM, Murthy GS (2013) Effects of environmental factors and nutrient availability on the biochemical composition of algae for biofuels production: a review. Energies 6:4607–4638

Kang NK, Kim EK, Kim YU, Lee B, Jeong WJ, Jeong BR, Chang YK (2017) Increased lipid production by heterologous expression of AtWRI1 transcription factor in Nannochloropsis salina. Biotechnol Biofuels 10:1–14

Katiyar R, Gurjar BR, Bharti RQ, Kumar A, Biswas S, Pruthi V (2017) Heterotrophic cultivation of microalgae in photobioreactor using low cost crude glycerol for enhanced biodiesel production. Renew Energy 113:1359–1365

Khademi F, Yıldız İ, Yıldız AC, Abachi S (2015) Advances in algae harvesting and extracting technologies for biodiesel production. In: Progress in clean energy, vol 2. Springer, Cham, pp 65–82

Khozin-Goldberg I, Cohen Z (2006) The effect of phosphate starvation on the lipid and fatty acid composition of the fresh water eustigmatophyte *Monodus subterraneus*. Phytochemistry 67:696–701

Klaus D, Ohlrogge JB, Neuhaus HE, Dormann P (2004) Increased fatty acid production in potato by engineering of acetyl-CoA carboxylase. Planta 219:389–396

Klok AJ, Lamers PP, Martens DE, Draaisma RB, Wijffels RH (2014) Edible oils from microalgae: insights in TAG accumulation. Trends Biotechnol 32:521–528

Koller M (2015) Design of closed photobioreactors for algal cultivation. In: Prokop A, Bajpai RK, Zappi ME (eds) Algal biorefineries, products and refinery design, vol 2. Springer, Cham, pp 139–186

Koller M, Salerno A, Tuffner P, Koinigg M, Böchzelt H, Schober S et al (2012) Characteristics and potential of micro algal cultivation strategies: a review. J Clean Prod 37:377–388

Lal A, Das D (2016) Biomass production and identification of suitable harvesting technique for Chlorella sp. MJ 11/11 and Synechocystis PCC 6803. 3. Biotech 6:41–51

Laurens LML, Chen-Glasser M, McMillan JD (2017) A perspective on renewable bioenergy from photosynthetic algae as feedstock for biofuels and bioproducts. Algal Res 24:261–264

Lee OK, Lee EY (2016) Sustainable production of bioethanol from renewable brown algae biomass. Biomass Bioenergy 92:70–75

Lenka SK, Carbonaro N, Park R, Miller SM, Thorpe I, Li Y (2016) Current advances in molecular, biochemical, and computational modeling analysis of microalgal triacylglycerol biosynthesis. Biotechnol Adv 34:1046–1063

Leyva L, Bashan Y, de-Bashan L (2015) Activity of acetyl-CoA carboxylase is not directly linked to accumulation of lipids when Chlorella vulgaris is co-immobilised with Azospirillum brasilense in alginate under autotrophic and heterotrophic conditions. Ann Microbiol 65:339–349

Li P, Miao X, Li R et al (2011) In situ biodiesel production from fast-growing and high oil content Chlorella pyrenoidosa in rice straw hydrolysate. J Biomed Biotechnol 1:8

Li Y, Han D, Yoon K, Zhu S, Sommerfeld M, Hu Q (2013) Molecular and cellular mechanisms for lipid synthesis and accumulation in microalgae: biotechnological implications. Handbook of microalgal culture. Wiley, Chichester, pp 545–565

Li J, Han D, Wang D, Ning K, Jia J, Wei L et al (2014a) Choreography of transcriptomes and lipidomes of nannochloropsis reveals the mechanisms of oil synthesis in microalgae. Plant Cell 26:1645–1665

Li J, Stamato M, Velliou E, Jeffryes C, Agathos SN (2014b) Design and characterization of a scalable airlift flat panel photobioreactor for microalgae cultivation. J Appl Phycol 27:75–86

Liang Z, Liu Y, Ge F, Liu N, Wong M (2015) A pH-dependent enhancement effect of co-cultured Bacillus licheniformis on nutrient removal by Chlorella vulgaris. Ecol Eng 75:258–263

Liao JC, Mi L, Pontrelli S, Luo S (2016) Fuelling the future: microbial engineering for the production of sustainable biofuels. Nat Rev Microbiol 14(5):288–304

Li-Beisson Y, Shorrosh B, Beisson F, Andersson MX, Arondel V, Bates PD et al (2013) Acyl-lipid metabolism. Arabidopsis Book 11:e0161

Li-Beisson Y, Beisson F, Riekhof W (2015) Metabolism of acyl-lipids in *Chlamydomonas reinhardtii*. Plant J 82:504–522

Linares P, Perez-Arriaga IJ (2013) A sustainable framework for biofuels in Europe. Energy Policy 52:166–169

Liu B, Benning C (2012) Lipid metabolism in microalgae distinguishes itself. Curr Opin Biotechnol 24:300–309

Liu ZY, Wang GC, Zhou BC (2008) Effect of iron on growth and lipid accumulation in Chlorella vulgaris. Bioresour Technol 99(11):4717–4722

Liu J, Han D, Yoon K, Hu Q, Li Y (2016) Characterization of type 2 diacylglycerol acyltransferases in *Chlamydomonas reinhardtii* reveals their distinct substrate specificities and functions in triacylglycerol biosynthesis. Plant J 86:3–19

Lohman EJ, Gardner RD, Pedersen T, Peyton BM, Cooksey KE, Gerlach R (2015) Optimized inorganic carbon regime for enhanced growth and lipid accumulation in Chlorella vulgaris. Biotechnol Biofuels 1:13

Lu Y, Zhai Y, Liu M, Wu Q (2010) Biodiesel production from algal oil using cassava (Manihot esculenta Crantz) as feedstock. J Appl Phycol 22:573–578

Lundquist T, Woertz I, Quinn N, Benemann J (2010) A realistic technology and engineering assessment of algae biofuel production. Energy Bio-Sci Inst:1–178

Luo Y, Le-Clech P, Henderson RK (2016) Simultaneous microalgae cultivation and wastewater treatment in submerged membrane photobioreactors: a review. Algal Res 24:425–437

Lv H, Qu G, Qi X, Lu L, Tian C, Ma Y (2013) Transcriptome analysis of Chlamydomonas reinhardtii during the process of lipid accumulation. Genomics 101:229–237

Mackay S, Gomes E, Holliger C, Bauer R, Schwitzguébel JP (2015) Harvesting of Chlorella sorokiniana by co-culture with the filamentous fungus Isariafumosorosea: a potential sustainable feedstock for hydrothermal gasification. Bioresour Technol 185:353–361

Marbella L, Bilad MR, Passaris I, Discart V, Bañadme D, Beuckels A et al (2014) Membrane photobioreactors for integrated microalgae cultivation and nutrient remediation of membrane bioreactors effluent. Bioresour Technol 163:228–235

Markou G, Georgakakis D (2011) Cultivation of filamentous cyanobacteria (blue-green algae) in agro-industrial wastes and wastewaters: a review. Appl Energy 88:3389–3401

Maroneze MM, Barin JS, Menezes CR, Queiroz MI, Zepka LQ, Jacob-Lopes E (2014) Treatment of cattle-slaughterhouse wastewater and the reuse of sludge for biodiesel production by microalgal heterotrophic bioreactors. Sci Agric 71:521–524

Maroneze MM, Siqueira SF, Vendruscolo RG, Wagner R, Menezes CR, Zepka LQ et al (2016) The role of photoperiods on photobioreactors – a potential strategy to reduce costs. Bioresour Technol 219:493–499

Mascolo MC, Pei Y, Ring TA (2013) Room temperature co-precipitation synthesis of magnetite nanoparticles in a large pH window with different bases. Materials 6:5549–5567

Masojidek J, Torzillo G (2008) Mass cultivation of fresh water microalgae. Encyclopedia of Ecology:2226–2235

McGinn PJ, Dickinson KE, Park KC, Whitney CG, MacQuarrie SP, Black FJ et al (2014) Assessment of the bioenergy and bioremediation potentials of the microalga Scenedesmus sp. AMDD cultivated in municipal wastewater effluent in batch and continuous mode. Algal Res 1:155–165

McKendry P (2002) Energy production from biomass (Part 2): conversion technologies. Bioresour Technol 83(1):47–54

Miao X, Wu Q (2006) Biodiesel production from heterotrophic microalgal oil. Bioresour Technol 97(6):841–846

Miller R, Wu G, Deshpande RR, Vieler A, Gartner K, Li X et al (2010) Changes in transcript abundance in *Chlamydomonas reinhardtii* following nitrogen deprivation predict diversion of metabolism. Plant Physiol 154:1737–1752

Molina-Grima E, Belarbi EH, Fernandez AFG, Robles MA, Chisti Y (2003) Recovery of microalgal biomass and metabolites: process options and economics. Biotechnol Adv 20:491–515

Münkel R, Schmid-Staiger U, Werner A, Hirth T (2013) Optimization of outdoor cultivation in flat panel airlift reactors for lipid production by Chlorella vulgaris. Biotechnol Bioeng 110:2882–2893

Muñoz R, Guieysse B (2006) Algal-bacterial processes for the treatment of hazardous contaminants: a review. Water Res 40(15):2799–2815
Naik SN, Goud VV, Rout PK, Dalai AK (2010) Production of first and second generation biofuels: a comprehensive review. Renew Sust Energ Rev 14(2):578–597
Nayak M, Karemore A, Sen R (2016) Sustainable valorization of flue gas CO2 and wastewater for the production of microalgal biomass as a biofuel feedstock in closed and open reactor systems. RSC Adv 6:91111–91120
Neuling U, Kaltschmitt M (2017) Review of biofuel production- feedstock, processes and markets. J Oil Palm Res 29(2):137–167
Niu YF, Zhang MH, Li DW, Yang WD, Liu JS, Bai WB et al (2013) Improvement of neutral lipid and polyunsaturated fatty acid biosynthesis by overexpressing a Type 2 diacylglycerol acyltransferase in marine diatom *Phaeodactylum tricornutum*. Mar Drugs 11:4558–4569
Ooms MD, Dinh CT, Sargent EH, Sinton D (2016) Photon management for augmented photosynthesis. Nat Commun 7:1
Park JBK, Craggs RJ, Shilton AN (2013) Investigating why recycling gravity harvested algae increases harvestability and productivity in high rate algal ponds. Water Res 47:4904–4917
Pereira H, Gangadhar KN, Schulze PSC, Santos T, Sousa CB, Schueler L et al (2016) Isolation of a euryhaline microalgal strain, Tetraselmis sp. CTP4, as a robust feedstock for biodiesel production. Sci Rep 6:1–11
Pierobon SC, Cheng X, Graham PJ, Nguyen B, Karakolis EG, Sinton D (2018) Emerging microalgae technology: a review. Sustain Energy Fuels 2:13–38
Pittman JK, Dean AP, Osundeko O (2011) The potential of sustainable algal biofuel production using wastewater resources. Bioresour Technol 102(1):17–25
Pleissner D, Lam WC, Sun Z, Lin CSK (2013) Food waste as nutrient source in heterotrophic microalgae cultivation. Bioresour Technol 137:139–146
Pruvost J, Le Borgne F, Artu A, Legrand J (2017) Development of a thinfilm solar photobioreactor with high biomass volumetric productivity (AlgoFilm©) based on process intensification principles. Algal Res 21:120–137
Ra CH, Kang CH, Kim NK, Lee CG, Kim SK (2015) Cultivation of four microalgae for biomass and oil production using a twostage culture strategy with salt stress. Renew Energy 80:117–122
Radakovits R, Jinkerson RE, Darzins A, Posewitz MC (2010) Genetic engineering of algae for enhanced biofuel production. Eukaryot Cell 9:486–501
Radakovits R, Eduafo PM, Posewitz MC (2011) Genetic engineering of fatty acid chain length in Phaeodactylum tricornutum. Metab Eng 13:89–95
Radakovits R, Jinkerson RE, Fuerstenberg SI, Tae H, Settlage RE, Boore JL et al (2012) Draft genome sequence and genetic transformation of the oleaginous alga Nannochloropis gaditana. Nat Commun 3:686
Raheem A, Azlina WAKG, Taufiq-Yap YH, Danquah MK, Harun R (2015) Thermochemical conversion of microalgal biomass for biofuel production. Renew Sust Energ Rev 49:990–999
Ramzan N, Ashraf A, Naveed S, Malik A (2011) Simulation of hybrid biomass gasification using Aspen plus: a comparative performance analysis for food, municipal solid and poultry waste. Biomass Bioenergy 35(9):3962–3969
Rasool U, Hemalatha S (2016) A review on bioenergy and biofuels: sources and their production. Braz J Biol 3(5):3–21
Razzak SA, Hossain MM, Lucky RA, Bassi AS, de Lasa H (2013) Integrated CO2 capture, wastewater treatment and biofuel production by microalgae culturing – a review. Renew Sustain Energy Rev 27:622–653
Resurreccion EP, Colosi LM, White MA, Clarens AF (2012) Comparison of algae cultivation methods for bioenergy production using a combined life cycle assessment and life cycle costing approach. Bioresour Technol 126:298–306
Richmond A, Cheng-Wu Z, Zarmi Y (2003) Efficient use of strong light for high photosynthetic productivity: interrelationships between the optical path, the optimal population density and cell-growth inhibition. Biomol Eng 20:229–236

Rodolfi L, Zittelli GC, Bassi N, Padovani G, Biondi N, Bonini G et al (2008) Microalgae for oil: strain selection, induction of lipid synthesis and outdoor mass cultivation in a low-cost photobioreactor. Biotechnol Bioeng 102(1):100–112

Ruiz-Marin A, Mendoza-Espinosa LG, Stephenson T (2010) Growth and nutrient removal in free and immobilized green algae in batch and semi-continuous cultures treating real wastewater. Bioresour Technol 101:58–64

Saifullah AA, Karim MA, AhmadYazid A (2014) Microalgae: an alternative source of renewable energy. AJER 3(3):330–338

Salam KA, Velasquez-Orta SB, Harvey AP (2016) A sustainable integrated in situ transesterification of microalgae for biodiesel production and associated co-product-a review. Renew Sust Energ Rev 65:1179–1198

Salim S, Bosma R, Vermuë MH, Wijffels RH (2011) Harvesting of microalgae by bioflocculation. J Appl Phycol 23(5):849–855

San Pedro A, González-López CV, Acién FG, Grima EM (2014) Outdoor pilot-scale production of Nannochloropsis gaditana: influence of culture parameters and lipid production rates in tubular photobioreactors. Bioresour Technol 169:667–676

Sanchez-Silva L, López-González D, Garcia-Minguillan AM, Valverde JL (2013) Pyrolysis, combustion and gasification characteristics of Nannochloropsis gaditana microalgae. Bioresour Technol 130:321–331

Santos AM, Deprá MC, Santos AM, Zepka LQ, Jacob-Lopes E (2015) Aeration energy requirements in microalgal heterotrophic bioreactors applied to agroindustrial wastewater treatment. Curr Biotechnol 4:249–254

Santos AB, Fernandes AS, Wagner R, Jacob-Lopes E, Zepka LQ (2016a) Biogeneration of volatile organic compounds produced by Phormidium autumnale in heterotrophic bioreactor. J Appl Phycol 28(3):1561–1570

Santos TD, Martín JLM, Fernández FGA, Molina E, Costa JAV, Heaven S (2016b) Optimization of carbon dioxide supply in raceway reactors: influence of carbon dioxide molar fraction and gas flow rate. Bioresour Technol 212:72–81

Sathish A, Sims RC (2012) Biodiesel from mixed culture algae via a wet lipid extraction procedure. Bioresour Technol 118:643–647

Savchenko O, Xing J, Yang X, Gu Q, Shaheen M, Huang M et al (2017) Algal cell response to pulsed waved stimulation and its application to increase algal lipid production. Sci Rep 7:1–13

Schenk P, Thomas-Hall S, Stephens E, Marx U, Mussgnug J, Posten C et al (2008) Second generation biofuels: high-efficiency microalgae for biodiesel production. Bioenergy Res 1(1):20–43

Schuhmann H, Lim DKY, Schenk PM (2016) Perspectives on metabolic engineering for increased lipid contents in microalgae perspectives on metabolic engineering for increased lipid contents in microalgae. Biofuels 3:71–86

Schulze PS, Barreira LA, Pereira HG, Perales JA, Varela JC (2014) Light emitting diodes (LEDs) applied to microalgal production. Trends Biotechnol 32:422–430

Scranton MA, Ostrand JT, Fields FJ, Mayfield SP (2015) Chlamydomonas as a model for biofuels and bio-products production. Plant J 82:523–531

Shao MF, Ning FY, Zhao JW et al (2011) Preparation of Fe_3O_4/SiO_2 layered double hydroxide core–shell microspheres for magnetic separation of proteins. J Am Chem Soc 134:1071–1077

Sharma A, Gupta MN (2006) Ultrasonic pre-irradiation effect upon aqueous enzymatic oil extraction from almond and apricot seeds. Ultrason Sonochem 13:529–534

Sharma KK, Schuhmann H, Schenk PM (2012) High lipid induction in microalgae for biodiesel production. Energies 5:1532–1553

Shen Y (2014) Carbon dioxide bio-fixation and wastewater treatment via algae photochemical synthesis for biofuels production. RSC Adv 4:49672–49722

Shuping Z, Yulong W, Mingde Y, Kaleem I, Chun L, Tong J (2010) Production and characterization of bio oil from hydrothermal liquefaction of microalgae Dunaliella tertiolecta cake. Energy 25(12):5406–5411

Sialve B, Bernet N, Bernard O (2009) Anaerobic digestion of microalgae as a necessary step to make microalgal biodiesel sustainable. Biotechnol Adv 27(4):409–416

Simionato D, Basso S, Giacometti GM, Morosinotto T (2013) Optimization of light use efficiency for biofuel production in algae. Biophys Chem 182:71–78

Singh J, Saxena RC (2015) An introduction to microalgae: diversity and significance. In: Kim S-K (ed) Handbook of marine microalgae, 2nd edn. Academic Press/Elsevier, Amsterdam, pp 11–24

Singh RN, Sharma S (2012) Development of suitable photobioreactor for algae production – a review. Renew Sust Energ Rev 16:2347–2353

Singh SP, Singh P (2015) Effect of temperature and light on the growth of algae species: a review. Renew Sustain Energy Rev 50:431–444

Singh P, Gupta SK, Guldhe A, Rawat I, Bux F (2015) Microalgae isolation and basic culturing techniques. In: Kim S-K (ed) Handbook of marine microalgae. Academic, San Diego, pp 43–54

Singh P, Kumari S, Guldhe A, Misra R, Rawat I, Bux F (2016) Trends and novel strategies for enhancing lipid accumulation and quality in microalgae. Renew Sust Energ Rev 55:1–16

Singh SP, Pathak J, Sinha RP (2017) Cyanobacterial factories for the production of green energy and value-added products: an integrated approach for economic viability. Renew Sust Energ Rev 69:578–595

Sirajunnisa AR, Surendhiran D (2016) Algae – a quintessential and positive resource of bioethanol production: a comprehensive review. Renew Sust Energ Rev 66:248–267

Suganya T, Varman M, Masjuki HH, Renganathan S (2016) Macroalgae and microalgae as a potential source for commercial applications along with biofuels production: a biorefinery approach. Renew Sust Energ Rev 55:909–941

Suh IS, Lee CG (2003) Photobioreactor engineering: design and performance. Biotechnol Bioprocess Eng 8(6):313–321

Surendhiran D, Vijay M (2014) Effect of various pretreatment for extracting intracellular lipid from Nannochloropsis oculata under nitrogen replete and depleted conditions. ISRN Chem Eng 2004:1–9

Surendhiran D, Vijay M, Sirajunnisa M (2014) Biodiesel production from marine microalga Chlorella salina using whole cell yeast immobilized on sugarcane bagasse. J Environ Chem Eng 2:1294–1300

Sutherland DL, Turnbull MH, Craggs RJ (2014) Increased pond depth improves algal productivity and nutrient removal in wastewater treatment high rate algal ponds. Water Res 53:271–281

Tafreshi AH, Shariati M (2006) Pilot culture of three strains of Dunaliella salina for beta-carotene production in open ponds in the central region of Iran. World J Microbiol Biotechnol 22:1003–1006

Takeda S, Nishijima S (2006) The development of new magnet carrier method for the waste water treatment by the magnetic separation method. Chem Eng 51:614–619

Thompson SP (2013) Biofuel patent trends: 2012 in review. Ind Biotechnol 9(1):13–16

Torzillo G, Zittelli GC, Chini Zittelli G (2015) Tubular photobioreactors. In: Prokop A, Bajpai RK, Zappi ME (eds) Algal biorefineries volume 2: products and refinery design. Springer, Cham, pp 187–212

Trentacoste EM, Shrestha RP, Smith SR, Glé C, Hartmann AC, Hildebrand M (2013) Metabolic engineering of lipid catabolism increases microalgal lipid accumulation without compromising growth. PNAS 110:19748–19753

Trivedi J, Aila M, Bangwal DP, Kaul S, Garg MO (2015) Algae based biorefinery – how to make sense? Renew Sust Energ Rev 47:295–307

Tuantet K, Temmink H, Zeeman G, Janssen M, Wijffels RH, Buisman CJN (2014) Nutrient removal and microalgal biomass production on urine in a short light-path photobioreactor. Water Res 55:162–174

Ugwu CU, Ogbonna J, Tanaka H (2002) Improvement of mass transfer characteristics and productivities of inclined tubular photobioreactors by installation of internal static mixers. Appl Microbiol Biotechnol 58(5):600–607

Ugwu CU, Aoyagi H, Uchiyama H (2008) Photobioreactors for mass cultivation of algae. Bioresour Technol 99(10):4021–4028

Vadiveloo A, Moheimani NR, Cosgrove JJ, Bahri PA, Parlevliet D (2015) Effect of different light spectra on the growth and productivity of acclimated *Nannochloropsis* sp. (Eustigmatophyceae). Algal Res 8:121–127

Valipour M, Mousavi SM, Valipour R et al (2012) Air, water, and soil pollution study in industrial units using environmental flow diagram. J Basic Appl Sci Res 2(12):12365–12372

Vandamme D, Foubert I, Muylaert K (2013) Flocculation as a low-cost method for harvesting microalgae for bulk biomass production. Trends Biotechnol 31(4):233–239

Vieler A, Wu G, Tsai CH, Bullard B, Cornish AJ, Harvey C et al (2012) Genome, functional gene annotation, and nuclear transformation of the heterokont oleaginous alga Nannochloropsis oceanica CCMP1779. PLoS Genet 8:e1003064

Vonlanthen S, Dauvillée D, Purton S (2015) Evaluation of novel starch-deficient mutants of Chlorella sorokiniana for hyper-accumulation of lipids. Algal Res 12:109–118

Wan M, Liu P, Xia J, Rosenberg JN, Oyler GA, Betenbaugh MJ et al (2011) The effect of mixotrophy on microalgal growth, lipid content, and expression levels of three pathway genes in Chlorella sorokiniana. Appl Microbiol Biotechnol 91:835–844

Wan C, Alam MA, Zhao XQ, Zhang XY, Guo SL, Ho SH et al (2015) Current progress and future prospect of microalgal biomass harvest using various flocculation technologies. Bioresour Technol 184:251–257

Wang B, Li Y, Wu N, Lan C (2008) CO2 bio-mitigation using microalgae. Appl Microbiol Biotechnol 79(5):707–718

Wang SK, Hu YR, Wang F, Stiles MR, Liu CZ (2014) Scale-up cultivation of Chlorella ellipsoidea from indoor to outdoor in bubble column bioreactors. Bioresour Technol 156:117–122

Wang X, Liu YH, Hu DX, Balamurugan S, Lu Y, Yang WD et al (2015a) Identification of a putative patatin-like phospholipase domain-containing protein 3 (PNPLA3) ortholog involved in lipid metabolism in microalga Phaeodactylum tricornutum. Algal Res 12:274–279

Wang Y, Yang Y, Ma F, Xuan L, Xu Y, Huo H et al (2015b) Optimization of Chlorella vulgaris and bioflocculant-producing bacteria co-culture: enhancing microalgae harvesting and lipid content. Lett Appl Microbiol 60:497–503

Wang C, Li Y, Lu J, Deng X, Li H, Hu Z (2017) Effect of overexpression of LPAAT and GPD1 on lipid synthesis and composition in green microalga Chlamydomonas reinhardtii. J Appl Phycol 1:9

Wei L, Xin Y, Wang Q, Yang J, Hu H, Xu J (2017) RNAi-based targeted gene knockdown in the model oleaginous microalgae Nannochloropsis oceanica. Plant J 89:1236–1250

Wu C, Xiong W, Dai J, Wu Q (2015) Genome-based metabolic mapping and 13C flux analysis reveal systematic properties of an oleaginous microalga Chlorella protothecoides. Plant Physiol 167:586–599

Wulf C, Thormann L, Kaltschmitt M (2017) Biohydrogen production: sustainability of current technology and future perspective (Anoop Singh and Dheeraj Rathore eds). New Delhi: Springer

Xie SX, Sun S, Dai SY, Yuan JS (2013) Efficient coagulation of microalgae in cultures with fi lamentous fungi. Algal Res 2:28–33

Xu X, Shen Y, Chen J (2015) Cultivation of *Scenedesmus dimorphus* for C/N/P removal and lipid production. Electron J Biotechnol 18:46–50

Yang H, He Q, Rong J, Xia L, Hu C (2014) Rapid neutral lipid accumulation of the alkali-resistant oleaginous Monoraphidiumdybowskii LB50 by NaCl induction. Bioresour Technol 172:131–137

Yang IS, Salama ES, Kim JO, Govindwar SP, Kurade MB, Lee M, Roh HS, Jeon BH (2016) Cultivation and harvesting of microalgae in photobioreactor for biodiesel production and simultaneous nutrient removal. Energy Convers Manag 117:54–62

Yavuz CT, Mayo JT, Yu WW et al (2006) Low-field magnetic separation of monodisperse Fe_3O_4 nanocrystals. Science 314:964–967

Yoon K, Han D, Li Y, Sommerfeld M, Hu Q (2012) Phospholipid: diacylglycerol acyltransferase is a multifunctional enzyme involved in membrane lipid turnover and degradation while synthesizing triacylglycerol in the unicellular green microalga Chlamydomonas reinhardtii. Plant Cell 24:3708–3724

Zeng F, Huang J, Meng C, Zhu F, Chen J, Li Y (2016) Investigation on novel raceway pond with inclined paddle wheels through simulation and microalgae culture experiments. Bioprocess Biosyst Eng 39:169–180

Zhang Y, Su H, Zhong Y, Zhang C, Shen Z, Sang W et al (2012) The effect of bacterial contamination on the heterotrophic cultivation of Chlorella pyrenoidosa in wastewater from the production of soybean products. Water Res 46:5509–5516

Zhang R, Li L, Tong D, Hu C (2016) Microwave-enhanced pyrolysis of natural algae from water blooms. Bioresour Technol 212:311–317

Zheng Y, Chi Z, Lucker B, Chen S (2012) Two-stage heterotrophic and phototrophic culture strategy for algal biomass and lipid production. Bioresour Technol 103:484–488

Zhou W, Min M, Hu B, Ma X, Liu Y, Wang Q et al (2013) Filamentous fungi assisted bioflocculation: a novel alternative technique for harvesting heterotrophic and autotrophic microalgal cells. Sep Purif Technol 107:158–165

Zhu L, Wang Z, Shu Q, Takala J, Hiltunen E, Feng P, Yuan Z (2013a) Nutrient removal and biodiesel production by integration of freshwater algae cultivation with piggery wastewater treatment. Water Res 47:4294–4302

Zhu L, Wang Z, Takala J, Hiltunen E, Qin L, Xu Z et al (2013b) Scale-up potential of cultivating Chlorella zofingiensis in piggery wastewater for biodiesel production. Bioresour Technol 137:318–325

Zhu L, Xu Z, Qin L, Wang Z, Hiltunen E, Li Z (2016) Oil production from pilot-scale microalgae cultivation: an economics evaluation. E Energ Sour Part B: Economics, Plan Pol 11:11–17

Zhu L, Nugroho YK, Shakeel SR, Li Z, Martinkappi B, Hiltunen E (2017) Using microalgae to produce liquid transportation biodiesel: what is next? Renew Sust Energ Rev 78:391–400

Recent Update on Biodiesel Production Using Various Substrates and Practical Execution

5

S. J. Geetha, Saif Al-Bahry, Yahya Al-Wahaibi, and Sanket J. Joshi

Abstract

Worldwide fossil fuel (liquid crude oil-based) demand is projected to exceed 100 million barrel per day by 2040, with diesel dominating by more than 5 million barrel per day. However, dwindling resource, severe ups and downs in crude oil prices, and harmful environmental impact are compelling scientific communities and different countries to look for better and long-lasting substitute renewable and environment-friendly energy fuel. Biodiesel is one such alternative liquid fuel which is almost similar to petro-based diesel and has some critical advantages over it: eco-friendly; suitable to normal diesel engines (either as 100% or as a blend in commercial diesel); and can be synthesized using different renewable resources – vegetable oils (edible and nonedible), waste frying oil, waste animal fat, algal and microbial oil. It could be produced by almost anyone locally, using either local oil produce or wastes, with almost no major consumables and chemicals, and also easy to separate and store. Based on the characteristics and type of feedstock used, overall production costs are quite low as compared to diesel production. Since it could also be produced from waste oily feedstocks

S. J. Geetha
Department of Biology, College of Science, Sultan Qaboos University, Muscat, Oman

S. Al-Bahry
Department of Biology, College of Science, Sultan Qaboos University, Muscat, Oman

Oil & Gas Research Center, Sultan Qaboos University, Muscat, Oman

Y. Al-Wahaibi
Petroelum and Chemical Engineering Department, Sultan Qaboos University, Muscat, Oman

S. J. Joshi (✉)
Oil & Gas Research Center, Sultan Qaboos University, Muscat, Oman

Central Analytical and Applied Research Unit, College of Science, Sultan Qaboos University, Muscat, Oman
e-mail: sanket@squ.edu.om

N. Srivastava et al. (eds.), *Substrate Analysis for Effective Biofuels Production*, Clean Energy Production Technologies, https://doi.org/10.1007/978-981-32-9607-7_5

and algal/microbial oils, biodiesel production does not suffer from "food versus fuel" dilemma. Worldwide, it is successfully produced both small scales and commercially, and used as a substitute for diesel (pure or blends). Overall, EU and the USA are leading producers and consumers of biodiesel and other countries (Asia, Middle East) are also showing tremendous growth opportunities in the coming years, and may surpass others. Apparent environmental, economic, and social benefits of biodiesel are encouraging both policy makers, stakeholders, and the common public to look forward with optimism that it could be the future of liquid energy fuel.

Keywords
Diesel · Biodiesel · Vegetable oil · Catalyst · Alcohol

5.1 Background: Diesel and Diesel Engines

Worldwide fossil fuels account for major energy sources, which include crude oil (derivatives such as petroleum, diesel, kerosene, etc.) and coal. However, fossil fuels are nonrenewable, incur higher production costs, susceptible to unstable market conditions (demand and supply), and lead to harmful environmental effects and therefore have stirred immense interest in different types of renewables – biofuels (Hill et al. 2006). Such renewable sources of energy should be environment and eco-friendly (less-polluting), economically viable (e.g., should be directly used in the transportation sector without considerable modifications in existing set-up), and easily available in sufficient quantity (Ingale et al. 2014, 2019). Solar, wind, hydrothermal, nuclear, and bio-based fuels (bioethanol, biodiesel, biomass) are such energy resources touted to replace fossil fuels in the near future.

In 1890, German inventor Rudolf Diesel developed an automobile engine, known as the diesel engine (from the inventor), which was initially run using vegetable oils (such as peanut oil). Although diesel engines could be run using untreated vegetable oils, it could pose certain challenges: low volatile nature, high viscosity, and other issues at lower temperatures. At present diesel engines are preferred for various heavy-duty usages, such as agriculture (farming), building new complexes, industrial boilers and generators, and heavy-duty trucks and transport sector (Van Gerpen 2010). Petroleum-derived diesel is comparatively more durable, provides better torque, and better fuel efficiency, as compared to other derivatives. The diesel engines are notably different from petroleum engines (which are spark-ignited ones), where fuel and air mixture are injected into the engine compartment, compressed, and then ignited by a spark. Whereas in a diesel engine (compression-ignited engine), first air (compressed at high pressure and temperatures) enters the engine compartment, followed by spray of finely atomized diesel at high velocity. When diesel comes into contact with pressurized air it leads to ignition. As such spark plug is not required in diesel engines, but some engines have the provision of electrically heated plugs to help start the engine during cold weathers. Some of the issues in diesel engines are the formation of soot particles and harmful emission of oxides of nitrogen (NOx). Though those are big environmental concerns, the diesel engine–based passenger cars and trucks are still widespread in Asia (Indian subcontinent), Europe, and other parts of the world. However, as it is evident that crude oil-based fuels (such as petroleum and diesel) are becoming more and more expensive, and there are several environmental concerns (with respect to emissions, global warming, etc.), biofuels (such as biodiesel and bioethanol) are the future of our energy needs and an alternative for such fossil fuels.

5.2 What Is/Why Biodiesel?

Due to those associated issues with using petro-derived diesel or using vegetable oil as such in diesel engines, further research was focused on various triglyceride derivatives, such as biodiesel. Biodiesel could be a substitute to petro-based diesel, mainly because of its environment-friendly nature and also as a renewable resource. Important landmarks in the history of advances of biodiesel industry starting from first experiments by Rudolf Diesel to recent past could be found from a review article by Silitonga et al. (2011). Biodiesel could simply be defined as alkyl esters of fatty acids derived from lipid feedstocks. It could be synthesized by a catalyzed transesterification of vegetable oil (or animal fat) with an alcohol and acid/alkali. As an alternative fuel, biodiesel can be used as such (B100) or blended with petro-based diesel, and same diesel engine could be used without any modifications. Biodiesel could be synthesized from a variety of vegetable (edible and nonedible) oils, animal fats, microbial oils, and even waste frying oil. Different benefits and challenges for biodiesel uses are mentioned below (Manzanera et al. 2008; Van Gerpen 2005; Bozbas 2008):

5.2.1 Benefits of Biodiesel

- It is clear, renewable, and energy-efficient, which does not need any additional lubricant to be added.
- Generally, it is used as a blend with diesel; concentrations up to 5–20% (depending on the concentration of biodiesel it is known as B5 or B20) are the most commonly used, which provides very good balance between performance and compatibility at lower temperatures, better cost-effectiveness and lower emissions, with almost no or few modifications.
- It is a green fuel which also reduces global warming gas emissions.
- In general, it is cheaper than diesel and can be quite practical to produce and utilize on-site.
- It has a higher cetane number (>100), making it more efficient fuel.
- As it could be produced from waste materials, overall environmental burden is reduced and provides better participation in the circular economy.
- Production is easier and more energy-efficient, as compared to normal petro-based diesel (which needs several energy-intensive steps to recover and produce).

- Everyone can produce it in their yard or available space, without any need to import fuels and looming costs associated with it.
- Pure biodiesel and blends with petro-based diesel, are biodegradable, making it more desirable during any spillage while storage, transport, or applications.
- B100 and different blends already been tested (for efficiency and emissions) in accordance with major requirements for various countries, and are reported to effectively reducing particulate matters (shoot), hydrocarbons, and carbon monoxide emissions.
- It has higher flash point (>100 °C), as compared to other petro-based fuels (kerosene or diesel), making it less dangerous and provides ease of transportation and storage.
- Overall it is regarded as safe and better fuel than conventional diesel.

5.2.2 Challenges for Biodiesel

- It has lower calorific value and volatility, and higher cloud point and pour point than conventional diesel.
- Biodiesel blends could freeze at lower temperatures.
- When it is used for the first time in the diesel engine run on diesel for a long time, it could release those hydrocarbon deposits, leading to blocking of filters, and leads to frequent need of changing filters.
- One major constraint for biodiesel production could be the availability of feedstocks, mainly of agricultural origin.
- As the viscosity is higher than diesel, it could affect engine performance and may need some modifications in the fuel injection systems.
- As the oxygen content is higher, it leads to relatively higher NOx emissions.
- It is more prone to oxidation, thus could lead to problems during storage and, it could lead to various oxidation products harmful to the engine.
- It should be visibly devoid of any undissolved water. As it is more hygroscopic, so during storage and transport contact with air and moisture could lead to water droplets.
- As the production process is not standardized throughout the world, inappropriate procedure could lead to corrosion (of copper or brass parts and galvanized parts), fuel system blockage, engine (rubber) seal failures, filter blockage, and buildup of particles.
- A modified refueling and distribution framework is needed to trade biodiesel, further increasing the cost of investment.
- As it is biodegradable, it is also more prone to microbial degradation than petro-based diesel. A positive trait in case of accidental spills, but a serious concern if storage tanks are contaminated, which can lead to blockage of fuel dispensers and filters. This could be avoided by proper housekeeping and maintenance of the quality by avoiding water in contact with biodiesel during post-processing to usage steps.
- It is not compatible with some type of plastics and rubber parts, and may infiltrate or dissolve parts made of that material over time.

5.3 Different Substrates for Biodiesel Production

In general, biodiesels are classified into three groups based on the feedstock from which they are produced. Typically, it was initially produced from vegetable oils (edible) such as cottonseed oil, palm oil, sunflower oil, and rapeseed oil (Table 5.1). At a later stage, different research groups started experimenting biodiesel production from nonedible vegetable oils such as Ethiopian mustard, babassu tree, neem, or jatropha oil (Table 5.1). Among those feedstocks, ones with a higher oil content/extraction possibility were deemed to be more suitable for commercial processes. Those nonedible oil sources showed quite prominent advantages as compared to edible sources: no competition with edible food resources, comparatively reduced production cost, no specific requirement of cultivation land (as most of these tree grows without much specific care or requirements), and more ecofriendly nature. However, the major disadvantage of such nonedible oil feedstocks are higher viscosity, unable to meet the demand for scaled-up processes, and more alcohol required for the biodiesel production process. Currently, different feedstocks are explored as suitable feedstocks for economical biodiesel production: algae/microalgae-based oil, used/waste vegetable cooking oil/yellow grease, animal tallow oils/fats, bacteria/microbial oil, and others (Table 5.1). These feedstocks have several advantages as compared to other edible or nonedible oils: faster and higher growth rate and productivity of microbial cells containing a higher percentage of oil, almost no effect/competition with food resources/chain, and fewer burdens for farming land. However, those feedstocks also pose certain challenges, such as the requirement of sunlight/additional nutrients, maintenance of temperature/environmental parameters, difficulties in oil extraction, challenge for industrial-scale production, and needs huge CAPEX. Overall, cheaper and easily available feedstock (such as waste products) will lead to the economic production of biodiesel at commercial levels.

5.4 Biodiesel Production Methods

Several methods are reported for the production and applications of biodiesel from various feedstocks: blending vegetable oils with diesel or dilution, micro-emulsion, pyrolysis, and transesterification (Manzanera et al. 2008; Koh and Ghazi 2011).

5.4.1 Blending Vegetable Oils with Diesel or Dilution

The vegetable oil can be mixed or diluted with normal diesel to run conventional engines. In 1980, a report from Brazil highlighted 90:10 mixture of diesel:vegetable oil showed no loss of power in the engine, without changing anything in the engines (Singh and Singh 2010). The other report also showed the ratio of diesel:vegetable oil (sunflower oil) up to 75:25 to be effective (Ziejewski et al. 1986). The main goal of blending is to specifically reduce the viscosity of vegetable oil-based fuel to enhance the engine performance which improves with a higher percentage of diesel.

Table 5.1 Different feedstocks for biodiesel production

Vegetable oil (edible)	Vegetable oil (non-edible)	Other substrates/feedstocks
Barley oil (*Hordeum vulgare*)	Abyssinian mustard, Ethiopian mustard, Abyssinian cabbage, Ethiopian rapeseed (*Brassica carinata*)	Algae/microalgae (*Phormidium autumnale, Dunaliella salina, Chlorella vulgaris, Chlorella protothecoides, Chaetoceros calcitrans, Skeletonema costatum, Phaeodactylum tricornutum, Chlamydomonas reinhardtii, Dunaliella teriolecta, Scenedesmes obliquus, Neochloris oleabundans, Chromonas salina, Nannochloropsis oculata, Isochrysis galbana, Botryococcus braunii, Cylindrotheca sp., Dunaliella primolecta, Monallanthus salina*)
Cashew nut oil (*Anacardium occidentale*)	Almond (Prunus dulcis)	Animal tallow oil
Castor oil (*Ricinus communis*)	Andiroba (*Carapa guianensis*)	Biomass pyrolysis oil
Cotton seed oil (*Gossypium hirsutum*)	Artichoke thistle or globe artichoke (*Cynara cardunculus*)	Chicken fat oil
Groundnut oil (*Arachis hypogaea*)	Babassu tree (*Attalea speciosa*)	Fish (waste sardine) oil
Hazelnut oil (*Corylus avellana*)	Camelina/false flax (*Camelina sativa*)	Lard
Mustard oil (*Sinapis hirta, Brassica juncea, B. nigra*)	Crambe oil (*Crambe abyssinica*)	Latexes
Oat oil (*Avena sativa*)	Fendler's bladderpod/popweed (*Lesquerella fendleri*)	Microbial oil
Palm oil (*Erythea salvadorensis*)	Indian mallow (*Abutilon muticum*)	Pine and kapok oil
Pistachio oil (*Pistacia vera*)	Indian soapberry or washnut (*Sapindus mukorossi*)	Poultry fat oil
Radish oil (*Raphanus sativus*)	Jatropha (*Jatropha curcus*)	Tarpenes
Rapeseed oil/ canola oil (*Brassica napus*)	Jojoba (*Simmondsia chinesis*)	Waste cooking oil (yellow grease)

(continued)

Table 5.1 (continued)

Vegetable oil (edible)	Vegetable oil (non-edible)	Other substrates/feedstocks
Rice bran oil (*Oryza sativa*)	Karanja (*Pongamia pinnata*)	
Sorghum oil	Kokum (*Garcinia indica*)	
Soyabean oil (*Glycine max*)	Linseed (*Linum Usitatissimum* L.)	
Sunflower oil (*Helianthus annuus*)	Mahua (*Mahua indica*)	
Tigernut oil (*Cyperus esculentus*)	Milk bush (*Euphorbia tirucalli*)	
Walnut oil (*Juglans regia*)	Moringa (*Moringa oleifera*)	
Wheat oil (*Triticum aestivum*)	Nag Champa (*Magnolia champaca*)	
Winter rapeseed oil	Neem (*Azadirachta indica*)	
	Othalanga seed (*Cerbera odollam*)	
	Petroleum nut (*Pittosporum resiniferum*)	
	Polanga (*Calophyllum inophyllum*)	
	Rubber seed (*Hevea brasiliensis*)	
	Silk cotton tree (*Bombax ceiba*)	
	Tall oil	
	Tobacco (*Nicotiana tabacum*)	
	Tung tree seeds (*Aleurites fordii*)	
	Yellow oleander or lucky nut (*Thevettia peruviana*)	

Adapted from: Sakthivel et al. (2018), Nomanbhay and Ong (2017), Siqueira et al. (2016), Sharma et al. (2008), Hoekman et al. (2012), Ambat et al. (2018), Banković-Ilić et al. (2014), Tasić (2016), Borugadda and Goud (2012), Singh and Singh (2010), Manzanera et al. (2008)

5.4.2 Micro-emulsification

The vegetable oils' viscosity could also be lowered by microemulsion formation. Microemulsions are clear fluids containing both oil and aqueous phase stabilized by surface-active agents – surfactants. This formation further improves the spray characteristics by explosive vaporization of the low boiling constituents in the micelles. It is reported that different microemulsions formed by higher "C" alcohols could help the viscosity reduction for diesel engines (Jain and Sharma 2010). One report also showed that there was no difference between microemulsion of 53% sunflower oil and the 75:25 mixture of diesel:sunflower oil (Ziejewski et al. 1983). Even though it is comparatively a simpler process, issues such as high viscosity and poor volatility/stability are there.

5.4.3 Pyrolysis

Pyrolysis is the process of thermal decomposition of substances (into the liquid and gaseous fuels and char) using a catalyst(s) in the absence of oxygen (Jahirul et al. 2012). Overall, this process is quite simple, yet effective, eco-friendly, and does not generate much waste. Different types of feedstocks could be utilized for pyrolysis: biomass, edible/nonedible vegetable oils, animal fats, fatty acids, and methyl ester of fatty acids (Jahirul et al. 2012; Jain and Sharma 2010). The limitations of this process are that it requires high temperature for operation, needs expensive equipment, and the purity of produced biodiesel is generally low due to the remaining amount of carbon residues.

5.4.4 Transesterification

Transesterification (alcoholysis) is the reaction of triglycerides (fat or oil) with alcohol (usually methanol) to form esters and glycerol, at a specific temperature (Fig. 5.1). Generally, different types of catalysts are used to enhance the reaction rate and overall yield%. The alkaline catalysts are usually preferred over acidic ones, because of the higher reactivity and the lower temperature requirement. As it is a reversible reaction, usually excess alcohol is used to shift the equilibrium toward the product. The byproduct of this reaction is glycerol/glycerin – which should be recovered, as it is also a value-added product having several applications.

Different alcohols such as methanol, ethanol, propanol, butanol, and amyl alcohol are reported to be used in the transesterification method, but methanol is the mainly used one specifically due to lower cost and other advantages. However, methanol or its vapor could lead to an explosion due to its low boiling point and hence it should be carefully handled during storage, transport, and biodiesel production process. At present different types of transesterification processes are reported (Koh and Ghazi 2011; Singh and Singh 2010; Mahmudul et al. 2017; Aransiola et al. 2014):

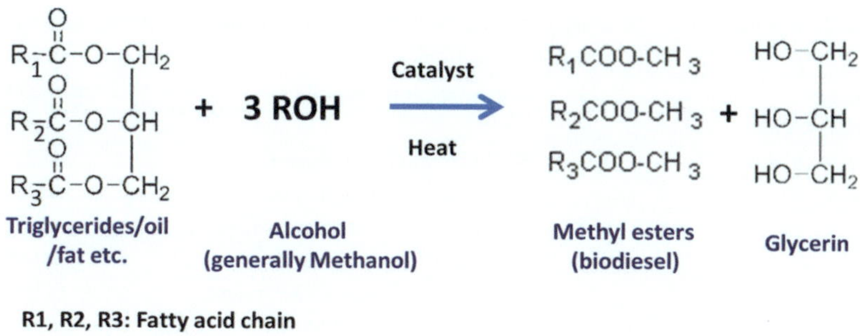

Fig. 5.1 Alkali-mediated transesterification reaction of triglycerides (oil or fat) with an alcohol (usually methanol) to form methyl esters (biodiesel) and glycerol, at specific temperatures

- Alkali based
- Acid based
- Enzyme based
- Noncatalytic supercritical alcohol based
- Ultrasound assisted
- Microwave assisted

Among those methods, alkali-catalyzed transesterification processes are more favored and reported in several papers (Manzanera et al. 2008; Ambat et al. 2018; Singh and Singh 2010), as it is more suitable for commercial-scale production, and thus produced biodiesel is comparable to normal diesel. However, overall conversion efficiency is low, cost of the process is high, and reusability of catalyst are some of the issues. In current chapter, we would like to emphasize more on "enzyme-catalyzed transesterification" process. Enzymes such as lipases are utilized in this process, and it is similar to alkali transesterification, with catalyst–solvent ratio, and duration of the reaction is different, where lipase is used as a catalyst. However, there are some challenges while using lipases to produce biodiesel: The cost of enzymes is quite high, it is a lengthy process (reaction could take 48 hours or more), and the enzyme showed very sluggish efficiency with methanol, and preferred alcohols are long-chain fatty alcohols (Manzanera et al. 2008). Various studies have been reported for transesterification by lipases of microbial and plant origins, for different alcohols and feedstocks under different conditions (Table 5.2), with either free enzymes and/or immobilized enzymes. When microbes capable to produce lipase were used to produce biodiesel using different substrates such as soybean oil, jatropha oil, nonedible oils, grease, rapeseed oil, waste cooking oil, in the presence of methanol or ethanol, efficiencies in the range of ~78–94% were reported (Qin et al. 2008; Ban et al. 2001; Tamalampudi et al. 2008; Huang et al. 2012; Yan et al. 2012a, b, 2014). Whereas, for free-enzymes or immobilized lipases, the conversion efficiencies were observed in the range of ~15–100% (Du et al. 2005; Halim and Kamaruddin 2008; Hernández-Martín and Otero 2008; Hsu et al. 2004; Jang et al. 2012; Lai et al. 2005; Li et al. 2006, 2007, 2009; Modi et al. 2006, 2007; Salis et al. 2005; Watanabe et al. 2002). Some reports also highlighted efficient reaction and

Table 5.2 Biodiesel production by biocatalysts using microbial cells or microbial enzymes (free or immobilized lipases) and plant enzymes (lipases)

Biocatalyst	Feedstock and alcohol	Yield
Microbial cells		
Rhizopus chinensis	Soybean oil and methanol	86%
R. oryzae	Soybean oil and methanol	91.1%
	Jatropha curcas oil and methanol	80%
	Soybean oil and methanol	90.9%
	Nonedible oil from *Calophyllum inophyllum* and methanol	87%
ROL-surface displayed on *S. cerevisiae*	Soybean oil and methanol	78.3%
RML-surface displayed on *P. pastoris*		83.14%
P. pastoris cells co-displaying CALB and TLL		95.4%
S. cerevisiae cells with intracellular overproducing of ROL		71%
Recombinant *E. coli* cells with intracellular co-expression of CALB and TLL	Grease (15–40% of FFA) and methanol	93%
A. oryzae cells with the expression of *Fusarium heterosporum* lipase	Rapeseed oil and ethanol (produced by fermentation of brown rice with ~8% of water)	94%
P. pastoris cells with intracellular overexpression of TLL	Waste cooking oil and methanol	82%
Microbial (free) enzyme		
Novozym 435	Dewaxed/degummed rice bran oil (85% of FFA) and methanol	96%
	Waste cooking palm oil (0.2% of FFA) and methanol	88%
	Waste Tuna-FFA (98.5% of FFA) methanol	97.7%
	Sunflower oil and methanol or ethanol	79–82%
	Soybean oil and methyl acetate	92%
	Jatropha sunflower oils and 2-propanol	92.8–93.4%
	Jatropha sunflower oils and ethyl acetate	91.3–92.7%
	Cotton seed oil and methanol	97%
	Acid oil (77.9% free fatty acids) and methanol	>90
	Waste edible oil (2.5% free fatty acids) and methanol	>90
Novozym-435 preincubated 0.5 h in ethyl oleate	Soybean oil and methanol	97%
Various commercial lipases	Triolein and linear and branched alcohols, fusel oil-like alcohol mixture	~100%
	Microalgae oil and long-chain alcohols	–

(continued)

Table 5.2 (continued)

Biocatalyst	Feedstock and alcohol	Yield
Mixture of Lipozyme TL IM and Novozym 435	Waste oil (70% of FFA) and methanol	95%
	Soybean oil deodorizer distillate (28% of FFA) and methanol	97%
	Vegetable oils and methanol/ethanol	~100%
Powder lipase from *C. cylindracea* (*C. rugosa*)	Waste-activated bleaching earth (35% of oil) and methanol	100%
Lipase from Candida sp. 99–125	Oil from microalga *Chlorella protothecoides* (4.5% of FFA) and methanol	98.15%
Candida antarctica lipase	Soybean (crude) oil and methanol	93.8%
M. miehei (Lipozyme IM60) *C. Antarctica* (SP435) *M. miehei* (Lipozyme IM60) *M. miehei* (Lipozyme IM60)	Tallow, soybean, rapeseed oil and primary alcohols; Secondary alcohols (a-methanol, ethanol, propanol, butanol, and isobutanol; b-Isopropanol and 2-butanol)	19.4–98.5%
J. cepacia (lipase PS-30) + *C. anturclica* (lipase SP435)	Recycled restaurant grease and ethanol	85.4%
M. miehei (Lipozyme IM-20)	Mowrah, mango, kernel, sal and C, -C, alcohols	86.8–99.2%
Pseudomonas fluorescens (Amano AK)	Sunflower oil and methanol	>90%
I. cepuciu (lipase PS-30)	Palm kernel; oil and methanol; ethanol	15%; 72%
Microbial (immobilized) enzyme		
Immobilized lipase from *P. expansum*	Waste oil (27% of FFA) and methanol	92.8%
Lipase from *T. lanuginosus* immobilized in a phyllosilicate sol–gel matrix	Recycled restaurant greases (6.8% of FFA) and methanol	89%
Co-immobilized lipases from *R. oryzae* and *C. rugosa*	Crude canola oil (0.64% of FFA) and methanol	88.9%
Candida sp. lipase (immobilized)	Vegetable oils and methanol	93–96%
Candida sp. lipase (immobilized)	Microalgae oil and methanol	98%
Plant enzymes		**Yield activity in (U/g)**
Germinated rape (*Brassica napus*) seeds lipase	Oleic acid and N-butanol	~96%
	Oleic acid and oleic alcohol	~94
	Methyl oleate and N-butanol	93%
Black cumin (*Nigella sativia*) seeds lipase	Vinyl acetate and N-octanol	99%
Germinated linseed (*Linum usitatissimum*) seeds lipase	Acetic acid and ethanol	63.8%
Germinated Jatropha (*Jatropha curcas*) seeds lipase	Triolein and lauric acid	~50%
Castor (communis) beans lipase	Oleic acid and methanol or ethanol	~80%
Frangipani (*Plumeria rubra*) latex lipase	Lauric acid and butanol	>90%
Babaco (*Carica pentagona*) latex lipase	Sunflower seed oil and ethanol or propanol or methanol or butanol	12.3–68%

Adapted from: Aguieiras et al. (2015), Robles-Medina et al. (2009), Mounguengui et al. (2013), Singh and Singh (2010) yield (~12–99%) using plant-derived lipases (Cambon et al. 2006, 2009; Jachmanián and Mukherjee 1996; Liaquat 2010; Su et al. 2010; Tuter et al. 2003).

5.5 Biodiesel Production Standard Processing Steps (Small Scale, Using Alkali-Based Transesterification)

General layout for biodiesel production using alkali-based transesterification and methanol as alcohol is given in Fig. 5.2. Currently, this is the protocol we are following for biodiesel production in our research lab, using waste cooking oil in initial trials (Fig. 5.3).

- First collect feedstock (waste frying oil) from on-campus kitchens and restaurants.
- Check oil for water content and filter to remove settled food particles.
- Before proceeding for the reaction process, first check the waste oil for free-fatty acid content using titration to determine the amount of alkali catalyst (potassium hydroxide in our case) needed for transesterification step.
- Fill the reaction chamber with waste cooking oil and start the heating (preferably using electric heating and not gas-based heating).
- Simultaneously mix sodium hydroxide with methanol, and dissolve it completely. Be careful at this step as it is quite corrosive and dangerous to work with alkali and methanol.

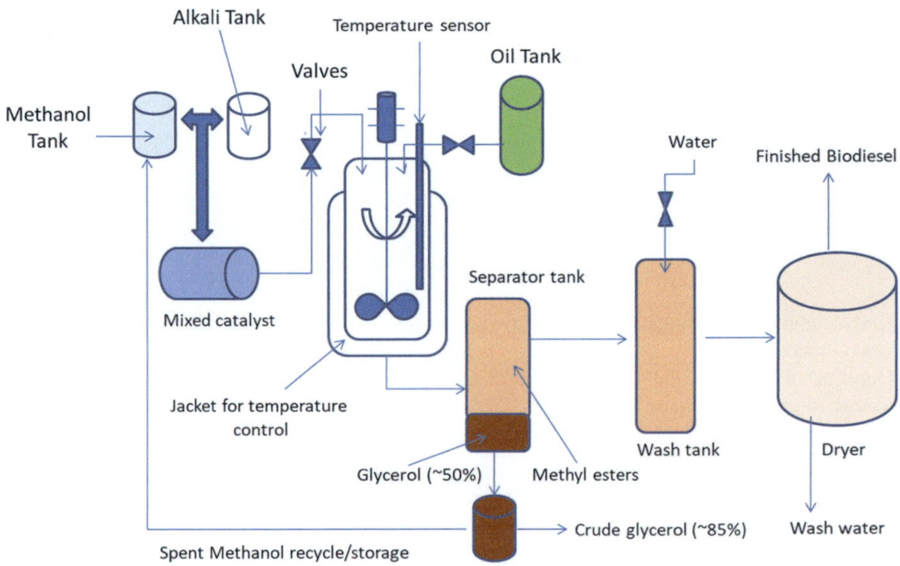

Fig. 5.2 A generalized scheme for the processes involved in biodiesel production by alkali-based transesterification

Fig. 5.3 Laboratory-scale biodiesel production results

- When oil has reached the reaction temperature, add this mixture of sodium hydroxide and methanol to the reaction chamber and mix for up to 2 h, and then it to settle. In this step, carefully monitor the reaction temperature, and do not exceed it beyond the boiling point of methanol.
- After overnight settling time, drain glycerol at the bottom of the vessel and collect it for further processing or disposal. This partially purified glycerol will still contain unused methanol and should be handled with a lot of care.
- Transfer the top layer of biodiesel and wash it several times with tap water.
- Completely separate water from biodiesel either by settling it or via some processes.
- Further filter the finished product and store into the drums/storage containers.
- Perform various analytical tests for different set parameters and record the observations.
- Separate methanol from spent glycerol before further processing or disposal.

5.6 Analytical Methods to Determine Biodiesel and Its Specifications Based on Standards Followed by Different Countries

The basic instruments needed for characterization of the biodiesel is rheometer, cetane number analyzer, flash point/cloud point/pour point analyzers, elemental analyzer, and other consumables. Other instruments used for chemical characterization are gas chromatography-mass spectroscopy (GC-MS) or flame ionization detector (GC-FID), high-performance liquid chromatography (HPLC), Fourier transformed infrared (FTIR) spectrometer, etc. Thin layer chromatography (TLC) was initially used for qualitative analysis of fatty acid esters and glycerides; however, it has limitations like low accuracy and sensitivity to moisture in air/environment.

The GC-MS or GC-FID is highly accurate and most followed methods for quantitative analysis of minor and major compounds (Thangaraj and Piraman 2016; Feyzi and Shahbazi 2015; Teo et al. 2016; Bet-Moushoul et al. 2016; Xue et al. 2006; Wang et al. 2007; Karmee and Chadha 2005; Sarin et al. 2007). Mainly biodiesel is analyzed as the percentage of fatty acid methyl esters (FAMEs). HPLC can also be used to measure the total amount of glycerides and to characterize biodiesel from more than one different feedstock (Liu et al. 2012; Veljković et al. 2006; Sahoo et al. 2007; Meher et al. 2006). More recently nuclear magnetic resonance (NMR) technique is reported for the chemical characterization of biodiesel by determining the blend level (Kaur and Ali 2011; Thangaraj and Piraman 2016; Knothe 2001; Meher et al. 2006). FTIR spectroscopic analysis provides details of biodiesel and triglycerides simultaneously (Correa and Arbilla 2008; Sharma et al. 2008). The biodiesel quality is measured with the help of various standards based on its physical and chemical properties. The biodiesel specification and required properties based on standards from different countries are shown in Table 5.3.

5.7 Production Economics for Biodiesel

Biodiesel production economics are influenced by several factors such as the methodology, commercial plant scale, kind and cost of feedstock, CAPEX, and OPEX. Several studies were carried out to clearly lay out economics, each differs with respect to their approach in calculating the associated costs, but generally all agreed that the cost of raw materials (feedstocks) is mainly responsible for the major share of the operating cost (Gebremariam and Marchetti 2018; Karmee et al. 2015; Skarlis et al. 2012). For example, in India, vegetable oils generally cost almost two to four times more than the cost of diesel, which makes it difficult to be economically feasible and cannot compete economically with diesel (Koh and Ghazi 2011). The calculation of the total production cost and the unit production cost is summarized by several researchers (Sinnott 1999; Apostolakou et al. 2009; Gebremariam and Marchetti 2018; Karmee et al. 2015; Skarlis et al. 2012). Operating cost calculation (for a production capacity of 50 kt/year) were divided into capital investment, other fixed costs (maintenance, 1.43%; operating labor/personnel, 0.91%; laboratory expenses, 0.18%; supervision, 0.18%; overheads, 0.45%; capital charges,

Table 5.3 The specification of biodiesel to be used as fuel in different countries

	ASTM D6751 (United States)		EN 14214 (Europe)		IS15607:2005 (India)		Australian Biodiesel Standard			Provisional Brazilian Biodiesel Standard ANP (Agência Nacional do Petróleo) 255		Provisional South African Biodiesel Standard	
Property (unit)	Test method	Limits	Test method	Limits	Test method	Limits	Test method	Limits	Effective date	Test method	Limits	Limits	Test method
Flash point (°C)	D 93	130.0 min	EN ISO 3679	120	–	120	ASTM D93	120	Sept. 18, 2003	ISO/CD 3679	100 min	120	ISO 3679
Water (and sediment) (% v or mg/kg)	D 2709	0.050 max	EN ISO 12937	500	D 2709; ISO 3733; ISO 6296	500	ASTM D2709	0.5	Sept. 18, 2003	D 2709	0.02 max	0.05	ISO 12937
Kinematic viscosity, 40 °C (mm^2/s)	D 445	1.9–6.0	EN ISO 3104 ISO 3105	3.5–5	ISO 3104	2.5–6.0	ASTM D445	3.5–5	Sept. 18, 2003	D 445; EN/ ISO 3104	ANP 310[b]	3.5–5.0	ISO 3104
Sulfated ash (% mass or %m/m)	D 874	0.020 max	ISO 3987	0.02	ISO 6245	0.02	ASTM D 874	0.2	Sept. 18, 2003	D 874; ISO3 987[a]	0.02 max	0.02	ISO 3987
Sulfur content (% mass or %m/m)	D 5453	0.0015 max	EN ISO 20846/ EN ISO 20884	10	D 5453	50	ASTM D5453	50/10	Sept. 18, 2003/ Feb. 1, 2006	D 5453; EN/ ISO 14596	0.001 max	10	ISO 20846, ISO 20884

(continued)

Table 5.3 (continued)

	ASTM D6751 (United States)		EN 14214 (Europe)		IS15607:2005 (India)		Australian Biodiesel Standard			Provisional Brazilian Biodiesel Standard ANP (Agência Nacional do Petróleo) 255		Provisional South African Biodiesel Standard	
Property (unit)	Test method	Limits	Test method	Limits	Test method	Limits	Test method	Limits	Effective date	Test method	Limits	Limits	Test method
Copper strip corrosion (3 h, 50 °C or other specified conditions)	D 130	No. 3 max	EN ISO 2160	1	1S0 2160	1	ASTM D130	No. 3	Sept. 18, 2003	D 130; EN/ ISO[a] 2160	No. 1 max	1	ISO 2160
Cetane number	D 613	47 min	EN ISO 5165	51	ISO 5156	51	EN ISO 5165/ ASTM D613	51	Sept. 18, 2004	D 613; EN/ ISO 5165	45 min	51	ISO 5165
Cloud point (° C)	D 2500	Report	–	–	–	–	–	–	–	D 6371[a]	ANP 310[b]	–	–
Carbon residue (100% sample) (% mass)	D 4530	0.050 max	–	–	D 4530 ISO 10370	0.05	ASTM D4530	0.05	Sept. 18, 2003	D 4530; EN/ ISO 10370	0.05 max	–	–
Acid number (value) (mg KOH/g)	D 664	0.80 max	EN 14104	0.5	–	0.5	ASTM D664	0.8	Sept. 18, 2003	D 664; EN 14104	0.80 max	0.5	EN 14104
Free glycerin (glycerol) (% mass or %m/m)	D 6584	0.020 max	EN 14105, EN 14106	0.02	D 6584	0.02	ASTM D6584	0.02	Sept. 18, 2004	D 6854; EN 14105–6	0.02 max	0.02	EN 14105, EN 14106
Total glycerin (glycerol) (% mass or %m/m)	D 6584	0.240 max	EN 14105	0.25	D 6584	0.25	ASTM D6584	0.25	Sept. 18, 2004	D 6854; EN 14105	0.38 max	0.25	EN 14105

Phosphorus content (% mass or mg/kg)	D 4951	0.001 max	EN 14107	10	D 4951	10	ASTM D4951	10	Sept. 18, 2003	D 4951; EN 14107	10 max	10	EN 14107
Distillation temperature, atmospheric equivalent temperature, 90–95% recovered (T90) (° C)	D 1160	360 max	–	–	–	–	ASTM D1160	360	Sept. 18, 2003	D 1160	360 max		
Ester content (% (m/m)	–	–	EN 14103	96.5	EN 14103	96.5	EN 14103	96.5	Sept. 18, 2003	–	–	96.5	EN 14103
Density; 15 °C (kg/m^3)	–	–	EN ISO 3675/ EN ISO 12185	860–900	ISO 3675; 1S0 12,185; D 4052	860–900	ASTM D1298/EN ISO 3675	860–890	Sept. 18, 2003	–	–	860	ISO 3675, ISO 12185
Carbon residue (10% dist. Residue) (% mass or %m/m)	–	–	EN ISO 10370	0.3	–	–	EN ISO 10370	0.3	Sept. 18, 2003	–	–	0.3	ISO 10370
Total contamination (mg/kg)	–	–	EN 12662	24	EN 12662	24	EN 12662/ ASTM D5452	24	Sept. 18, 2004	–	–	24	EN 12662

(continued)

Table 5.3 (continued)

Property (unit)	ASTM D6751 (United States)		EN 14214 (Europe)		IS15607:2005 (India)		Australian Biodiesel Standard			Provisional Brazilian Biodiesel Standard ANP (Agência Nacional do Petróleo) 255		Provisional South African Biodiesel Standard	
	Test method	Limits	Test method	Limits	Test method	Limits	Test method	Limits	Effective date	Test method	Limits	Limits	Test method
Oxidative stability, 110 °C (h)	–	–	EN 14112	6	EN 14112	6	EN 14112/ ASTM D2274 (as relevant to biodiesel)	6	Sept. 18, 2004	EN 14112	6 min	6	EN 14112
Iodine value/ number (g iodine/100 g)	–	–	EN 14111	120	EN 14104	To report	–	–	–	EN 14111	Take note	140	EN 14111
Linolenic acid content (% (m/m))	–	–	EN 14103	12	–	–	–	–	–	–	–	12	EN 14103
Content of FAME with ≥4 double bonds (% (m/m))	–	–		1	–	–	–	–	–	–	–	–	–
Methanol content (% m/m)	–	–	EN 14110	0.2	EN 14110	0.20	–	–	–	–	–	0.2	EN 14110
Monoglyceride content (% m/m)	–	–	EN 14105	0.8	–	–	–	–	–	D 6584; EN 14105	1.00 max	0.8	EN 14105
Diglyceride content (% m/m)	–	–	EN 14105	0.2	–	–	–	–	–	D 6584; EN 14105	0.25 max	0.2	EN 14105
Triglyceride content (% m/m)	–	–	EN 14105	0.2	–	–	–	–	–	D 6584; EN 14105	0.25 max	0.2	EN 14105

Alkali metals: Group I (Na + K) (mg/kg)	–	–	EN 14108, EN 14109	5	EN 14108 and EN 1410 9	To report	EN 14108/ EN 14109	5	Sept. 18, 2004	EN 14108–9	10 max	5	EN 14108, EN 14109
Earth alkali metals:Group II (ca + mg) (mg/kg)	–	–	prEN 14,538	5	–	To report	prEN 14,538	5	Sept. 18, 2004	–	–	5	prEN 14,538
Cold-filter plugging point	–	–	–	–	–	–	TBA		Sept. 18, 2004	–	–	–	–
Specific gravity[a]	–	–	–	–	–	–	–	–		D 1298/4052	ANP 310[b]	–	–
Alcohol (%, m/m)	–	–	–	–	–	0.02	–	–		EN 14110	0.50 max	–	–
Cold filter plugging point (CFPP)	–	–	–	–	–	–	–	–		–	–		EN 116
Winter, °C, max	–	–	–	–	–	–	–	–		–	–	–4	
Summer, °C, max	–	–	–	–	–	–	–	–		–	–	3	

Adapted from: Indian Standard (2005), Australian Biodiesel Standard (2003), Robles-Medina et al. (2009), Jain and Sharma (2010), Silitonga et al. (2011), Borugadda and Goud (2012), Knothe et al. (2010)

2.13%; insurance, local taxes and royalties, 0.57%), variable costs (feedstocks, 87.86%; miscellaneous materials, 0.14%; utilities, 1.38%), and general overheads and basic research and development accounts for 4.80%. The estimated production cost of such capacity was reported to be ~1.15$/L (~4.353$/gal). This may vary from country to country, based on availability and costs of feedstocks, utility charges, government taxes (or exemptions), scale of commercial plant, etc.

5.8 Current Status of Practical Execution in Different Countries

In the late 2000, robust increase in world ethanol and biodiesel output was observed, mainly due to different governing policies, which eventually stimulated biofuel market potential. The major markets for biodiesel are the EU, the USA, South America (mainly Brazil and Argentina), and Asia (Indonesia, India, and other countries). Indonesia (in the Asian context), and Western region (in Europe) will continue to be the foremost players in biodiesel production by year 2027 (OECD/FAO 2018). It is suggested that if global crude oil prices would increase it would further stimulate worldwide biodiesel production.

If rightly put forward, policies rather than the global demand will decisively influence international biodiesel production and usage. Several benefits of using biodiesel such as eco-friendly and environment-friendly nature, and cost competitiveness as compared to diesel, encouraged worldwide policymakers to develop and support policies which lead to an increasing trend for biodiesel demand. Global biodiesel production is anticipated to increase around 9%, from 36 billion L in 2017 to 39 billion L by year 2027 (OECD/FAO 2018). It is expected that with this increased demand and production, the prices would drastically decrease by almost 14% by 2029 or before that. In the near future, palm oil–based biodiesel production would not be permitted and the major share of thus produced biodiesel would be produced using waste cooking oils and other similar feedstocks.

In the United States, current second major biodiesel producer, an increase in production from 6.9 billion L in 2017 to 7.2 billion L could be seen in 2019, but then it could decrease in future, due to anti-subsidy duties imposed on imported biodiesel and the government policies. In South American countries, such as Brazil and Argentina, the 10% biodiesel mandate is expected to be met by year 2020, where the government tax exemptions would further boost the local biodiesel industry. Brazil would further become leading biodiesel producer and could support almost 50% of global production, reaching to almost 5.6 billion L. In Asia, the Indonesian government already reached a 7% biodiesel mandate. In Thailand, the government has set targets for biodiesel use of 5.1 billion L by 2036. Other Asian countries such as India, Malaysia, and the Philippines will also continue expanding biodiesel production and fulfill domestic demands. Most of the oil-producing Middle Eastern countries are still in the early stages to implement policies for extensive usage of biodiesel and other biofuels as such. However, they are also reported to substantially supporting investments in research and developments of such renewable technologies. Back in 2015, it was reported that Dubai was to become the first city in the world to formally adopt biodiesel made from waste cooking oil for use in local municipal vehicles (https://www.weforum.org/agenda/2015/02/the-first-city-running-its-vehicles-on-waste-cooking-oil/). Neutral Fuels, the UAE-based biodiesel producer responsible for this commercialization, was also the first company in the world to commercialize an enzymatic biodiesel in the UAE and wider region. Neutral Fuels (http://www.neutral-fuels.com/) has for the last 6 years been producing biodiesel using used cooking oil, with some of its most high-profile clients including McDonalds UAE and Dubai's Gulf Indian High School, which runs a biodiesel-fuelled fleet of logistics trucks and school busses through its contract with Neutral Fuels. In Oman, Coeja Eco Solutions (http://www.coeja.com/projects.htm) in partnership with Neutral Fuels, UAE, initiated a trial for used cooking oil from McDonald's Oman restaurants that was collected and used to produce biodiesel, which in turn was used for their transportation sector.

As reported (https://www.statista.com/statistics/271472/biodiesel-production-in-selected-countries/), worldwide biodiesel production by different countries was (billion L): the USA, 6; Brazil, 4.3; Germany, 3.5; Argentina, 3.3; Indonesia, 2.5; France, 2.3; Thailand, 1.4; Spain, 1.3; China, 1; Poland, 1; Columbia, Canada, the Netherlands, and India around 0.2–0.6, in year 2017. Overall biodiesel production in year 2017, as percentage contribution by different countries, is shown in Fig. 5.4.

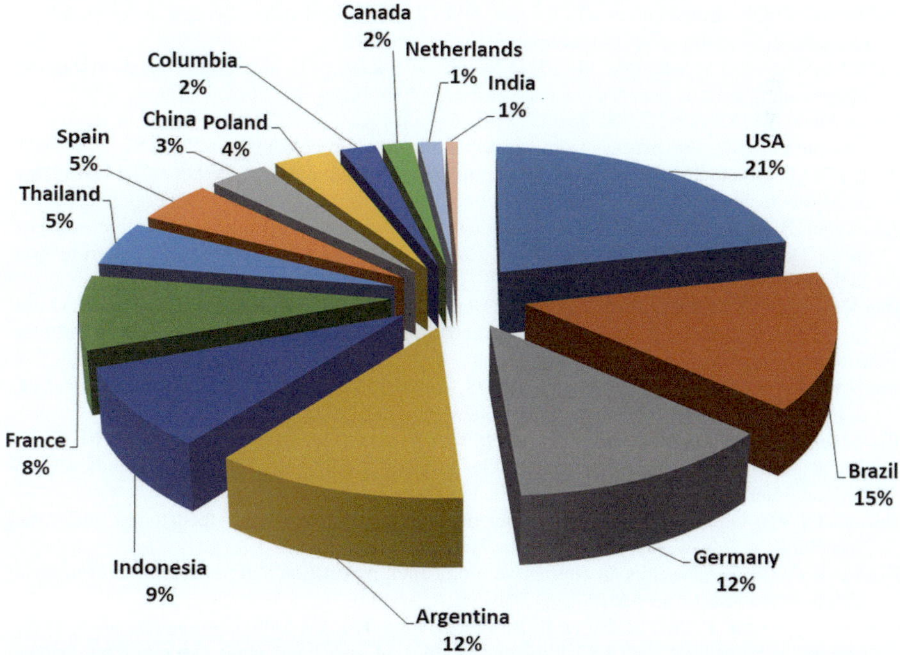

Fig. 5.4 Leading biodiesel producing countries in 2017, with their percentage contribution (https://www.statista.com/statistics/271472/biodiesel-production-in-selected-countries/)

5.9 Conclusion and Future Outlook

At present, fossil fuels lead and suffice most of our energy needs; however, several issues associated with it such as being a nonrenewable source and environmental burden encourages research, development, and further implementation of alternative and renewable fuels such as biodiesel. Those biofuels have several advantages over fossil fuels such as being environment friendly, being comparatively cheaper, and reducing greenhouse gases. Some reports are expecting to see almost 39% surge in world consumption of such biofuels by 2021. Biofuel markets for bioethanol or biodiesel are directly impacted by government policies, taxation (relief), availability of feedstocks, economy/profits, and international crude oil (and local diesel) prices. Such policies are revised from time to time, and recent policies appear to be favorable to biofuels, with a focus on the substantially aiding greenhouse gas mitigation in industrial and transportation segments. Another important aspect is "public perception" to such biofuels, and it cannot be widely accepted without clear information passed to general public in scientific way. Thus, it is quite critical to encourage and support (by government policies) communities by properly communicating the benefits (both environmental and economic) of biofuels such as biodiesel.

References

Aguieiras EC, Cavalcanti-Oliveira ED, Freire DM (2015) Current status and new developments of biodiesel production using fungal lipases. Fuel 159:52–67

Ambat I, Srivastava V, Sillanpää M (2018) Recent advancement in biodiesel production methodologies using various feedstock: a review. Renew Sust Energ Rev 90:356–369

Apostolakou AA, Kookos IK, Marazioti C, Angelopoulos KC (2009) Techno-economic analysis of a biodiesel production process from vegetable oils. Fuel Process Technol 90(7–8):1023–1031

Aransiola EF, Ojumu TV, Oyekola OO, Madzimbamuto TF, Ikhu-Omoregbe DIO (2014) A review of current technology for biodiesel production: state of the art. Biomass Bioenergy 61:276–297

Australian Biodiesel Standard (2003) Biodiesel fuel quality standard – Australian government, Department of Environment and Energy. http://www.environment.gov.au/protection/fuel-quality/standards/biodiesel

Ban K, Kaieda M, Matsumoto T, Kondo A, Fukuda H (2001) Whole cell biocatalyst for biodiesel fuel production utilizing Rhizopus oryzae cells immobilized within biomass support particles. Biochem Eng J 8:39–43

Banković-Ilić IB, Stojković IJ, Stamenković OS, Veljkovic VB, Hung YT (2014) Waste animal fats as feedstocks for biodiesel production. Renew Sust Energ Rev 32:238–254

Bet-Moushoul E, Farhadi K, Mansourpanah Y, Nikbakht AM, Molaei R, Forough M (2016) Application of CaO-based/Au nanoparticles as heterogeneous nanocatalysts in biodiesel production. Fuel 164:119–127

Borugadda VB, Goud VV (2012) Biodiesel production from renewable feedstocks: status and opportunities. Renew Sust Energ Rev 16(7):4763–4784

Bozbas K (2008) Biodiesel as an alternative motor fuel: production and policies in the European Union. Renew Sust Energ Rev 12(2):542–552

Cambon E, Gouzou F, Pina M, Barea B, Barouh N, Lago R et al (2006) Comparison of the lipase activity in hydrolysis and acyl transfer reactions of two latex plant extracts from babaco (Vasconcellea x Heilbornii Cv.) and Plumeria rubra: effect of the aqueous microenvironment. J Agric Food Chem 54:2726–2731

Cambon E, Bourlieu C, Salum TFC, Piombo G, Dubreucq E, Villeneuve P (2009) Ability of Vasconcellea x heilbornii lipase to catalyse the synthesis of alkyl esters from vegetable oils. Process Biochem 44:1265–1269

Correa SM, Arbilla G (2008) Carbonyl emissions in diesel and biodiesel exhaust. Atmos Environ 42(4):769–775

Du W, Xu YY, Liu D, Li Z (2005) Study on acyl migration in immobilized Lipozyme TL-catalyzed transesterification of soybean oil for biodiesel production. J Mol Catal B Enzym 37:68–71

Feyzi M, Shahbazi E (2015) Catalytic performance and characterization of Cs–Ca/SiO_2–TiO_2 nanocatalysts for biodiesel production. J Mol Catal A Chem 404:131–138

Gebremariam SN, Marchetti JM (2018) Economics of biodiesel production. Energy Convers Manag 168:74–84

Halim SFA, Kamaruddin AH (2008) Catalytic studies of lipase on FAME production from waste cooking palm oil in a tert-butanol system. Process Biochem 43:1436–1439

Hernández-Martín E, Otero C (2008) Different enzyme requirements for the synthesis of biodiesel: Novozym® 435 and Lipozyme®TL IM. Bioresour Technol 99:277–286

Hill J, Nelson E, Tilman D, Polasky S, Tiffany D (2006) Environmental, economic, and energetic costs and benefits of biodiesel and ethanol biofuels. Proc Natl Acad Sci 103(30):11206–11210

Hoekman SK, Broch A, Robbins C, Ceniceros E, Natarajan M (2012) Review of biodiesel composition, properties, and specifications. Renew Sust Energ Rev 16(1):143–169

Hsu A-F, Jones KC, Foglia TA, Marmer WN (2004) Transesterification activity of lipases immobilized in a phyllosilicate sol–gel matrix. Biotechnol Lett 26:917–921

Huang D, Han S, Han Z, Lin Y (2012) Biodiesel production catalyzed by Rhizomucor miehei lipase-displaying Pichia pastoris whole cells in an isooctane system. Biochem Eng J 63:10–14

Indian Standard, ICS 75.160.20, IS 15607:2005, Bio-diesel (B 100) blend stock for diesel fuel – specification (2005), Bureau of Indian Standards

Ingale S, Joshi SJ, Gupte A (2014) Production of bioethanol using agricultural waste: banana pseudo stem. Braz J Microbiol 45(3):885–892

Ingale S, Parnandi VA, Joshi SJ (2019) Bioethanol production using *Saccharomyces cerevisiae* immobilized in calcium alginate–magnetite beads and application of response surface methodology to optimize bioethanol yield. In: Srivastava N, Srivastava M, Mishra P, Upadhyay S, Ramteke P, Gupta V (eds) Sustainable approaches for biofuels production technologies, Biofuel and biorefinery technologies, vol 7. Springer, Cham, pp 147–181

Jachmanián I, Mukherjee K (1996) Esterification and interesterification reactions catalyzed by acetone powder from germinating rapeseed. J Am Oil Chem Soc 73:1527–1532

Jahirul MI, Rasul MG, Chowdhury AA, Ashwath N (2012) Biofuels production through biomass pyrolysis—a technological review. Energies 5(12):4952–5001

Jain S, Sharma MP (2010) Prospects of biodiesel from Jatropha in India: a review. Renew Sust Energ Rev 14:763–771

Jang MG, Kimb DK, Park SC, Lee JS, Kim SW (2012) Biodiesel production from crude canola oil by two-step enzymatic processes. Renew Energy 42:99–104

Karmee SK, Chadha A (2005) Preparation of biodiesel from crude oil of Pongamia pinnata. Bioresour Technol 96(13):1425–1429

Karmee SK, Patria RD, Lin CSK (2015) Techno-economic evaluation of biodiesel production from waste cooking oil – a case study of Hong Kong. Int J Mol Sci 16(3):4362–4371

Kaur M, Ali A (2011) Lithium ion impregnated calcium oxide as nano catalyst for the biodiesel production from karanja and jatropha oils. Renew Energy 36(11):2866–2871

Knothe G (2001) Determining the blend level of mixtures of biodiesel with conventional diesel fuel by fiber-optic near-infrared spectroscopy and ^{1}H nuclear magnetic resonance spectroscopy. J Am Oil Chem Soc 78(10):1025–1028

Knothe G, Krahl J, Van Gerpen J (ed) (2010) The biodiesel handbook, 2nd ed. Academic/AOCS Press, ISBN: 9781893997622

Koh MY, Ghazi TIM (2011) A review of biodiesel production from Jatropha curcas L. oil. Renew Sust Energ Rev 15(5):2240–2251

Lai C, Zullaikah S, Vali SR, Ju Y (2005) Lipase-catalyzed production of biodiesel from rice bran oil. J Chem Technol Biotechnol 80:331–337

Li L, Du W, Liu D, Wang L, Li Z (2006) Lipase-catalyzed transesterification of rapeseed oils for biodiesel production with a novel organic solvent as the reaction medium. J Mol Catal B Enzym 43:58–62

Li X, Xu H, Wu Q (2007) Large-scale biodiesel production from microalga *Chlorella protothecoides* through heterotrophic cultivation in bioreactors. Biotechnol Bioeng 98:764–771

Li NW, Zong MH, Wu H (2009) Highly efficient transformation of waste oil to biodiesel by immobilized lipase from *Penicillium expansum*. Process Biochem 44:685–688

Liaquat M (2010) Optimized synthesis of (Z)-3-hexen-1-yl caproate using germinated rapeseed lipase in organic solvent. J Mol Catal B Enzym 68:59–65

Liu H, Su L, Shao Y, Zou L (2012) Biodiesel production catalyzed by cinder supported CaO/KF particle catalyst. Fuel 97:651–657

Mahmudul HM, Hagos FY, Mamat R, Adam AA, Ishak WFW, Alenezi R (2017) Production, characterization and performance of biodiesel as an alternative fuel in diesel engines–a review. Renew Sust Energ Rev 72:497–509

Manzanera M, Molina-Muñoz ML, González-López J (2008) Biodiesel: an alternative fuel. Recent Pat Biotechnol 2(1):25–34

Meher LC, Dharmagadda VS, Naik SN (2006) Optimization of alkali-catalyzed transesterification of Pongamia pinnata oil for production of biodiesel. Bioresour Technol 97(12):1392–1397

Modi MK, Reddy JRC, Rao BVSK, Prasad RBN (2006) Lipase-mediated transformation of vegetable oils into biodiesel using propan-2-ol as acyl acceptor. Biotechnol Lett 28:637–640

Modi MK, Reddy JRC, Rao BVSK, Prasad RBN (2007) Lipase-catalyzed mediated conversion of vegetable oils into biodiesel using ethyl acetate as acyl acceptor. Bioresour Technol 98:1260–1264

Mounguengui RWM, Brunschwig C, Baréa B, Villeneuve P, Blin J (2013) Are plant lipases a promising alternative to catalyze transesterification for biodiesel production? Prog Energy Combust Sci 39(5):441–456

Nomanbhay S, Ong M (2017) A review of microwave-assisted reactions for biodiesel production. Bioengineering 4(2):57

OECD/FAO (2018) OECD-FAO Agricultural Outlook 2018–2027, OECD Publishing, Paris/ Food and Agriculture Organization of the United Nations, Rome https://doi.org/10.1787/agr_outlook-2018-en

Qin H, Yan X, Yun T, Dong W (2008) Biodiesel production catalyzed by whole-cell lipase from Rhizopus chinensis. Chin J Catal 29:41–46

Robles-Medina A, González-Moreno PA, Esteban-Cerdán L, Molina-Grima E (2009) Biocatalysis: towards ever greener biodiesel production. Biotechnol Adv 27(4):398–408

Sahoo PK, Das LM, Babu MKG, Naik SN (2007) Biodiesel development from high acid value polanga seed oil and performance evaluation in a CI engine. Fuel 86(3):448–454

Sakthivel R, Ramesh K, Purnachandran R, Shameer PM (2018) A review on the properties, performance and emission aspects of the third generation biodiesels. Renew Sust Energ Rev 82:2970–2992

Salis A, Pinna M, Monduzzi M, Solinas V (2005) Biodiesel production from triolein and short chain alcohols through biocatalysis. J Biotechnol 199:291–299

Sarin R, Sharma M, Sinharay S, Malhotra RK (2007) Jatropha–palm biodiesel blends: an optimum mix for Asia. Fuel 86(10–11):1365–1371

Sharma YC, Singh B, Upadhyay SN (2008) Advancements in development and characterization of biodiesel: a review. Fuel 87(12):2355–2373

Silitonga AS, Atabani AE, Mahlia TMI, Masjuki HH, Badruddin IA, Mekhilef S (2011) A review on prospect of Jatropha curcas for biodiesel in Indonesia. Renew Sust Energ Rev 15(8):3733–3756

Singh SP, Singh D (2010) Biodiesel production through the use of different sources and characterization of oils and their esters as the substitute of diesel: a review. Renew Sust Energ Rev 14(1):200–216

Sinnott RK (1999) Coulson & Richardson's chemical engineering, 2nd ed. Chemical engineering design, vol 6. Elsevier Butterworth Heinemann

Siqueira SF, Francisco ÉC, Queiroz MI, de Menezes CR, Zepka LQ, Jacob-Lopes E (2016) Third generation biodiesel production from microalgae Phormidium autumnale. Braz J Chem Eng 33(3):427–433

Skarlis S, Kondili E, Kaldellis JK (2012) Small-scale biodiesel production economics: a case study focus on Crete Island. J Clean Prod 20(1):20–26

Su EZ, Zhou Y, You PY, Wei DZ (2010) Lipases in the castor bean seed of Chinese varieties: activity comparison, purification and characterization. J Shanghai Univ (English Edition) 14(2):137–144

Tamalampudi S, Talukder MR, Hama S, Numata T, Kondo A, Fukuda H (2008) Enzymatic production of biodiesel from Jatropha oil: a comparative study of immobilized-whole cell and commercial lipases as a biocatalyst. Biochem Eng J 39:185–189

Tasić MB, Pinto LFR, Klein BC, Veljković VB, Maciel Filho R (2016) Botryococcus braunii for biodiesel production. Renew Sust Energ Rev 64:260–270

Teo SH, Islam A, Taufiq-Yap YH (2016) Algae derived biodiesel using nanocatalytic transesterification process. Chem Eng Res Des 111:362–370

Thangaraj B, Piraman S (2016) Heteropoly acid coated ZnO nanocatalyst for *Madhuca indica* biodiesel synthesis. Biofuels 7(1):13–20

Tuter M, Secundo F, Riva S, Aksoy HA, Ustun G (2003) Partial purification of *Nigella sativa* L. seed lipase and its application in transesterification reactions. J Am Oil Chem Soc 80(1):43–48

Van Gerpen J (2005) Biodiesel processing and production. Fuel Process Technol 86(10):1097–1107

Van Gerpen J (2010) The basics of diesel engines and diesel fuels. In: The biodiesel handbook. AOCS Publishing, pp 24–32

Veljković VB, Lakićević SH, Stamenković OS, Todorović ZB, Lazić ML (2006) Biodiesel production from tobacco (*Nicotiana tabacum* L.) seed oil with a high content of free fatty acids. Fuel 85(17–18):2671–2675

Wang Y, Ou S, Liu P, Zhang Z (2007) Preparation of biodiesel from waste cooking oil via two-step catalyzed process. Energy Convers Manag 48(1):184–188

Watanabe Y, Shimada Y, Baba T, Ohyagi N, Moriyama S, Terai T et al (2002) Methyl esterification of waste fatty acids with immobilized *Candida antarctica* lipase. J Oleo Sci 51:655–661

Xue F, Zhang X, Luo H, Tan T (2006) A new method for preparing raw material for biodiesel production. Process Biochem 41(7):1699–1702

Yan J, Li A, Xu Y, Ngo TPN, Phua S, Li Z (2012a) Efficient production of biodiesel from waste grease: one-pot esterification and transesterification with tandem lipases. Bioresour Technol 123:332–337

Yan Y, Xu L, Dai M (2012b) A synergetic whole-cell biocatalyst for biodiesel production. RSC Adv 2:6170–6173

Yan J, Zheng J, Li S (2014) A novel and robust recombinant Pichia pastoris yeast whole cell biocatalyst with intracellular overexpression of a *Thermomyces lanuginosus* lipase: preparation, characterization and application in biodiesel production. Bioresour Technol 151:43–48

Ziejewski MZ, Kaufman KR, Pratt GL. (1983) Vegetable oil as diesel fuel, USDA, Argic Rev Man, ARM-NC-28, pp 106–111

Ziejewski M, Goettler H, Pratt GL (1986) International congress and exposition. Detroit, MI; 24–28 February [Paper No. 860301]

Cellulose Nanofibers from Agro-Wastes of Northeast India for Nanocomposite and Bioenergy Applications

6

Suvangshu Dutta and Rashmi Rekha Devi

Abstract

Cellulose is the most profuse biopolymer ever existing on earth. Isolation of cellulose in nanodimension known as cellulose nanocrystals or cellulose nanofibers (CNF) from the relatively unexplored agro-wastes of Northeast India is being focused. Besides, the authors of the chapter report the utilization of *Mesua ferea* L. (locally known as Nahar) seed buckles and the de-oiled seeds as two potent sources of CNFs for the first time. The strong sulfuric acid hydrolysis technique and cellulase-based green enzymatic approaches are generally found to be mostly employed for CNF extraction. In addition to greenness, the later method encompasses better yield and quality. Characterization of the isolated nanofibers has been discussed in details with reference to the standard techniques like FTIR, XRD, TGA, SEM, TEM, and AFM. Research works procured on CNFs derived from selected agro-waste-based renewable resources and their nanocomposites are being highlighted. The chapter concludes that cellulose degradable CNFs isolated from agro-wastes of Northeast India has tremendous application potential as nanocomposites ranging from packaging to biomedical, biotechnological, and other bioenergy applications. Getting edge in the development of such biodegradable bionanomaterials in contemporary research is imperative from environment sustainability viewpoints.

Keywords

Cellulose nanofiber · Agro-waste · Nanocomposite · Cellulase · Biodegradation · Bioenergy · Sustainability

S. Dutta (✉)
Department of Chemistry, D.R. College, Golaghat, Assam, India

R. R. Devi
H.P.B. Girls College, Golaghat, Assam, India

Department of Chemical Engineering, IIT Guwahati, Guwahati, Assam, India

N. Srivastava et al. (eds.), *Substrate Analysis for Effective Biofuels Production*,
Clean Energy Production Technologies, https://doi.org/10.1007/978-981-32-9607-7_6

6.1 Introduction

Cellulose, the versatile biopolymer, was discovered officially way back in 1839 by the French chemist Anselme Payen (Prasad et al. 2017). The polymeric structure of cellulose [$(C_6H_{10}O_5)_n$] was determined by Hermann Staudinger in 1920. In nature, it occurs along with hemicellulose, lignin, pectin, and waxes in plant fibers. Cellulose is a linear isotactic homopolymer of high molecular weight consisting of β-D-glucopyranose units joined via β-1,4-glycosidic linkage and every other monomer is rotated 180° (Silverio et al. 2013) with respect to its neighboring unit (Fig. 6.1). More the aggregation of the microfibrils in an ordered manner, more will be the crystallinity of cellulose. On the other hand, the regions with irregular arrangement of cellulose chains are generally called amorphous. Cellulose, the key chemical component responsible for the mechanical properties of the natural fiber, is the main ingredient of several natural fibers such as cotton, flax, hemp, coconut, jute, sisal, sugarcane bagasse, and several other agro-wastes (Sonia and Dasan 2013; Kumar et al. 2014; Miao and Hamad 2013). Cellulose is also synthesized by algae, tunicates, and some bacteria (Henrique et al. 2013). In general, cellulose accounts for ca. 35–50% by mass of most of the lignocellulosic agro-wastes, whereas hemicelluloses account for ca. 20–35% and lignin ca. 5–30% (Zabeda et al. 2016). Cellulose nanofibers (CNF) are the nanosized cellulose which have at least one dimension in the nanometer range. Since its official introduction in 1980s, the conversion of cellulose into cellulose nanostructures is continuously getting edge in research as it imparts certain remarkable and unique features such as low density, low cost, biocompatibility, well-defined size and morphology, controlled surface chemistry, high specific strength, high modulus, and more environmental sustainability (Prasad et al. 2017). CNF can improve the biopolymer properties like toughness, mechanical strength, thermal stability, barrier properties, biodegradability, and optical properties via nanocomposite formation. In recent times, the development of nanocomposites derived from agro-waste-based renewable resources reinforced with CNF becomes hot spot of the scientists in this field.

Cellulase is a group of enzymes (endoglucanases, exoglucanases, or cellobiohydrolases and β-glucosidases) known to be produced by fungi and bacteria (Thota et al. 2017; Srivastava et al. 2018). The key function of this class of enzymes is to hydrolyze cellulose to glucose (sugar) molecules. Investigation on the isolation of

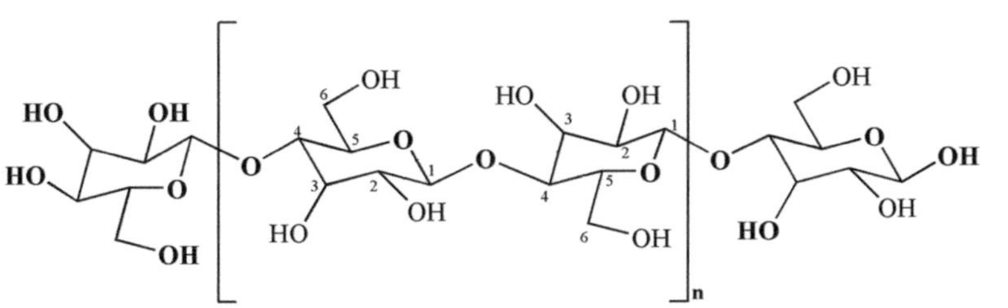

Fig. 6.1 Chemical structure of cellulose

cellulase producing microorganisms from agro-waste-based renewable resources paves the way to get novel cellulases with specific properties. Studies have shown that production of fermentable sugars by such hydrolysis may lead to the formation of the green fuel "bioethanol" (Balat 2011; Singhania et al. 2014; Cherian et al. 2015). Production of large-scale cellulase from agro-wastes using cost-effective techniques like solid-state fermentation adds boost to bioethanol production. However, the yield of sugar production depends strongly on the pretreatment of the biomass from where enzyme is to be isolated. Furthermore, research is underway for the production of enzymes with high enzymatic activity for improvement in technology to produce better cellulase enzyme systems (Garvey et al. 2013). Literature has evidenced that the cost of pretreatment can be drastically reduced and efficiency of the cellulase enzyme can be improved immobilizing enzymes on a covalent support. Various organic molecules like chitin, chitosan, polyvinyl alcohol, nylon, etc., have been used as supports for immobilized cellulase (Cherian et al. 2015). Such immobilization makes the enzyme thermostable, recyclable, and easily separable from the reaction mixture.

Northeast India is a hub of vast varietal agricultural biomass and renewable resources. Although the literature indicates the extensive preparation of CNFs from various sources of plants, fruits, vegetables, etc., however, there are no data indicating the utilization of many unexplored biomass of the region, particularly the waste generated from the plant resources. The chapter focuses on the development of CNFs from certain such agro-waste-based renewable resources of Northeast India worked out by the authors in recent times. Being inspired from the high cellulose content (Guna et al. 2017) of these resources, CNFs were developed by alkali treatment, bleaching, and acid hydrolysis techniques followed by sonication. Enzymatic approaches and their advantages are also discussed. Physical, chemical, morphological, and thermal characterization of the CNFs is done by standard techniques like FTIR, XRD, TGA, SEM, TEM, and AFM. These CNFs have tremendous scope to be utilized as prospective nano-reinforcing agents for the production of eco-friendly nanobiomaterials along with other versatile applications. After application, they are not going to create any environmental hazard as they are cellulase degradable. In fact, the degraded products can be reconverted into materials of worth importance as discussed.

6.2 Agro-Wastes of Northeast India for Cellulose Production

One of the major fruits consumed by the people of Northeast India is banana. The peel of the banana fruit is largely obtained as a byproduct generated from banana-processing industries during food or juice production, hotel, restaurant, and households. *Musa* ABB (locally known as kachkal), the only culinary variety of banana, is an important and cheap vegetable used by the people of the region. After proper utilization of the fruit part of banana, the peel becomes waste material. Deka and Khawas isolated CNFs successfully from this underutilized agro-waste and studied its potential application as an effective reinforcing material to be used as bionanocomposites which adds high value to culinary banana (Deka and Khawas 2016). Another agro-waste-based renewable resource of Northeast India is the pineapple leaf fiber which exhibits high specific strength and stiffness. Pineapple is a fibrous plant largely available in tropical countries. Pineapple leaf is a waste product of pineapple cultivation and hence can be obtained easily without any additional cost. Mahardika et al. showed CNF production from pineapple leaf which may be an effective reinforcing material for biocomposites (Mahardika et al. 2018; Cherian et al. 2010) having the wide range of biomedical and biotechnological applications like tissue engineering, medical implants, drug delivery, etc. Lately, Dai et al. developed nanocellulose from pineapple peel and evaluated the reinforcement capability for gellan gum film.

Two relatively unexplored biomass of Northeast India, as explored by the authors of this book chapter, were used as raw material as shown in Scheme 6.1. *Mesua ferrea* (*M. ferrea*) L. locally known as nahar belongs to the family *Guttiferae*. Owing to its antimicrobial efficacy, the essential oil extracted from the seeds of the plant is used for medicinal purpose. The nahar seed oil has also been reported to be utilized for the preparation of biodiesel and various versatile polymeric materials (Dutta and Karak 2005 2006; Dutta et al. 2010a, b). However, after debuckling the seeds for essential oil extraction, the hard buckles of nahar seeds are of no use and become agro-waste. Again, the seeds were powdered and then vacuum dried at 40 °C. The powdered seed was then subjected to solvent extraction in petroleum ether for a week to extract the oil. The process was repeated to confirm the de-oiled process. The raw de-oiled cake or seed powders are of no use and are agricultural wastes (Fig. 6.2). Dutta et al. also worked on pond hyacinth (locally known as Meteka), which is an aquatic plant available in Northeast India. This plant is an agro-waste, creates problems by blocking water and is of very limited use (Dutta 2018). Ramesh and Sundari studied on water hyacinth (*Eichhornia crassipes*)-based CNFs and found that the weed rich in cellulose content has no major differences in the percentage of cellulose content in the shoot and root (Ramesh and Sundari 2012).

Devi et al. proposed an innovative idea to fabricate CNFs from agricultural compost of Northeast India. Compost, an agro-based biomass feedstock, was obtained from mixing water hyacinth (*Eichhornia crassipes*), cow dung, and saw

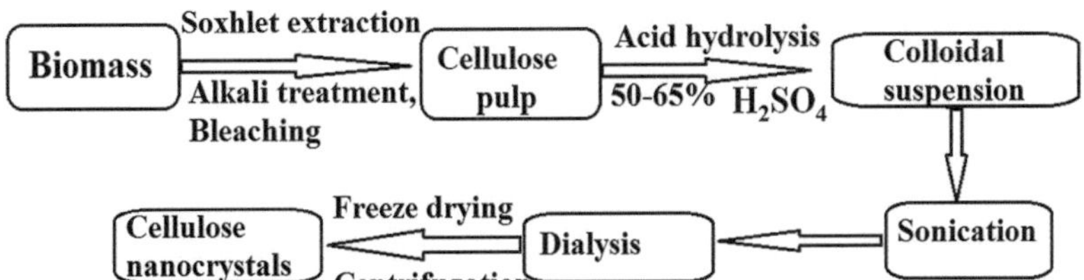

Scheme 6.1 Technique of isolation of cellulose nanocrystals/fibers (CNFs) from agro-wastes

Fig. 6.2 Two unexplored agricultural biomass of rich cellulose content

dust in the ratio 8:1:1 (Devi et al. 2015). The mixture was then utilized for the extraction of cellulose and then nanocellulose (CNF). The study reported a cost-effective and feasible approach of utilizing compost, an agro-waste-based renewable resource, for production of value-added products which could find potential application in the fields of healthcare, biomedical engineering, and packaging industry. Sugarcane bagasse is a major agro-industrial biomass of the sugarcane industry and its cultivation is found to be ever-growing in Northeast India. In Brazil, sugarcane bagasse is reported to generate bioethanol. Ferreira et al. isolated CNFs from this important biomass and modified the nanocellulose with adipic acid. They showed that unmodified CNFs can act as a potential reinforcing agent for matrices of hydrophilic polymers whereas modified CNFs for hydrophobic matrices (Ferreira et al. 2018). Besides the above-mentioned agro-waste-based renewable resources, there are still some underutilized resources available in Northeast India (Fig. 6.3) which has great potential for cellulose production (Table 6.1). The authors of this book chapter are now undertaking a project on the unexplored agro-wastes of the region namely soya seed buckle and jatropha seed (known for biodiesel production) buckle for CNF extraction.

Fig. 6.3 Some underutilized biomass available for cellulose extraction

Table 6.1 Selected agro-wastes of Northeast India with CNF isolation potential

Agro-waste	Scientific name of the plant	Moisture content (%)	Cellulose (%)	Hemicellulose (%)	Lignin (%)	References
Culinary banana peel	*Musa paradiciaca* L.	6.7	16.62	7.34	4.01	Deka and Khawas (2016)
Pineapple leaf	*Ananas comosus*	10.52	81.27	12.31	3.46	Mahardika et al. (2018)
Sugarcane bagasse	*Saccharum officinarum*	7.0	46.5	26	24	Ferreira et al. (2018)
Nahar seed buckle	*Mesua ferrea* L.	16.8	19.7	16.2	63.9	Rastrogi et al. (2016)
Water hyacinth	*Eichhornia crassipes*	–	24.5	34.1	8.6	Kumar et al. (2018)
Jatropha seed buckle	*Jatropha curcas* L.	8.7	20.7	24.6	54.7	Rastrogi et al. (2016)
Soya seed buckle	*Glycine max*	11.0	42.4	43.4	13.6	Rastrogi et al. (2016)
Ipomoea	*Ipomoea fistulosa*	9.72	32.5	47.6	14.7	Kataki et al. (2016)
Jute	*Corchurus olitorious* L.	1.1	64.4	12	11.8	Saravanan et al. (2016)
Onion skin	*Allium cepa* L.	–	41.1	38.9	16.2	Rhim et al. (2015)

6.3 Techniques of Cellulose Production

Deka and Khawas utilized culinary banana peel flour for CNF isolation. They digested the raw material with a solution of 20% (w/v) sodium hydroxide and 0.1% anthraquinone to solubilize the non-cellulosic part. Further delignification was carried out with 1% (w/v) sodium chlorite (at pH 5 adjusted with 10% (v/v) acetic acid) at 70 °C (Deka and Khawas 2016). Bleaching was performed repeatedly to leach out of phenolic compounds and lignins. The insoluble part was then subjected to acid hydrolysis via treatment with solution of 1% (v/v) sulfuric acid at 80 °C for 1 h. This caused hydrolysis of the amorphous cellulose and eased in getting the required CNFs. All the steps were followed by centrifugation @10,000 rpm at 4 °C for 20 min. The acid-hydrolyzed CNFs were subjected to size reduction by giving high-intensity ultrasonic treatment in ice water bath for 30 min before freeze drying. Deka and Khawas studied the effect of ultrasonic intensity on nanofibrilation of chemically purified cellulose (Deka and Khawas 2016). Cherian et al. isolated CNFs from pineapple fibers by steam explosion method. After bleaching with NaOH and acetic acid mixture followed by treatment with 11% oxalic acid solution, the pressure of the autoclave was released and the fibers were taken out. The solution was washed until $KMnO_4$ solution is no longer decolorized to ensure the washings free from acid (Cherian et al. 2010). The resulting CNFs were subjected to centrifugation @8000 rpm for 4 h.

Dutta et al. prepared CNFs by first dewaxing the dried nahar seed buckles (biomass 1) in a soxhlet apparatus with 2:1 (v/v) mixture of toluene–ethanol for 6 h. The dewaxed material was subjected to alkali treatment, bleaching and then acid hydrolysis using 64 wt% sulfuric acid in water at 45 °C. Centrifugation and sonication were performed until CNFs in the form of stable gel were obtained (Fig. 6.4). The authors are now trying to isolate CNF from the de-oiled cake (biomass 2) of nahar seed. Dutta et al. also prepared CNFs from alkali-treated and bleached fibers of pond hyacinth by hydrolysis treatment using concentrated sulfuric acid solution (60 wt %) at a temperature of 45 ± 3 °C under mechanical stirring conditions followed by sonication @7500 rpm at room temperature. The resulting colloidal suspension was dialyzed against water for 3 days until constant p^H is reached to remove nonreactive sulfate groups, salts, and soluble sugars. The suspension was again subjected to dialysis for further dispersion by constant stirring in a magnetic stirrer to obtain the nanocellulose suspensions (CNF) (Dutta 2018).

In another work, well-tattered pond hyacinth was mixed with cowdung and sawdust together in a definite ratio to obtain compost as reported by Devi et al. (2015). The mixed combinations were then subjected for composting for 20 days using a batch scale rotary drum composter (Singh and Kalamdhad 2013). In a separate study, a rotary drum composter of 550 L capacity was used as adopted by Singh

(Alkali Treated) (Bleached) (CNF Gel)

Fig. 6.4 CNF extraction from nahar seed buckles

et al. (2009). The compost was then alkali-treated, bleached and finally acid hydrolyzed using 40–50% sulfuric acid solution. The resulting CNF suspension was washed by centrifugation @8000 rpm at room temperature for 10 min repeatedly before the suspension was dialyzed using cellulose acetate dialysis membrane having 12,000–14,000 Da as the cut-off molecular weight against deionized water for several days until constant pH in the range of 5–6 was attained (Devi et al. 2015). The nanocellulose solution was then sonicated for 30 min before freeze-drying. A general and common scheme may be projected (Scheme 6.1) as evidenced from the latest literature for extraction of cellulose nanocrystals (CNF) from agro-waste-based plant biomass. Borkotoky et al. employed a similar technical method for isolation of CNFs from bamboo pulp. Bleaching of the pretreated bamboo pulp was done with H_2O_2 (2 wt%) and hypochlorite treatment at 90 °C for 2 h. The pulp was filtered, washed with Milli-Q at each step repeatedly and dried at 80 °C. CNFs were obtained by hydrolyzing the cellulose pulp with sulfuric acid (64 wt%) for 2 h, at room temperature. The resulting CNF suspension was centrifuged @10,000 rpm for 20 min and dialyzed until the pH reached 7 as reported elsewhere (Borkotoky et al. 2018). Ferreira et al. synthesized CNFs following the acid hydrolysis (65% v/v at 45 °C) technique from bleached sugarcane bagasse pulp and functionalized with adipic acid. Their study evidenced that the crystallinity index and aspect ratio of the CNFs were higher for unmodified samples than in the modified ones. They proposed significant decrease in nanocrystal dimensions due to surface modification with adipic acid arising from (i) removal of the amorphous region, (ii) changes in electrostatic repulsion, and (iii) hydrophilic affinity of CNFs (Ferreira et al. 2018). Pure cellulose microfibers were obtained from water hyacinth, the weed plant, by bleaching, alkaline, and sodium chlorite reactions (Ramesh and Sundari 2012). The fibers obtained from the stems in micron size were cryocrushed with liquid nitrogen and sonicated to nanofibriliate the fibers (CNFs).

Lately, cellulase-based enzymatic approaches are getting edge for nanocellulose production mainly due to two reasons. Firstly, it is a green technology which may reduce the pollution generated by traditional chemical processes (acid hydrolysis) and secondly, the low yield obtained from traditional acid hydrolysis technique may be overcome. In addition, enzymatic preparation of CNFs provides improved quality of final product compared to pure chemical processes. The first attempt in this regard was initiated by Zhu et al. (2011) who used commercial endoglucanase and purified exoglucanase from commercial cellulase. Beltramino et al. reported well optimized enzymatic conditions giving up to ca. 82% yield of CNFs which is much higher than the CNFs obtained from sulfuric acid hydrolysis techniques (Beltramino et al. 2018). Their study evidenced reduction in surface charge and increase in crystallinity of the CNFs without affecting other characteristics of cellulose. In recent times, a few studies have been carried out to utilize novel enzyme technologies for effective fractionation of agro-wastes into CNFs (Anderson et al. 2014; Teixeira et al. 2015). Studies have been initiated to utilize covalently immobilized cellulase to increase the applicability of the enzymatic approaches for a wide range of temperatures. Rehim et al. proposed formation of low cost immobilized cellulase in the form of polymeric gel disks that can be reused several times for CNF production from sugarcane bagasse. The report estimates retention of 85% of activity of immobilized cellulase even after six cycles (Rehim et al. 2019).

6.4 Characterization of CNFs

6.4.1 FTIR Characterization

FTIR is an analytical tool to study the changes in the chemical composition of the chemically treated fibers. The spectrum is generally recorded in the transmission mode in the range of 4000–400 cm^{-1}. The formation of nanocellulose fiber (CNF) can be well understood by comparing the FTIR spectrum of alkali-treated, bleached, and acid-treated agro-waste cellulose fibers. Peak at ca. 3400 and 2900 cm^{-1} is due to –OH and –CH stretching vibration which will be observed in all the samples. In the crude or alkali-treated fiber, the peak at ca. 1735–1740 cm^{-1} is observed due to either the acetyl and uronic ester linkage of carboxylic group of the ferulic and p-coumeric acids of lignin and hemicelluloses (Chen et al. 2011). This peak disappears gradually with treatments like bleaching, acid hydrolysis, and mechanical treatment. The peak at ca. 1515–1520 cm^{-1} observed in the crude or alkali-treated fiber indicates the C=C stretching of aromatic rings of lignin but in the treated one it is not likely to be observed due to the partial removal of lignin (Sain and Panthapulakkal 2006). The peak at ca. 1637–1650 cm^{-1} due to O–H bending mode of adsorbed water to cellulose is found to be decreased in the nanofibers with the removal of hemicelluloses. In the FTIR spectrum of CNFs, peaks at ca. 1030–1050 and 850–880 cm^{-1} are expected due to C–O–C pyranose ring skeletal vibration stretching and C–H rocking vibration of cellulose structure respectively. Increase in intensity of these two peaks indicates an increase in crystalline cellulose part. The FTIR spectra as observed for water hyacinth samples (Dutta 2018) are shown in Fig. 6.5. The figure indicates a diminishing trend of peaks at 1740 $cm^{-1}$and 1520 cm^{-1} in the spectrum of acid-treated final fiber which clearly evidences significant removal of hemicelluloses and mainly lignin, without destroying the cellulosic structure, by the purification process (alkali and bleaching treatments). Studies show that new peaks at ca. 1173 and ca. 677 cm^{-1} for CNFs is due to sulfate groups arising from sulfuric acid hydrolysis (Dai et al. 2018).

6.4.2 X-Ray Diffraction Study

The crystallinity of the cellulose fibers is evaluated by X-ray diffraction study. The crystallinity index (CrI) can be calculated by using Segal equation (Thygesen et al. 2005),

$$\text{C.I.} = \left[\left(I_{200} - I_{am}\right) / I_{200}\right] \times 100$$

where,

I_{200} = maximum intensity of the (200) diffraction peak close to $2\Theta = 22^0$
I_{am} = intensity of diffraction for amorphous part close to $2\Theta = 18^0$

Cellulose is found to exhibit characteristic X-ray diffraction peaks at $2\Theta = 16.1$–$18.5°$, $22.5°$ and $34.6°$, corresponding to the (110), (200), and (004) planes of the typical structure of cellulose I.

Dutta et al. recorded strong crystalline nature of the CNFs as reflected from sharp diffraction intensity at $2\Theta = 22^0$. The cellulose fibers obtained from the pond hyacinth samples showed the diffraction intensity at 22^0 and a shoulder in the region $2\Theta = 14$–16^0 and also exhibited peaks at some lower intensity such as 13.8^0, 17^0 and 34.2^0 (Shahabi-Ghahafarrokhi et al. 2015; Peng et al. 2013). The broad peaks of the alkali-treated fibers are due to the amorphous nature of the lignin whereas α-celluloses are crystalline in nature. The higher peak intensity of the nanocellulose indicates the complete removal of non-cellulosic materials during acid hydrolysis step (Fig. 6.6a). The additional peaks observed are expected to be arising from the rearrangement of fibrils in the interfibrillar regions leading to new crystalline regions (Dutta 2018). The crystallinity index (CI) as determined by Segal equation is shown in the bar diagram (Fig. 6.6b, Dutta 2018). Deka and Khawas studied the variation of CI of culinary banana peel-based CNF with ultrasonication intensity and found a maximum of 63.64% CI at 1000 W intensity. They concluded from the study that with an increase in the intensity of ultrasonication from 0 to 1000 W, the degree of homogenization and fragmentation increases which in turn increases the crystalline nature of CNFs (Deka and Khawas 2016).

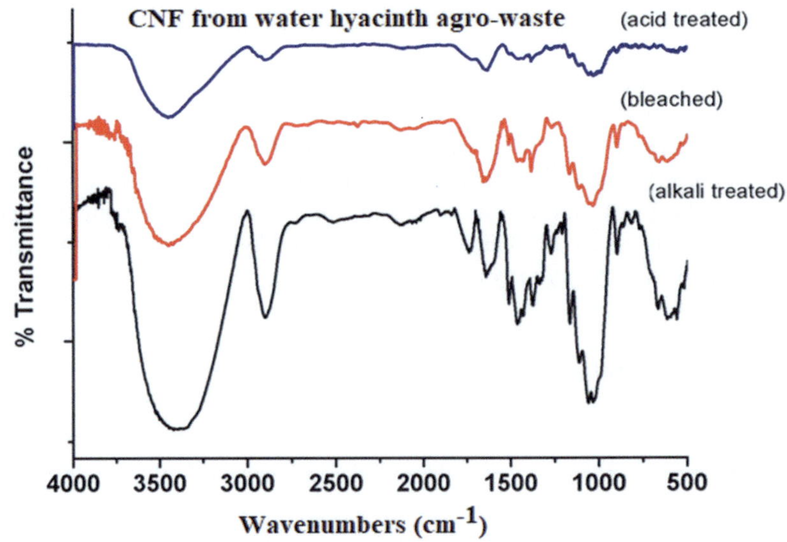

Fig. 6.5 Representing FTIR curves for CNF formation

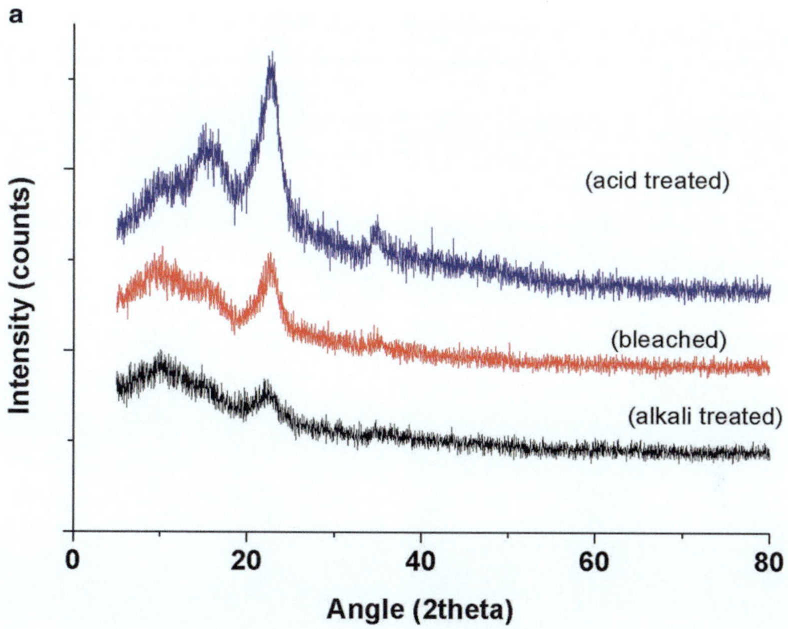

Fig. 6.6a XRD patterns of alkali-treated, bleached, and CNF samples

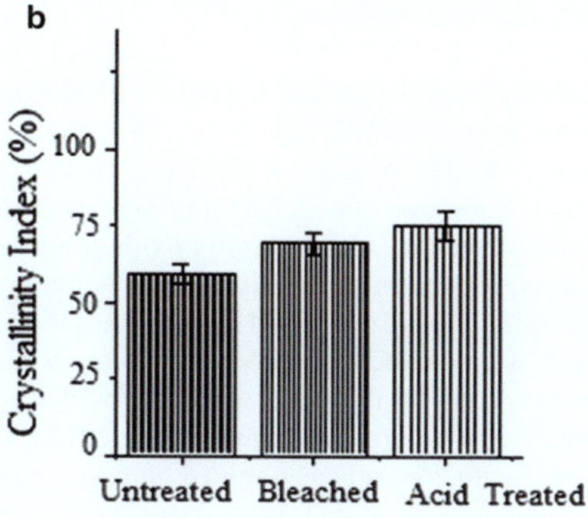

Fig. 6.6b Crystallinity index of alkali-treated, bleached, and CNF samples

6.4.3 SEM/TEM/AFM Analysis

The changes in the microstructure of cellulose nanofibers can be well evaluated after the chemical and ultrasonication treatments by SEM, TEM, and AFM analysis. Scanning electron microscope (SEM) image of the agro-waste before any treatment generally shows a rough and irregular microstructure (Fig. 6.7a), sometimes agglomerated, with some starch granules. During alkaline treatment, the

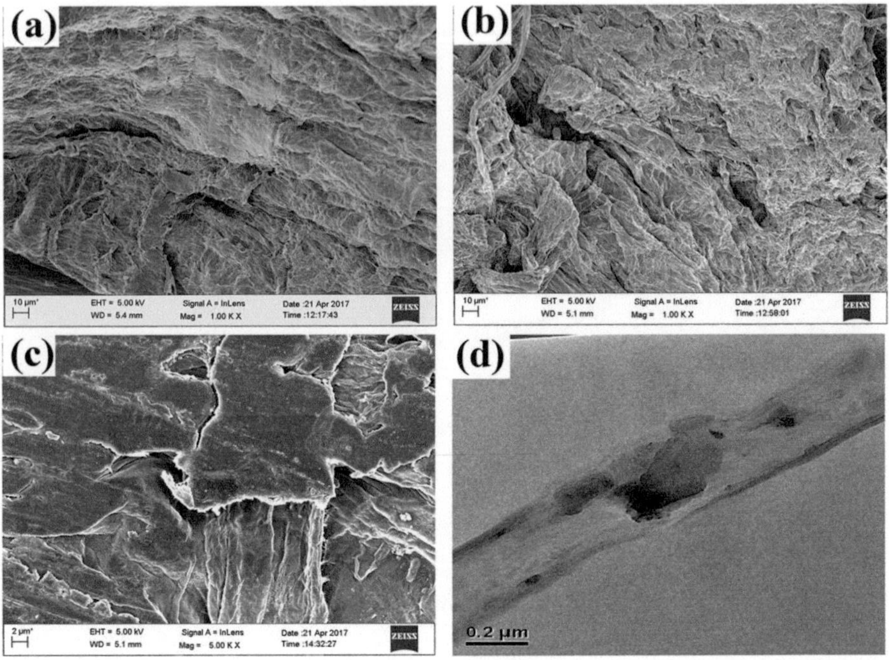

Fig. 6.7 SEM micrograph of (**a**) untreated, (**b**) bleached, (**c**) acid hydrolyzed, (**d**) TEM micrograph of acid hydrolyzed, water hyacinth samples

hemicellulose is partially hydrolyzed and becomes soluble in water. The lignin gets depolymerized and hence defibrillation occurs (Fig. 6.7b) due to the removal of the cementing materials. The process of bleaching makes the fibers whitened by oxidation reaction and removes most of the lignin present in the fiber and further defibrillation takes place (Fig. 6.7c). Lignin oxidation leads to lignin degradation and results in the formation of functional groups like hydroxyl, carbonyl, etc., which facilitate easier solubilization and separation of the lignin part (Cherian et al. 2010). Further, a close look at SEM images of water hyacinth CNFs (Fig. 6.7a–c) indicates a considerable reduction in size after acid hydrolysis due to the successful removal of amorphous phase (Dutta 2018). The loss of fibrous character during CNF formation indicates the etching and erosion nature of the acid hydrolysis process.

Transmission electron microscope (TEM) images generally illustrate a more clear picture which mostly suggests the spherical shapes and rod-like shapes of CNFs (Fig. 6.7d). However, the actual dimension of nanocellulose depends mainly on three factors: strength of the acid, reaction time, and temperature. The spheres are formed due to self-assembly of short cellulose rods/ chains via. intermolecular hydrogen bonds. Such strong hydrogen bonding among CNF chains overcomes the repulsion of negative charges on the surface, leading to the formation of self-assembled porous networks (Berg et al. 2007). High-intensity ultrasonic

Table 6.2 Nanodimensions and crystallinity index of selected CNFs

Agro-waste	Diameter in nm (SEM)	Diameter in nm (TEM)	% Crystallinity Index (XRD)	References
Water hyacinth	20–100	25	68.1	Ramesh and Sundari (2012) and Dutta (2018)
Nahar seed buckle	40–70	30	67.2	Dutta (2018)
Cassava peel	–	3–8	51.2	Widiarto et al. (2017)
Pineapple peel	<50	32.5	73.62	Cherian et al. (2010)
Pistachio shell	–	12	67.0	Marett et al. (2017)
Oil palm (empty fruit bunch)	–	5–40	69.0	Jonoobi et al. (2015)
Soy hull	–	20–120	69.6	Jonoobi et al. (2015)
Sugarcane bagasse	–	4–6	87.5	Jonoobi et al. (2015)

treatments result in large aggregates consisting of needle-like or wire-like shapes. Dutta et al. found that TEM studies of water hyacinth samples exhibited the approximate ranges for diameter of nanocelluloses as ca. 25 nm having length of ca. 210 nm (Table 6.2). The CNFs looked like spherical rod-shaped (Dutta 2018). Most of the agro-waste-based CNFs are found to possess aspect ratio of ca. 3–15. Such higher aspect ratio helps in creating a strong composite material as it increases the interface area and hence allows more load to be transferred to the reinforcing cellulose fiber (Marett et al. 2017). In a comparative TEM study on banana peel CNF, Tibolla et al. suggest that acid hydrolysis technique offers better CNFs than enzyme hydrolysis (Tibolla et al. 2014).

A far better insight into the cellulose nanostructure is provided by atomic force microscopy (AFM). AFM is generally recorded by two modes namely height image and 3D image. The height image mode demonstrates topography of the surface by tracking the surface with the probe while the 3D image mode demonstrates the dissimilarity between soft and hard moieties of the material. The bright section of the AFM image indicates crystallinity whereas dark sections represent the amorphous part in the direction of the fiber axis in the cellulose structure (Mandal and Chakrabarty 2011).

6.4.4 Thermogravimetric Analysis (TGA)

The thermogravimetric analysis (TGA) of the CNFs showed comparable and even higher thermal stability than the cellulose fibers obtained from other resources. The thermal stability is found to be progressively increased after every step of treatment from alkali refining to acid hydrolysis as evidenced from TG curves (Fig. 6.8). The sharp peak observed in the DTG curve of nanosized cellulose (Fig. 6.9) obtained by the present authors from nahar seed buckles gives a more clear picture of the

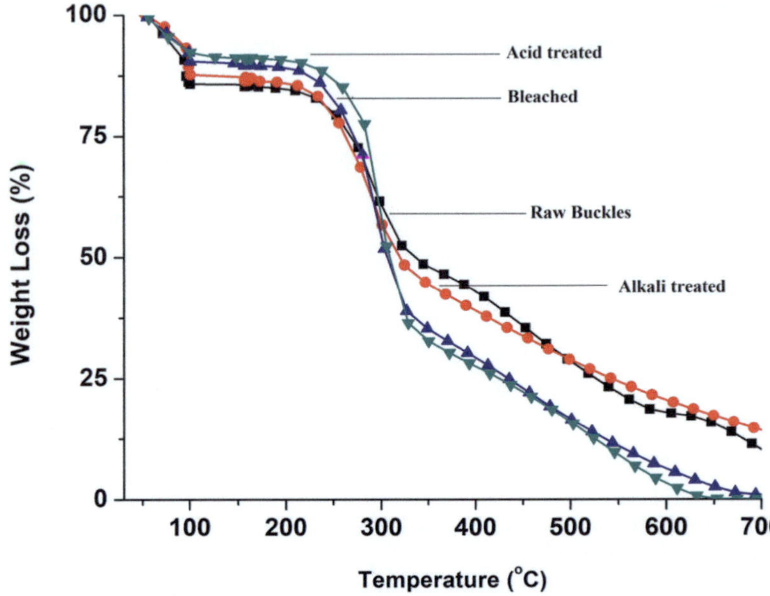

Fig. 6.8 TG curves for (**a**) raw (**b**) alkali-treated, (**c**) bleached, and (**d**) acid hydrolyzed nahar seed buckles

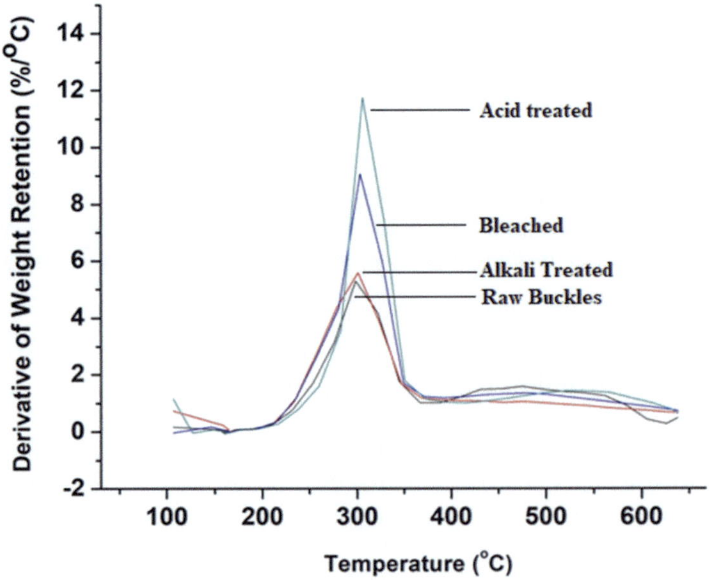

Fig. 6.9 DTG curves for (**a**) raw (**b**) alkali-treated, (**c**) bleached, and (**d**) acid hydrolyzed nahar seed buckles

process. Both TG and DTG curves indicate high cellulose content and improvement in homogeneity of the crystalline structure (Ghahafarrokhi et al. 2015).

In general, the TG curve shows an initial weight loss (first step or initial degradation) in the region of ca. 30–120 °C which may be attributed to the loss of moisture content. The major degradation is the second-step degradation which takes place in the range of ca. 210–300 °C. This step is obviously due to the breakage of β-1,4-glycosidic linkages of cellulose and depolymerization of hemicelluloses (Deepa et al. 2011). The final minor degradation (third and last step of degradation) is generally observed after ca. 400 °C which is due to the decomposition of the thermally stable aromatic lignin and wax part of the fiber. The degradation temperatures are always found to be higher for the acid hydrolyzed sample (CNF) than that of the alkali treated or bleached samples. Thus most of the agro-waste-based CNFs could find application in the field of nanocomposite as a potent reinforcing agent wherein the processing temperature for thermoplastic polymer rises above 210 °C (Saravanan et al. 2016; Deka and Khawas 2016).

6.5 Agro-Waste Based CNFs for Nanocomposite Applications

Nanocomposites are two-phase structural materials having at least one dimension in the nanometer range (1 nm = 10^{-9} m). They are much superior to conventional composite materials in terms of thermal, mechanical, barrier, and recyclable properties. Cellulose nanofibers extracted from agro-wastes/renewable resources have great potential to act as the reinforcing agents for a wide range of polymers. It has been noticed that CNFs possess Young's modulus as high as 134 GPa and tensile strength as high as 10 GPa (Lee et al. 2009) which make them unique in terms of stiffness. The strong interfiber H-bonds present in CNFs make them unique and helps in forming a dense network with the host polymers. The resulting nanocomposites become more versatile with enhanced performance characteristics. CNFs have some disadvantages also as they are not compatible with most of the hydrophobic polymers. Due to its moisture sensitivity or hydrophilic nature, the CNF nanoparticles become agglomerated via. strong H-bonding while trying to disperse in hydrophobic polymers. This limitation can however be overcome by surface modification of the CNFs by the chemical reaction of its hydroxyl groups with different organic reagents during processes like silylation and acetylation (Siqueira et al. 2010). CNFs have also been reported to be surface-modified by grafting of PLA (polylactic acid) or PCL (polycaprolactone) polymers on their surface (Kan et al. 2013). Surface modification can also be performed by physical adsorption of a surfactant as mentioned elsewhere (Petersson et al. 2007). They reported the development of PLA-based composites using cellulose nanocrystals modified with the surfactant Beycostat AB09. The surfactant-treated nanocrystals showed very good dispersion in the PLA matrix with enhanced thermal stability and storage modulus of the composites.

Two methods are found to be utilized in literature for the preparation of nanocomposites – solvent casting and melt compounding. In the solvent casting method, the CNF particles are suspended into a polymer solution with a water-miscible solvent or the surface-modified CNF particles are dissolved into an organic solvent (Dufresne 2013). After proper dispersion, the nanocomposites are fabricated on molds or glass plates by compression molding and then evaporating the solvent system. Murphy and Collins (2016) prepared polylactic acid (PLA)/microcrystalline cellulose composites by solvent casting followed by extrusion of the powdered films in a twin-screw extruder at an elevated temperature. They proved that 3 wt% and 5 wt% CNF/PLA composite films with both modified and natural fibers experienced a wider glass transition range and higher strength. Melt compounding method, on the other hand, is an industrial method for producing cellulose-based nanocomposites. The polymer melt is processed either in batches or extruded continuously depending on the requirement of product amount (Kargarzadeh et al. 2018). The CNFs can then be introduced into the polymer melt in the extrusion process either in dried or in solution form.

Dai et al. demonstrated that introduction of CNFs derived from pineapple peel into gellan gum solutions results in some composite films with enhanced thermal stabilities. They showed that 4% loading of CNF increases the tensile strength of gellan gum films by 48% (Dai et al. 2018). More than 150 publications reported nanocomposites of CNFs with PLA with improved mechanical stiffness arising from increase in strength and elastic modulus (Kargarzadeh et al. 2018). When PCL chains grafted onto CNFs are used as reinforcing filler for PCL matrices, then elongation at break is found to be decreased whereas Young's modulus and storage modulus are found to be increased drastically for the resulting nanocomposites (Goffin et al. 2011). Polyvinyl alcohol composites loaded with CNFs exhibited improved tensile and thermal properties (Lee et al. 2009).

6.6 CNFs-Based Nanocomposites for Photocatalytic and Biomedical Applications

Polymer nanocomposites have widely been used in food packaging applications as they offer an improved barrier for gases like O_2 and CO_2, harmful UV rays as well as heat. For food safety reasons, only nontoxic reinforcements are allowed as additives for packaging materials. Therefore, biobased and particularly the CNF-based bionanocomposites are getting importance in this regard (Dahman 2009). Agar/CNFs composite films were developed and the effects of CNFs on the mechanical and water vapor barrier properties were tested. Films with up to 1.98 GPa and very low water vapor permeability were obtained (Rhim et al. 2015) which provided scopes of the bionanocomposites for utilization in food-packaging applications. Borkotoky et al. fabricated CNF-based PLA nanocomposite foams having ability as a green replacement for the petroleum-derived synthetic foams and applicability in the field of packaging biomedical and drug delivery (Borkotoky et al. 2018). Owing to its very good moisture barrier properties, CNF-based nanocomposite films extend food shelf-life by serving as carriers for antioxidants and antimicrobials (Dufresne 2013). Innovative nanocomposites were prepared by Fortunati et al. by reinforcement of PLA matrix with a combined filler system of CNF and silver nanoparticles. Besides enhancing the mechanical and barrier properties of the PLA film, the synergism between silver and CNF fillers imparted antibacterial properties to the nanocomposites which could be used as food active packaging material (Fortunati et al. 2012).

Among other applications, the use of CNF-based hybrid composites in the field of wastewater treatment is relatively new. The small-scale commercially available water purification systems involve photocatalysis with artificial UV light. The photocatalysts capture the UV radiation from sunlight and breaks down the various organic and inorganic pollutants present in water. Titanium dioxide (TiO_2) is the most promising in this regard. Kwon et al. studied the photocatalytic activity of nano-TiO_2 and studied the action mechanism of contaminant degradation for environment applicability (Kwon et al. 2008). However, research works on ZnO photocatalysis are relatively low although they have tremendous potential in the field. In a recent study, it has been found that oil palm empty fruit bunch-based CNF with ZnO displays more photocatalytic activity than the pure ZnO nanostructures. This may be due to the interactions of the carboxylic acid groups present on the CNF surface and the bivalent zinc ions which facilitate the growth and dispersion of the ZnO nanoparticles. Lefatshe et al. tested ZnO nanocomposites and showed their efficiency for wastewater treatment by the degradation of methylene blue dye (Lefatshe et al. 2017). Our group studied *Mesua ferrea* L. buckle-based CNF capped ZnO for photocatalytic activity and it has been observed that the dye has been considerably degraded after 100 min of exposure to visible light (Fig. 6.10).

The study of ZnO nanocomposites obtained by Lefatshe et al. also demonstrated improved antibacterial activity against Gram-positive *Staphylococcus aureus* and

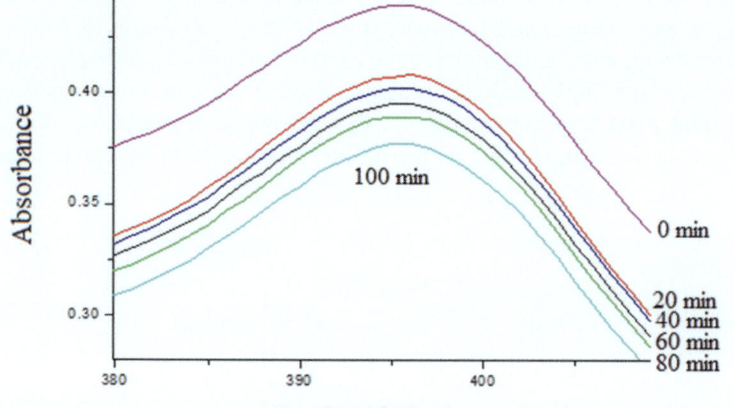

Fig. 6.10 Absorbance spectra for the degradation of MB dye by CNF/ZnO composite

Gram-negative *Escherichia coli*. The very common organic pollutant phenol in wastewater is found to be photodegraded by ZnS/CNF nanocomposites. The resulting nanocomposite was explored as a biomaterial for in vitro faster drug release and other antibacterial applications (Pathania et al. 2015). Synthesis of pharmacologically important nanocomposites with citrus waste-based cellulose can be exploited in wound dressing, antimicrobial and healing applications (Ali et al. 2016).

6.7 CNFs-Based Nanocomposites for Bioenergy Applications

Besides the myriads of application potency of the CNF-based nanocomposites, the bioenergy applications are of worth importance. Bioenergy has turned out to be a great alternative to energy security for sustainable development as it can counteract the vicious problems like depletion and price rise of petroleum resources, greenhouse gas emission, and the overall environment denudation. CNF-based polyurethane composites can be used for encapsulation of dye sensitized solar cells. Stability and efficiency of solar cells are expected to be enhanced by using such nanocomposites (Sulaiman et al. 2014). Ana et al. synthesized carbon nanodots by using green algae *Cladophora rupestis*-based CNF as precursors and showed its potential as photosensitizer in solar cells (Ana and Camacho 2019).

Two major pathways for biodiesel production are the esterification of free fatty acids and the transesterification of triglycerides (Scheme 6.2). Generally acid or base catalyst, enzyme catalysts are used during the process. It has been seen that heterogeneous acid catalysts with porous structure and high surface area lead to a high yield of biodiesel. Nanostructured catalysts like TiO_2, Fe_3O_4, SnO_2, ZnO, CaO, MgO, carbon, graphene, and fullerenes are in use for the production of biofuels (Antunes et al. 2017). Magnetic nanoparticle-based catalysts are extra special in this regard as they have the capability of easily getting separated from reaction medium which leads to more economical, industrial-scale biodiesel production (Gardy et al. 2018). In a work enunciated by Guan et al., 97.8% conversion of biodiesel was reported by using sulfonated multiwalled carbon nanotube from triglyceride transesterification. Such high catalytic efficiency may be attributed to the porous cyclic structure and marked Lewis acidity of the sulfonated nanocatalyst (Guan et al. 2017). Further, the hydrophobic nature of the carbon-based nanocatalyst facilitates

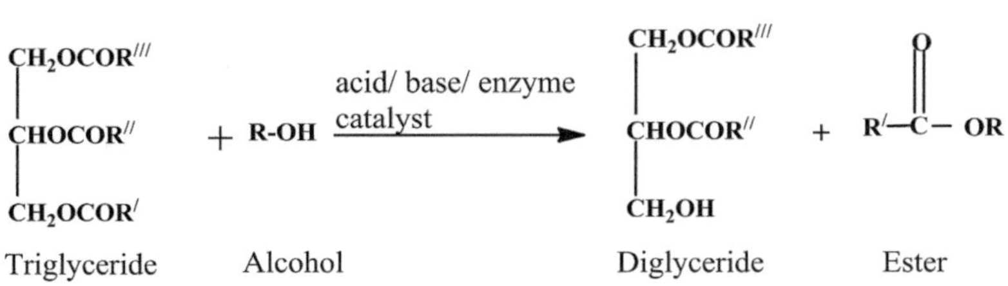

Scheme 6.2 Transesterification of triglycerides

transesterification of triglycerides. Oliveira et al. explained that free fatty acids like levulinic acid get adsorbed on the sulfonated carbon nanotube surface quite easily and converted into levulinate ester in high yields (Oliveira and da Silva 2014).

Inspired from the presence of sulfonic acid (–SO_3H) functional groups on the surface of CNFs, they could be expected as potential candidates as nanocatalysts for biodiesel production. The main advantages of using such nanomaterials are their low toxicity, naturally abundant raw materials, surface with strong acidity, increased solubility, and hydrophobic property of the surface. Stellwagen et al. reported acid functionalized CNFs as an efficient catalyst for transesterification of triolein and methanol (Stellwagen et al. 2013). They grafted aryl sulfonic acid groups additionally on the CNF surface and showed the increased stability and efficiency the cellulose-based catalyst. This study opens up new scopes for designing and developing functionalized CNF catalysts in the field of acid catalyzed biodiesel production. The nanosized cellulose developed by our group from the underutilized agro-wastes of northeastern region of India has now been introduced as a catalyst in the production of biodiesels. The study concludes that the mesoporous structure of CNF facilitates easy accessibility of their active sites to the triglyceride and alcohol molecules (Scheme 6.3). The catalytic performance was found to be increased with increase in –SO_3H groups on the surface of the catalyst.

Recently various carbon nanomaterial immobilized enzymes have been reported for bioethanol production. In a comparative study carried out by Lupoi and Smith,

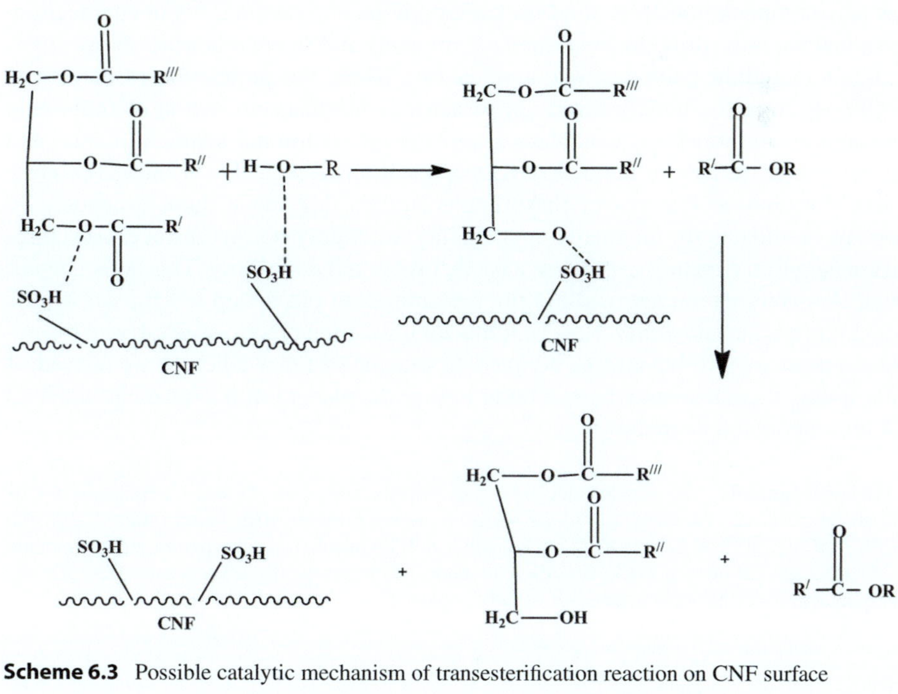

Scheme 6.3 Possible catalytic mechanism of transesterification reaction on CNF surface

it has been concluded that immobilized cellulase enzyme on silica nanoparticles results in a greater yield of bioethanol as compared to pristine enzyme (Lupoi and Smith 2011). These approaches of biofuel production utilizing nanomatarials or nanocomposites definitely offers a safe and economical pathway with enhanced yield from the cheapest renewable lignocellulosic materials and agro-wastes.

6.8 Conclusion

Inimitable mechanical properties, low density, biodegradability, and availability from agro-waste-based renewable resources make the ever-abundant cellulose increasingly important. The electron rich hydroxyl group of celluloses/CNFs play key role in forming both intermolecular and intramolecular H-bonds which make them unique in properties and helps in forming a dense network with wide range polymeric matrices. This chapter highlights the potential of development of cheap, renewable, and environment-friendly CNFs from agro-wastes of Northeast India for nanocomposite applications. The resulting bionanomaterials have twofold advantages. First of all, they can be synthesized by enzymatic hydrolysis approach that possesses very less or no threat to the environment as compared to the traditional acid hydrolysis technique. Secondly, they are biodegradable leading to simpler glucose units. The study further demonstrates that CNFs isolated from agro-wastes of Northeast India are cellulase degradable and has tremendous application potential as nanocomposites such as food packaging, tissue engineering, organ engineering, surgical wounds, drug delivery, medical implants and many other bioenergy applications including photocatalytic applications. Thus, the production of nanosized cellulose from the underutilized agro-wastes of northeastern region of India may emerge as a competitive candidate to replace the traditional synthetic, toxic, and hazardous materials for nanocomposite applications especially for biofuel production. The unique features of these nanomaterials that make them exceptionally potent candidates for bioenergy applicability are highly porous surface, crystallinity, adsorption capability, stability, easy recovery, and reusability. This book chapter will definitely encourage and aid the researchers to put a step toward worldwide environment sustainability. New ventures for utilization of agro-waste-based renewable resources will have to be adopted as sources of renewable energy instead of discarding them here and there in order to save the planet Earth from environmental deterioration and degradation.

Acknowledgments The authors sincerely acknowledge University Grants Commission for its financial assistance for carrying out the research project through MRP Grant (No. F.5-2015-16 (MRP/NERO)/505). R.D. thankfully acknowledges IIT Guwahati, India, for providing Institute Post-Doctoral Fellowship. NEIST (CSIR Laboratory) Jorhat is greatly acknowledged for offering helping hands during various analyses of the samples.

References

Ali A, Ambreen S, Maqbool Q, Naz S, Shams MF, Ahmad M, Phul AR, Zia M (2016) Zinc impregnated cellulose nanocomposites: synthesis, characterization and applications. J Phys Chem Solids 98:174–182

Ana JCS, Camacho DH (2019) Influence of precursor size in the hydrothermal synthesis of cellulose based carbon nanodots and its applications towards solar cell sensitisation. Mater Chem Phys 228:187–193

Anderson SR, Esposito D, Gillette W, Zhu JY, Baxa U, McNeil SE (2014) Enzymatic preparation of nanocrystalline and microcrystalline cellulose. TAPPI J 13:35–42

Antunes FAF, Gaikwad S, Ingle AP, Pandit R, dos Santos JC (2017) Bioenergy and biofuels: nanotechnological solutions for sustainable production. In: Rai M, da Silva SS (eds) Nanotechnology for bioenergy and biofuel production, green chemistry and sustainable technology. Springer International Publishing, Cham, pp 3–18

Balat M (2011) Production of bioethanol from lignocellulosic materials via the biochemical pathway: a review. Energy Convers Manag 52:858–875

Beltramino F, Roncero MB, Vidal T, Valls C (2018) A novel enzymatic approach to nanocrystalline cellulose preparation. Carbohydr Polym 189:39–47

Berg O, Capadona JR, Weder C (2007) Preparation of homogeneous dispersions of tunicate cellulose whiskers in organic solvents. Biomacromolecules 8:1353–1357

Borkotoky SS, Dhar P, Katiyar V (2018) Biodegradable poly (lactic acid)/cellulose nanocrystals (CNCs) composite microcellular foam: effect of nanofillers on foam cellular morphology, thermal and wettability behaviour. Int J Biol Macromol 106:433–446

Chen W, Haipeng Y, Liu Y, Hai Y, Zhang M, Chen P (2011) Isolation and characterization of cellulose nanofibers from four plant cellulose fibers using a chemical-ultrasonic process. Cellulose 18:433–442

Cherian BM, Leao AL, de Souza SF, Thomas S, Pothan LA, Kottaisamy M (2010) Isolation of nanocellulose from pineapple leaf fibres by steam explosion. Carbohydr Polym 81:720–725

Cherian E, Dharmendirakumar M, Baskar G (2015) Immobilization of cellulase onto MnO_2 nanoparticles for bioethanol production by enhanced hydrolysis of agricultural waste. Chin J Catal 36:1223–1229

Dahman Y (2009) Nanostructured biomaterials and biocomposites from bacterial cellulose nanofibers. J Nanosci Nanotechnol 9:5105–5122

Dai H, Ou S, Huang Y, Huang H (2018) Utilisation of pineapple peel for production of nanocellulose and film application. Cellulose 25:1743–1756

Deepa B, Abraham E, Cherian BM, Bismarck A, Blaker JJ, Pothan LA, Leao AL, de Souza SF, Kottaisamy M (2011) Structure, morphology and thermal characteristics of banana nano fibers obtained by steam explosion. Bioresour Technol 102:1988–1997

Deka S, Khawas P (2016) Isolation and characterization of cellulose nanofibers from culinary banana peel using high-intensity ultrasonication combined with chemical treatment. Carbohydr Polym 137:608–616

Devi RR, Dhar P, Kalamdhad A, Katiyar V (2015) Fabrication of cellulose nanocrystals from agricultural compost. Compost Sci Util 23:104–116

Dufresne A (2013) Nanocellulose: a new ageless bionanomaterial. Mater Today 16:220–227

Dutta S (2018) Pond hyacinth based nanocellulose for eco-friendly nanobiomaterial applications. Genius 6:298–307

Dutta S, Karak N (2005) Synthesis, characterization of poly(urethane amide) resins from nahar seed oil for surface coating applications. Prog Org Coat 53:147–152

Dutta S, Karak N (2006) Effect of the NCO/OH ratio on the properties of *Mesua ferrea* L. seed oil modified polyurethane resins. Polym Int 55:49–56

Dutta S, Karak N, Baruah S (2010a) Jute fiber reinforced polyurethane green composites based on *Mesua ferrea* L. seed oil. J Appl Polym Sci 115:843–850

Dutta S, Karak N, Saikia JP, Konwar BK (2010b) Biodegradation of epoxy and MF modified polyurethane films derived from a sustainable resource. J Polym Environ 18:167–176
Ferreira FG, Mariano M, Rabelo SC, Lona LMF (2018) Isolation and surface modification of cellulose nanocrystals from sugarcane bagasse waste: from a micro- to a nano-scale view. Appl Surf Sci 436:1113–1122
Fortunati E, Armentano I, Zhou Q, Ianmoni A, Saino E, Visai L, Berglund L, Kenny JM (2012) Multifunctional binanocomposite films of poly(lactic acid), cellulose nanocrystals and silver nanoparticles. Carbohydr Polym 87:1596–1605
Gardy J, Osatiashtiani CO, Hassanpour A, Lai X, Lee AF, Wilson K, Rehan M (2018) A magnetically seperable SO_4/Fe-Al-TiO_2 solid acid catalyst for biodiesel production from waste cooking oil. Appl Catal B Environ 234:268–278
Garvey M, Klose H, Fischer R, Lambertz C, Commandeur U (2013) Cellulases for biomass degradation: comparing recombinant cellulase expression platforms. Trends Biotechnol 31:581–593
Ghahafarrokhi IS, Khodaiyan F, Mousavi M, Yousefi H (2015) Preparation and characterization of nanocellulose from beeri industrial residues using acid hydrolysis/ultrasound. Fibers Polym 16:529–536
Goffin AL, Raquez JM, Duquesne E, Siqueira G, Habibi Y, Dufresne A (2011) Poly(ε-caprolactone)- based nanocomposites reinforced by surface-grafted cellulose nanowhiskers via extrusion processing: morphology, rheology, and thermo-mechanical properties. Polymer 52:1532–1538
Guan Q, Li Y, Chen Y, Shi Y, Gu J, Li B, Miao R, Chen Q, Ning P (2017) Sulfonated multi-walled carbon nanotubes for biodiesel production through triglecerides transesterification. RSC Adv 7:7250–7258
Guna VK, Ilangovan M, Anantha Prasad MG, Reddy N (2017) Water hyacinth: a unique source for sustainable materials and products. ACS Sustain Chem Eng 5:4478–4490
Henrique MA, Silverio HA, Neto WP, Pasquini D (2013) Effect of incorporating cellulose nanocrystals from corncob on the tensile, thermal and barrier properties of. Poly(vinyl alcohol) nanocomposites. J Environ Manag 121:202–209
Jonoobi M, Oladi R, Davoudpour Y, Oksman K, Dufresne A, Hamzeh Y, Davoodi R (2015) Different preparation methods and properties of nanostructured cellulose from various natural resources and residues: a review. Cellulose 22:935–969
Kan KH, Li J, Wijesekera K, Cranston ED (2013) Polymer-grafted cellulose nanocrystals as pH-responsive reversible flocculants. Biomacromolecules 14:3130–3139
Kargarzadeh H, Huang J, Lin N, Ahmad I, Mariano M, Dufresne A, Thomas S, Gałeski A (2018) Recent developments in nanocellulose-based biodegradable polymers, thermoplastic polymers, and porous nanocomposites. Prog Polym Sci 87:197–227
Kataki R, Goswami R, Jaehong S, Deka S, Kumari D, Kumar M (2016) Characterization of cadmium removal from aqueous solution by biochar produced from *Ipomoea fistulosa* at different pyrolytic temperatures. Ecol Eng 97:444–451
Kumar A, Choudhary V, Bhardwaj NK (2014) Characterization of cellulose nanocrystals produced by acid hydrolysis from sugarcane as agro-waste. J Mater Phys Chem 2:1–8
Kumar M, Kaur M, Sachdeva S, Puri SK (2018) Aquatic weeds as the next generation feedstock for sustainable bioenergy production. Bioresour Technol 251:391–402
Kwon S, Fan M, Cooper A (2008) Photocatalytic applications of micro and nano TiO_2 in environmental engineering. Crit Rev Environ Sci Technol 38:197–226
Lee SY, Mohan DJ, Kang I, Doh GH, Lee S, Han SO (2009) Nanocellulose reinforced PVA composite films: effects of acid treatment and filler loading. Fibers Polym 10:77–82
Lefatshe K, Muivaa CM, Kebaabetswe LP (2017) Extraction of nanocellulose and in-situ casting of ZnO/cellulose nanocomposite with enhanced photocatalytic and antibacterial activity. Carbohydr Polym 164:301–308
Lupoi JS, Smith EA (2011) Evaluation of nanoparticle immobilised cellulase for improved yield in simultaneous saccharification and fermentation reactions. Biotechnol Bioeng 108:2835–2843
Mahardika M, Abral H, Kasim A, Arief S, Asrof M (2018) Production of nanocellulose from pineapple leaf fibers via high-shear homogenization and ultrasonication. Fibers 6:28–39

Mandal A, Chakrabarty D (2011) Isolation of nanocellulose from waste sugarcane bagasse (SCB) and its characterization. Carbohydr Polym 86:1291–1299

Marett J, Aning A, Johan Foster E (2017) The isolation of cellulose nanocrystals from pistachio shells via acid hydrolysis. Ind Crop Prod 109:869–874

Miao C, Hamad WY (2013) Cellulose reinforced polymer composites and nanocomposites: a critical review. Cellulose 20:2221–2262

Murphy CA, Collins MN (2016) Microcrystalline cellulose reinforced polylactic acid biocomposite filaments for 3D printing. Polym Compos 39:1311–1320

Oliveira BL, da Silva TV (2014) Sulfonated carbon nanotubes as catalysts for the conversion of levulinic acid into ethyl levulinate. Catal Today 234:257–263

Pathania D, Kumari M, Gupta VK (2015) Fabrication of ZnS–cellulose nanocomposite for drug delivery, antibacterial and photocatalytic activity. Mater Des 87:1056–1064

Peng Y, Gardner DJ, Han Y, Kiziltas A, Cai Z, Tshabalala MA (2013) Influence of drying method on the material properties of nanocellulose I: thermostability and crystallinity. Cellulose 20:2379–2392

Petersson L, Kvien I, Oksman K (2007) Structure and thermal properties of poly (lactic acid)/cellulose whiskers nanocomposite materials. Compos Sci Technol 67:2535–2544

Prasad V, Bharimalla AK, Deshmukh SP, Vigneshwaran N, Patil PG (2017) Nanocellulose-polymer composites for applications in food packaging: current status, future prospects and challenges. Polym Plast Technol Eng 56:805–823

Ramesh A, Sundari MT (2012) Isolation and characterization of cellulose nanofibers from the aquatic weed water hyacinth – *Eichhornia crassipes*. Carbohydr Polym 86:1701–1705

Rastrogi A, Jha MK, Sarma AK (2016) A comparative study of kinetics for combustion versus pyrolysis of Mesua ferrea husk, soya husk, and Jatropha curcas husk using thermogravimmetry and different methods. Energy Sources Part A Recover Util Environ Eff 38:1355–1363

Rehim MHA, Mohamad AY, Abdul Aziz MG, Ahmad MG, Abdel MH (2019) Green synthesis of cellulose nanofibers using immobilized cellulase. Carbohydr Polym 205:255–260

Rhim JW, Reddy JP, Luo X (2015) Isolation of cellulose nanocrystals from onion skin and their utilization for the preparation of agar-based bio-nanocomposites films. Cellulose 22:407–420

Sain M, Panthapulakkal S (2006) Bioprocess preparation of wheat straw fibers and their characterization. Ind Crop Prod 23:1–8

Saravanan N, Sampath PS, Sukantha TA (2016) Extraction and characterisation of new cellulose fiber from the agro-waste of *lagenaria siceraria* (Bottle Guard). J Adv Chem 12:4382–4388

Shahabi-Ghahafarrokhi I, Khodaiyan F, Mousavi M, Yousefi H (2015) Preparation and characterization of nanocellulose from beer industrial residues using acid hydrolysis/ultrasound. Fibers Polym 16:529–536

Silverio HA, Neto WP, Dantas NO, Pasquini D (2013) Extraction and characterization of cellulose nanocrystals from agro-industrial residue–soy hulls. Ind Crop Prod 44:427–436

Singh J, Kalamdhad AS (2013) Assessment of bioavailability and leachability of heavy metals during rotary drum composting of green waste (Water Hyacinth). Ecol Eng 52:59–69

Singh YK, Kalamdhad AS, Ali M, Kazmi A (2009) Rotary drum composting of vegetable waste and tree leaves. Bioresour Technol 100:6442–6450

Singhania RR, Saini JK, Saini R, Adsul M, Mathur A, Gupta R, Tuli DK (2014) Bioethanol production from wheat straw via enzymatic route employing Penicillium janthinellum cellulases. Bioresour Technol 169:490–495

Siqueira G, Bras J, Dufresne A (2010) A. New process of chemical grafting of cellulose nanoparticles with a long chain isocyanate. Langmuir 26:402–411

Sonia A, Dasan KP (2013) Chemical, morphology and thermal evaluation of cellulose microfibers obtained from Hibiscus sabdariffa. Carbohydr Polym 92:668–674

Srivastava N, Srivastava M, Mishrab PK, Gupta VK, Molinaf G, Rodriguez-Coutog R, Manikantab A, Ramteke PW (2018) Applications of fungal cellulases in biofuel production: advances and limitations. Renew Sust Energ Rev 83:2379–2386

Stellwagen DR, van der Klis F, van Es DS, de Jong KP, Bitter JH (2013) Functionalised carbon nanofibers as solid acid catalysts for trans-esterification. ChemSusChem 6:1668–1672
Sulaiman S, Mokhtar MN, Naim MN, Baharuddin AS, Sulaiman A (2014) A review: potential usage of cellulose nanofibers (CNF) for enzyme immobilization via covalent interactions. Appl Biochem Biotechnol 175:1817–1842
Teixeira RSS, Da Silva AS, Jang JH, Kim HW, Ishikawa K, Endo T, Bon EPS (2015) Combining biomass wet disk milling and endoglucanase/β-glucosidase hydrolysis for the production of cellulose nanocrystals. Carbohydr Polym 128:75–81
Thota SP, Badiya PK, Yerram S, Vadlani PV, Pandey M, Golakoti NR, Belliraj SK, Dandamudi RB, Ramamurthy SS (2017) Macro-micro fungal cultures synergy for innovative cellulase enzymes production and biomass structural analyses. Renew Energy 103:766–773
Thygesen A, Oddershede J, Lilholt H, Thomsen AB, Stahl K (2005) On the determination of crystallinity and cellulose content in plant fibres. Cellulose 12:563–576
Tibolla H, Pelissari FM, Menegalli FC (2014) Cellulose nanofibers produced from banana peel by chemical and enzymatic treatment. LWT-Food Sci Technol 59:1311–1318
Widiarto S, Yuwono SD, Rochliadi A, Arcana IM (2017) Preparation and characterization of cellulose and nanocellulose from agro-industrial waste – Cassava Peel. Mater Sci Eng 176:012052. (IOP Conference Series 2016)
Zabeda H, Sahu JN, Boyce AN, Faruq G (2016) Fuel ethanol production from lignocellulosic biomass: an overview on feed-stocks and technological approaches. Renew Sust Energ Rev 66:751–774
Zhu JY, Sabo R, Luo X (2011) Integrated production of nano-fibrillated cellulose and cellulosic biofuel (ethanol) by enzymatic fractionation of wood fibers. Green Chem 13:1339–1344

Impact of Pretreatment Technologies for Biomass to Biofuel Production

7

Sanjay Sahay

Abstract

Lignocellulose biomass (LCB) is a sustainable resource and raw materials for various products of human uses (fuel, adhesive, bioplastic, biopolymer, bionanomaterials, phenolics, etc.), biofuel being on the top. LCB however is a complex mass of three very different types of chemicals such as lignin, hemicelluloses, and cellulose. A pretreatment technique is applied to weaken or break the linkages between and within these three components. An ideal pretreatment technology to be used must ensure obtainment of three components from LCB in the purest form, at an affordable cost without concurrent production of any toxic by-products. For the last many decades, extensive works have been carried out with the objective to find out nearly an ideal pretreatment method. As a consequence, a number of pretreatment techniques have been reported that can be grouped into five major classes such as physical, physicochemical, chemical, biological, and nanoscale methods. They have their own pros and cons, and thus search for that ideal technique or improvising an existing one is still going on. In the following section, a critical assessment of the reported techniques under the above said categories and their impact on breaking of recalcitrance of LCB and in turn possible impact on 2G bioethanol technology is given.

Keywords

Lignocellulosic biomass · Pretreatment · Hydrolyzable cellulose · Biorefinery compatible pretreatment · Cost-effective pretreatment

S. Sahay (✉)
Government Postgraduate College, Biaora, Rajgarh, Madhya Pradesh, India

N. Srivastava et al. (eds.), *Substrate Analysis for Effective Biofuels Production*,
Clean Energy Production Technologies, https://doi.org/10.1007/978-981-32-9607-7_7

7.1 Introduction

The world is almost unanimous about reducing or replacing fossil fuels as transport fuel because of their negative impact on country's energy security and foreign exchange and above all on global climate. Experiences gained with the use of bioethanol as transport fuel in Brazil and other countries have proved its suitability as petroleum extender in the existing vehicles and as pure fuel in vehicles with some engine modifications. Even in view of current aggressive drive to develop electric (battery)-driven vehicles, it is felt that these batteries may not be applicable for at least heavy vehicles such as lorries, ships, and aeroplanes. Bioethanol as transport fuel will thus not lose relevance in many years to come.

Currently, commercial production of bioethanol relies mainly on food crops (sugarcane molasses, corn, sugar beet, etc.) (1G bioethanol); with this technology, any increase in production would mean grow more crops for energy at the cost of crops to feed people as the area of cultivable land in a country is usually limited. To avoid such food vs fuel conflict, it is essential to go for 2G bioethanol or lignocellulosic bioethanol relying on agro-forest residues, municipal wastes, and dedicated energy plants; together, these constitute the lignocellulosic biomass (LCB).

Plant biomass or LCB is the product of the most important and extensive chemical reaction on the earth, i.e., carbon dioxide fixation (photosynthesis). Therefore, it is also the most abundant and renewable bioresource whose global yield is roughly estimated to be 1×10^{10}MT per annum (Sánchez and Cardona 2008). Moreover, the use of plant biomass itself has two major benefits, viz., (a) it would encourage further plantation facilitating absorption of more carbon dioxide from environment and (b) discourage the use of unsustainable processes such as application of fossil fuels in transport vehicles or burning of agricultural residues in the field. Other benefits associated with expansion of plantation area are:

(a) Maintenance of biodiversity since forests serve as shelter and breeding areas for enormous variety of life forms (about 80% of terrestrial biodiversity resides in forests)
(b) Ecological services such as control of flood, soil erosion, hungry herbivores, erroneous rainfall, etc.
(c) Social service as source of livelihood for indigenous and poor landless people living nearby (as per UN reports, more than 1.6 billion people draw their livelihoods from forest, 10 million are directly employed in forest-related activities, and about 1 percent of the global GDP or gross domestic product emanates from the sale of timber and non-timber products)
(d) Human health services by providing medicines, opportunities for recreational activity, and boost for meditation, mindfulness, and creativity

LCB includes both (a) agro-forestry residues and plant-based municipal wastes and (b) dedicated potential energy crops (switch grass, willow, populus, salix, spruce, etc.). There is growing interest all over the world to apply these bioresources for producing platform chemicals (cellulose, hemicelluloses, and lignin) potentially convertible to an enormous number of products (e.g., pentoses, hexoses, furfural, HMF, bioplastics, adhesives, various phenolics, etc.) of human uses (Isikgor and Becer 2015). Making use of LCB, however, poses one major challenge, i.e., its recalcitrance which is due to highly intimate relationship among its major chemical constituents, i.e., lignin, hemicelluloses, and cellulose. The challenge is further complicated by the varied nature of feedstocks (farmland residues such as wheat and paddy straws, maize cobs and stover, sugarcane leaves and bagasse, etc.; forest residues such as bark, off-cuts, saw dust, shavings, etc.; and dedicated energy crops belonging to angiosperm, i.e., hardwood, and gymnosperm, i.e., softwood) and varied proportion of three major chemicals in different organs of biomass (stem, leaf, cob, or kernel) and in various plant species (herbaceous and angiospermic wood or hardwood have generally less lignin content than woody plants and gymnospermic wood or softwood). Thus, any technique to make use of LCB must consider these facts into account. Because of recalcitrance of LCB, some pretreatment is essential for LCB to be subjected to. Ideally pretreatment is intended for getting the three major components of LCB (i.e., lignin, hemicelluloses, and celluloses) in the purest form so that they can be subsequently processed to obtain various useful products in cost-effective and environment-friendly ways. Moreover, the variation in chemical composition from LCB to LCB requires testing of the applicability of a pretreatment technique in many of the potential LCB types.

The quest for designing such an ideal pretreatment method has resulted into many techniques falling within five main classes, viz., physical, physicochemical, chemical, biological, and nanoscale methods ones having their own advantages and disadvantages. At the end of the days however all these techniques have to answer some common queries. Is it cost-effective? Major items of cost calculation are the cost of chemical(s), requirement of special vessel resistant to the chemical if any, amount of energy used, and remediation of discharges. Is it environment-friendly? Major concerns are whether the technique uses or produces by-products of toxic or non-biodegradable nature. Is it suitable for process development? For example, is it compatible with "single pot-process," i.e., pretreatment, cellulolysis, and fermentation occurring in the same pot?

The present chapter intends to compile the reported pretreatment techniques with emphasis on their impact (science and applicability) on LCB biorefinery technology.

7.2 LCB Composition

LCB is composed of three major constituents, viz., lignin, hemicelluloses, and cellulose, and many minor materials such as pectin, protein, extractives, and ash (Baruah et al. 2018). The proportion of these heterogenous chemicals however varies in different plant species (Table 7.1) (Sun and Cheng 2002) and also with ages and growth conditions (e.g., presence of stresses).

Table 7.1 Composition of some common LCBs. (Modified from Baruah et al. 2018)

Type of LCB	LCB source	Cellulose	Hemicellulose	Lignin
A. Agro-residues				
	Wheat straw	33–40	20–25	15–20
	Sugarcane bagasse	40–45	30–35	20–30
	Rice straw	38	32	12
	Barley straw	38	35	16
	Corn stover	38	23	20
	Corn cob	41	31	12
B. Potential energy plants				
a. Softwood	Poplar	44	20	29
	Silver birch	41	32.4	22.0
	Pinus sylvestris	40	28.5	27.7
b. Hardwood	Bamboo	45	24	20
	Eucalyptus	45	19	31
c. Herbaceous	Switch grass	31	24	18
	Napier grass	47	31	22
C. Municipal waste	Waste papers	65	13	1

7.3 Structure of LCB

The most possible structure of the biomass with respect to the three major constituents may be expressed as having deep-seated cellulose fibrils linked and overlaid by further linked hemicelluloses and lignin polymers (Fig. 7.1), lignin being linked with both cellulose and hemicelluloses. Lignin is also considered as matrix wherein lie hemicelluloses and cellulose embedded (Faulon et al. 1994). The ether and ester bonds link lignin with cellulose and hemicelluloses, while hydrogen bonds stabilize cellulose and hemicelluloses (Faulon et al. 1994). The details of the chemical characteristics are as follows.

7.3.1 Cellulose

Cellulose is a linear chain of (homopolymer) glucose (more accurately cellobiose or glucose pair) linked with β-1,4 glycosidic bonds (linkages); the number of glucose molecules forming cellulose (DP, i.e., degree of polymerization) is variable which may extend up to 17,000. DP of LCB cellulose however is generally of the order of 300–1700. Glucose monomers are connected by twisting around 180° with each other, a feature that facilitates the formation of intermolecular hydrogen bonding responsible for higher tensile strength and solvent insolubility of cellulose (Sjöström 1993). Structurally, cellulose is a straight chain whose surface has protruded hydroxides that facilitate linking of chains via hydrogen bonding to form group of parallel chains of different orders (microfibril and fibril). The structure of higher order is also responsible for crystallinity of the cellulose. Thus, cellulose has both amorphous (less organized) and crystalline (highly organized) regions; the crystalline

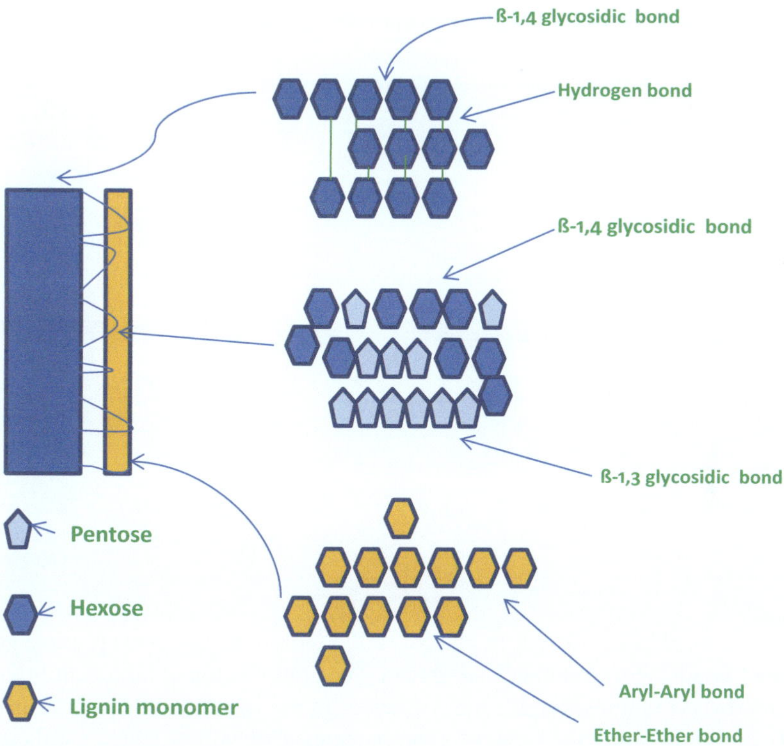

Fig. 7.1 Basic chemistry of three major components of LCB

region is relatively resistant to enzymatic degradation. It is hygroscopic in nature absorbing water up to 8–14% (Harmsen et al. 2010). At low or ambient temperature, it swells in water and dilutes acid (at low temperature) and remains insoluble. As temperature is raised, it starts turning soluble (at higher temperatures) since hydrogen bonds that maintain the crystallinity are broken down under this condition. Higher concentration of acid also makes it soluble but at the cost of heavy degradation of the polymer. In alkaline solutions, it is relatively insoluble, and only low molecular fractions (less than 200 glucose forming chain) are dissolved (Krassig and Schurz 2002).

7.3.2 Hemicelluloses

These are heteropolymers of pentoses, hexoses, and sugar acids. Pentoses mainly comprise D-xylose and L-arabinose, whereas hexoses mainly consist of mannose, glucose, and galactose. When obtained from different plant sources, they exhibit extremely variable structures and composition (Table 7.2). They may be classified as xylan (with backbone exclusively has xylose), mannan (with backbone may consist exclusively of mannan or of mannose and glucose in a nonrepeating pattern),

Table 7.2 Composition (%) of hemicelluloses of some representative plants

Source plant/part	Xylose	Arabinose	Galactose	Mannose	Glucose	Sugar acids	Refs.
Tobacco stalks Cluster bean seed Husk and esparto grass	100	–	–	–	–	–	Eda et al. (1976)
Birch wood (Roth)	89.3	1	–	–	1.4	8.3	Kormelink and Voragen (1993)
Wheat							
Arabinoxylan	65.8	33.5	0.1	0.1	0.3		Gruppen et al. (1992)
Corn fiber xylan	48–54	33–35	5–11	–	–	3–6	Doner and Hicks (1997)
Rice bran							
Neutral xylem	46	44.9	6.1	–	1.9	1.1	Shibuya and Iwasaki (1985)

xyloglucan (backbone consists of glucose with xylose linked to most glucose groups by a 6→1 bond), glucomannan, and glucan. The composition of hemicelluloses varies in hardwood (angiospermic) and softwood (gymnospermic); the former is made mainly of xylans while the latter of glucomannans (McMillan 1993). Furthermore, softwood heteroxylans exhibit esterification at their arabinofuranosyl with ferulic acids and p-coumaric acids (Mueller-Hartlev et al. 1986), but hardwood xylans exhibit acetylation at majority of their xylose residues (60–70%) (Timell 1967). The DP (degree of polymerization) is also much lower (~200) for hardwood- and (~100) for softwood-derived hemicelluloses (Ebringerová et al. 2005).

Unlike cellulose, hemicellulose chains cannot bind side-on as they have branched structure and protruded acetyl groups stemming from xylose residues, and thus a crystalline structure is not achieved (Kirk-Othmer 2001). At low temperature (>100 °C), they maintain insolubility in water, but they start dissolving at elevated temperature (100–200 °C). Even then, hemicelluloses dissolve at a temperature (100–200 °C) which is lower than that (>200 °C) at which cellulose dissolves (Thermowood handbook 2003). Furthermore, lowering the pH of hemicellulose suspension also catalyzes the dissolution of hemicellulose. The higher solubility of this biomaterial makes it a good candidate for such applications as in hydrogels, cosmetics, and drug carriers (Farhat et al. 2017).

7.3.3 Lignin

Lignin is a heteropolymer of phenolic monomers (sinapyl alcohol, coniferyl alcohol, and P-coumaryl alcohol) present in the cell wall of the plants. The three types of linkages such as aryl-aryl, alkyl-aryl, and alkyl-alky link the phenolic monomers. As binder, it is mainly responsible for binding the varied types of cell wall's chemicals forming a tough composite material. Lignin content is present in the lowest amount in the herbaceous plants (e.g., grasses) and in the highest amount in softwoods (Table 7.1). The composition of hardwood and softwood lignin is also different, while that of hardwood mostly consists primarily of coniferyl alcohol and p-coumaryl alcohol units; that of softwood is primarily made up of coniferyl and sinapyl alcohol residues in various proportions (Kirk-Othmer 2001). Lignin is relatively soluble in many organic solvents such as acetone, lower alcohols, dioxane, etc. At higher temperature, its depolymerization is accelerated in presence of both alkalies and acids (O'Connor et al. 2007).

7.3.4 Minor Chemical Components

Minor chemical components include extractives and metals. Extractives consist of diverse types of compounds which may be hydrophilic (e.g., polyphenol, etc.) or lipophilic ones (e.g., fatty acids, glyceride, waxes, etc.) or even neutral organic solvents (e.g., alcohols, ethyl acetate, etc.). In additions, various metal ions are also present in the biomass.

7.4 Pretreatment

7.4.1 Objectives

Since the plant biomass is composed of three utterly different chemical entities (lignin, hemicelluloses, and cellulose), an ideal pretreatment technology must ensure release of these in the purest forms so as to employ them in subsequent applications profitably. Although concept of biorefinery to utilize each and every component of biomass is gaining momentum, cellulose is currently the most important molecule, and thus all the pretreatment methods mainly target cellulose. One of the important reasons for this is the availability of mature techniques to hydrolyze (enzymatic) cellulose to glucose and ability of yeast (*Saccharomyces cerevisiae*, the invincible ethanalogen) to ferment glucose into bioethanol. Thus, the goals of pretreatment currently are:

(a) Loosening and increasing the surface area
(b) Structural modification and release of lignin
(c) Partial hydrolysis and release of hemicelluloses
(d) Decreasing the crystallinity of cellulose
(e) Least possible degradation of sugars/lignin-producing inhibitory compounds
(f) Highest possible recycling of pretreatment solvent(s)
(g) Cost-effectiveness in terms of investment in equipment, chemical, and energy
(h) Flexibility to accommodate varied feedstocks

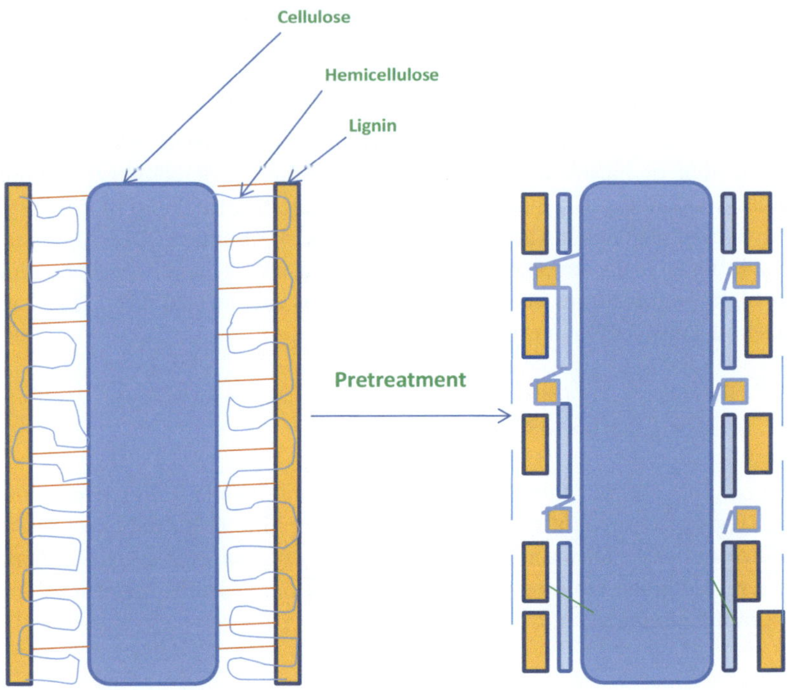

Fig. 7.2 Desired changes in LCB as a result of pretreatment

(i) Environmental friendliness (i.e., starting chemicals or by-products not to be toxic/non-biodegradable)

Ideally, a process accommodating maximum number of these objectives to get maximum recovery of pure cellulose in non-crystalline form is the end (Fig. 7.2).

7.4.2 Central Theme

Plant cell wall has acquired during evolution a structure as resistant as possible (recalcitrant) to parasitic (enzymes from pathogens) attack and other environmental adversaries. Matching level of pretreatment severity thus is required to deal with the level of LCB recalcitrance. Overend and Chornet (1987) experimenting on steam explosion method of pretreatment derived formula of severity index (R_0) as given below:

$$R_0 = \int_0^t \exp\left\{\frac{(T-100)}{14.75}\right\} \mathrm{d}t \tag{7.1}$$

In the equation, severity index (R_0) is expressed in terms of reaction time (t) and temperature (T). The formula is also relevant to understand and design other pretreatment processes. For example, in all the reported methods of pretreatment (except solvent methods carried out at room temperature), severity is in the center, though factors may include besides reaction temperature and residence (reaction or retention) time, pressure, and various other chemicals (acid, alkali, ammonia, CO_2, organic acids, etc.). The dimension of severity is then enhanced to incorporate multiple factors, and the phenomenon is renamed as "comprehensive severity." Even in case of acid and alkali, their strengths and concentrations have direct impact on the pretreatment efficiency. Higher severity, however, generally goes hand in hand with production of higher concentration of inhibitors.

7.4.3 Pretreated LCB and Enzyme Digestibility

Although a number of goals are assigned to pretreatment, the most important one is to achieve higher enzyme digestibility of cellulose with theoretically 100% cellulosic sugar yield. To this end, an experiment was carried out to demonstrate relative importance of three most important structural features of pretreated LCB, viz., lignin content, crystallinity, and acetyl (Chang et al. 1998). It was found that it was lignin that was responsible for extensive nonproductive adsorption of enzymes and thus a significant proportion of applied enzymes were deprived from accessing to their substrate (cellulose). Lignin content has thus been concluded to be the most important factor that affects digestibility. Other factors may delay the process, but not the final yield. In one report, attaining a lignin content <10% in the pretreated LCB has been found to be of utmost importance to achieve effective enzyme digestibility (Kim and Holtzapple 2006).

7.4.4 Pretreatment Methods

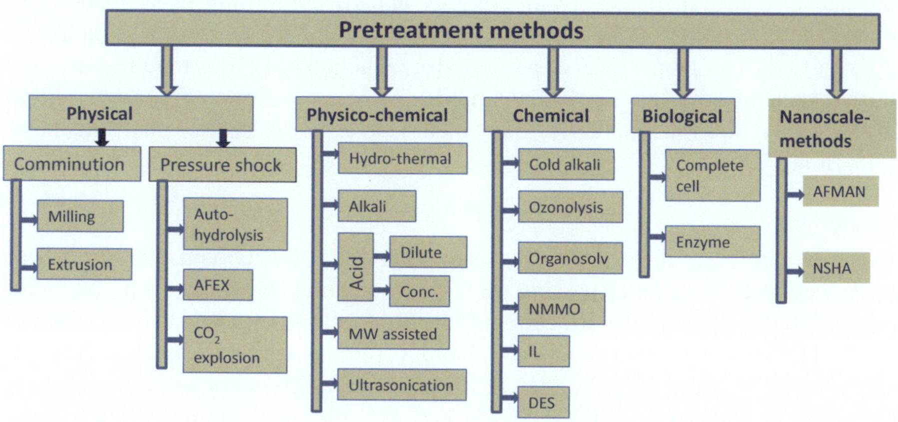

7.4.4.1 Physical (Mechanical) Pretreatment

Physical pretreatment aims at particle size reduction and feedstock-surface area increment. It may reduce crystallinity of cellulose in extreme case. Owing to its highly energy dependence, it becomes costly; the harder the LCB, the more is the energy or the cost involved. For example, energy requirement for size reduction of soft corn stover (11 kWh per metric ton) and hard poplar wood (118.5 kWh per metric ton) is substantially different (Rajendran et al. 2017). While some physical treatment (e.g., chipping) is common to all pretreatment methods, hardly any physical pretreatment alone is feasible as it would call for higher amount and types of costly enzymes (cellulases, hemicellulases, and pectinases) to be used subsequently.

7.4.4.1.1 Comminution

The common methods to achieve comminution are milling and extrusion.

Milling

Milling is used to reduce the size of feedstock as small as to 0.2 mm. Reduction in particle size is believed to accompany with destabilization of the chemical configuration of LCB, increased porosity and bulk density, reduction of cellulose crystallinity, and expansion of LCB surface area; the overall effect is enhanced exposure of cellulose to enzyme (Barakat et al. 2014). A reduction of particle size in LCB to 1–2 mm has been recommended to achieve maximum cellulose digestibility (Schell and Harwood 1994). A number of milling equipment are available, viz., ball, rod, two-roll, hammer, vibratory, colloid, and wet disk-type milling ones. Among planetary, ball, and attrition methods, planetary milling was found the best yielding highest amount of glucose and galactose (Kim et al. 2013).

Milling is deemed to be a suitable pretreatment method as it does not yield any toxic or inhibitory compounds. Due consideration is to be given to the extent of size reduction of LCB, since it has inverse relationship with machine operation cost, i.e., the smaller the particle size achieved by this method, the larger the machine operating and thus overall process cost. Another demerit of milling is its inability to remove lignin, while lignin removal is very important for enzyme and microbial activity. Only a moderate level of particle size reduction of LCB is required, and any further reduction has no significant impact on sugar recovery. Moreover, milling itself may not be sufficient pretreatment method to ensure release of platform chemicals from LCB.

Extrusion

Extrusion combines the effect of high temperature (>300 °C) and pressure (sheer forces) exerted on material passing through the rotating screw blades in the barrel disrupting the recalcitrant structure of the lignocelluloses (Kumar and Sharma 2017). The screw while moving pushes biomass through it, and in doing so, it builds and releases pressure of great amount. In one study with sweet sorghum bagasse, the optimum extrusion parameters were found as barrel temperature (100 °C), screw speed (200 rpm), and moisture content of feedstock (70%) that could release 70%

of the sugars from feedstock and give a final ethanol yield of about 200 ml kg^{-1} of bagasse (Heredia-Olea et al. 2015). The main constraint of the technique however is its energy intensiveness, an energy demand of 226–324 kwh per ton feedstock has been reported (Heredia-Olea et al. 2015).

7.4.4.1.2 Pressure Shock Mediated

LCB in a closed reactor is first subjected to high pressure steam, and then pressure is released suddenly. The technique may be used without (steam explosion) or with a catalyst (ammonia and CO_2 explosion).

Steam Explosion (Autohydrolysis)

In this method, LCB is first put under pressurized steam, and then the pressure is released abruptly causing the water molecules from LCB to escape in an explosive way. This results in highly decompressed and splitted LCB. Indeed, the method creates simultaneously three extreme physical conditions, viz., higher temperature (160–260 °C), pressure (0.69–4.83 MPa), and drastic pressure change. In addition, acids (organic acids) formed due to degradation of carbohydrate under these conditions do provide catalytic force. Overall, effects of steam explosion thus are decompression of LCB and nonspecific breakage of linkages within and between cellulose, hemicelluloses, and lignin. Steam explosion is also called autohydrolysis provided there is no extraneous catalyst (acid or alkali) added to this. The factors affecting the process are particle size of LCB, moisture content, reaction time, and temperature (Rabemanolontsoa and Saka 2016). Only a moderate level of particle size reduction of LCB is sufficient to obtain optimum level of sugar, and any further reduction would only enhance overall process cost. Low moisture condition of LCB is necessary as it permits vapor to enter inside and affect LCB thoroughly. High moisture condition keeps void spaces filled with moisture that prevents entry of steam at higher temperature. There is a direct relationship between temperature and residence time. Process at higher temperature requires lesser residence time (e.g., 270 °C and 1 min, respectively) or vice versa (190 °C and 10 min, respectively) (Duff and Murrayh 1996). Formation of acetic acid, formic acid, levulinic acid, etc. at higher temperature further catalyzes the process.

The process can be optimized by adjusting two variables (residence or reaction time and temperature) of severity factor as given in Eq. (7.1) (Overend and Chornet 1987). The log R_o is the severity factor whose given values fall in the range of 3.14–3.56 for steam explosion. In this equation, "T" represents the temperature in Celsius (°C), 100 °C is considered as reference temperature as there is no solubilization occurring at this temperature, "t" represents the reaction time in minute (min), and 14.75 represents the activation energy in the existing conditions. The equation (process) proceeds as per the first-order kinetics and the Arrhenius law (Overend and Chornet 1987).

This pretreatment technique is rated as environment-friendly because it involves limited chemical use and is efficient as it involves less energy use, no recycling costs, and higher sugar yield (Pielhop et al. 2016). The process has however been found more applicable to low lignin containing LCB such as hardwoods and agro-wastes. Softwood containing relatively more lignin is less responsive to this pretreatment. Another limitation of the process is incomplete digestion of lignin-hemicellulose linkages resulting into release of semi-digested complexes; the latter tend to condense and precipitate and thus become unavailable reducing overall sugar recovery to that extent. Further disadvantage of the process is the production of undesirable compounds such as furfural, hydroxymethyl furfural (due to degradation of sugars), and levulinic acid, vanillic acid, caproic acid, caprylic acid, pelargonic acid, and palmitic acid (due to degradation of lignin) which are inhibitory to enzymes and fermenting microbes.

Ammonia Fiber Explosion (AFEX)

In this process, the LCB in presence of liquid ammonia is subjected to high pressure in a reactor; the temperature and residence time are maintained at 60–100 °C and 5–30 min, respectively; and then the pressure is abruptly reduced (Shirkavand et al. 2016). Under these conditions, LCB swells up, its fibrous structure is disrupted, and crystallinity of cellulose is reduced. The AFEX pretreatment is affected by four factors, namely, temperature, blow down pressure, moisture content of LCB, and amount of ammonia used (El-Naggar et al. 2014). For instance, in case of corn stover, combination of ammonia to LCB loading of 5:1, moisture content of 70%, and temperature of 170 °C has been found optimal (Zhao et al. 2014). The good part of this process is almost no formation of inhibitors, while the bad part is higher operating cost mainly because ammonia itself is costly, its recycling process is costly, and maintenance of reactor handling corrosive ammonia is also cost-intensive. Moreover, the process has not been found effective in case of biomass containing higher lignin, for example, AFEX-treated newspaper and aspen chips with 25% lignin were found to yield only 40–50% sugars following saccharification (McMillan et al. 1994).

CO_2 Explosion

In this process, CO_2 is used instead of extraneous water or ammonia. LCB with CO_2 is kept in a closed reactor at high pressure, and suddenly pressure is released. CO_2 in gaseous or liquid form (dissolved CO_2 in cellular water forming carbolic acid) when explosively released from LCB causes the same disruptive effect on it as steam explosion. The process is affected by pretreatment pressure, reaction time, and temperature. The positive effect of supercritical CO_2 on subsequent enzymatic saccharification of LCB as in case of sugarcane bagasse (Benazzi et al. 2013), corn stover (Narayanaswamy et al. 2011), etc. has been reported. The advantages of the process are negligible production of inhibitors (vis-à-vis steam explosion), lower cost of CO_2, no corrosive effect on vessel, and no adverse effect on environment (vis-à-vis AFEX). The disadvantages are high maintenance cost to create higher pressure say 24 MPa (Narayanaswamy et al. 2011) and costly reaction set up (reactor) to handle pressurized CO_2.

Although in the center of all these processes, the driving forces are the sudden pressure release and a catalyst such as ammonia or CO_2 or organic acid (acetic acid), addition of additional catalyst, for example, water-ethanol as co-solvents in case of CO_2 explosion pretreatment of corn stover (Serna et al. 2016) or extruder in AFEX (called FIBEX or fiber extrusion), has been found to enhance efficiency of treatment significantly.

7.4.4.2 Physicochemical Pretreatment

Generally, physical methods in combination with chemical methods are found more effective to achieve the purposes of pretreatment, for example, chopping or milling of biomass before treating with any chemical (Sahay and Rana 2017), extrusion in combination with ammonia fiber explosion (AFEX) called FIBEX, and acid or alkali treatment under higher temperatures and pressure (Redding et al. 2011) or assisted with microwave (Gabhane et al. 2011).

7.4.4.2.1 Hydrothermal Method

This pretreatment method is known by other names also, viz., liquid hot water (LHW), hydrothermolysis, hydrothermal treatment, solvolysis, and so on (Mosier et al. 2005). This involves application of hot water (170–230 °C) to LCB. During the process lignin and hemicellulose are released, and the latter is removed because of the release of its acetyl groups (Zhuang et al. 2016). The process depends on water temperature and residence time, though controlled pH (4.0–7.0) has also been reported to limit inhibitor formation (Li et al. 2014). Thus, the optimum hydrothermal pretreatment conditions for sugarcane bagasse (180 °C and 30 min) (Hongdan et al. 2013) and corn cob (160 °C and 10 min) (Imman et al. 2018) have been reported.

During this process, water acts as both solvent and reagent. Water enters the biomass, acts on hemicelluloses and lignin, and removes them partially (He et al. 2016). Both exogenous and biomass-based water and organic acids generated via degeneration of hemicellulosic sugars (acetic acid) and lignin (e.g., phenolic acid) catalyze further dissolution of lignin and hemicelluloses (Galia et al. 2015). Hydrothermal pretreatment is highly studied method for its efficiency and comparatively lower cost (Galia et al. 2015). The process does not require costly size reduction of the LCB since cooking itself effectively creates interparticle separation. There are three possible ways to apply hot water to LCB, viz., concurrent, counter-current, and flow-through. Concurrent pretreatment process moves both LCB and hot water in the same direction; counter-current moves LCB and water in the opposite direction. The flow-through reactor on the other hand works by having stationary LCB bed and hot water flowing over it. The residence time is regulated by controlling the flow of water. A residence time of about 15 min has been found suitable at water temperature range of 200–230 °C, as under these conditions all hemicelluloses, 4–22% cellulose, and 35–60% of the lignin have been reported to be removed from LCB (Harmsen et al. 2010).

This pretreatment method has been found free from the problems of corrosion and need for neutralization steps (cf acid pretreatment). The method is simple and straightforward and involves a comparatively lesser number of process steps in the protocol, a feature that is advantageous for process integration and process development. Thus this pretreatment method has enough potential for commercialization. The process has however certain serious limitations such as higher consumption of water and energy, lower hemicelluloses and lignin dissolution, and high amount of inhibitor generation (Yang et al. 2018).

7.4.4.2.2 Alkali Methods

Technically alkalies cause saponification reaction in LCB and breakage of lignin-carbohydrate linkages (Tarkow and Feist 1969). Apart from the removal of lignin, the method helps to solubilize lignin and swells cellulose, thereby reducing its crystallinity and degree of polymerization. Alkali specifically acts on α- and β-aryl ethers present in lignin and glycosidic linkages present in carbohydrates (Lai 1991). A 10% or less of residual lignin in LCB is the target for successful pretreatment process (Kim and Holtzapple 2006). Alkali does remove acetyl groups and uronic acid substitutions in hemicelluloses in course of pretreatment (Maurya et al. 2015). All these have positive effect on the accessibility of the enzymes to the cellulose. These form the bases for alkali pretreatment, and the alkali may be NaOH, KOH, $Ca(OH)_2$ and NH_4OH, and NaOH (sodium hydroxide) having been reported to be better in action and availability.

Among alkalies, slake lime or $Ca(OH)_2$ is the cheapest and the most environment-friendly one. Its application is also very simple, involving slurrying of lime with water, sprinkling it on the pile of LCB, and incubating for some hours to weeks (Kumar et al. 2009). Slake lime pretreatment of LCB has been reported to remove lignin, thereby reducing nonproductive adsorption sites for enzymes, and thus results in a significant yield of sugars due to improved enzyme access and activity on pretreated LCB. The recovery of calcium from reaction mix is possible by converting it into $CaCO_3$ with cheaply available CO_2 (Kumar et al. 2009). Applying lime kiln technology, calcium hydroxide can subsequently be regenerated. Slake lime pretreatment has been found to dissolve amorphous components of LCB (lignin and hemicellulose), and thus proportion of crystalline substances (crystalline cellulose) increases in LCB (Kumar et al. 2009). The method becomes more effective if temperature is raised. Earlier, the method has been applied to pretreat wheat straw at 85 °C for 3 h (Chang et al. 1998), switch grass at 100°C for 13 h (Chang et al. 1997), and poplar wood at 150 °C for 6 h (Chang et al. 2001). In case of corn stover, slake lime pretreatment conditions have been optimized as having lime loading (0.075 g /g of dry LCB), a water loading (5 g/g of LCB), reaction temperature 120 °C, and residence time of 4 h (Kumar et al. 2009). Under these conditions, about 100% polysaccharide conversion has been reported, which is nine times more than the untreated corn stover (Kumar et al. 2009).

Another alkali, NaOH, may be used as dilute (0.5–4% w/v) or concentrated (6–20% w/v) solution (Mirahmadi et al. 2010); dilute NaOH process however is the more studied one. Generally, low concentration process requires higher temperature and vice versa. Dilute NaOH method is more effective than slake lime in causing cellulose swelling and breakage of α- and β-aryl ether bonds present in lignin and glycosidic linkages present in carbohydrates. Dilute NaOH has however found to be unsuitable for softwoods having lignin content >26% (Millet et al. 1976). Sometimes as in case of water hyacinth containing low lignin content, its pretreatment with dilute NaOH shows very small effect (Patil et al. 2011). Some modifications of dilute NaOH pretreatment methods have also been reported. For example, an experiment on the impact of presence or absence of oxygen at various temperatures at 25, 35, 45, and 55 °C on alkali pretreatment of corn stover revealed that delignification is highly dependent on temperature and presence of oxygen (Kim and Holtzapple 2006). Likewise, in case of switch grass, the alkali pretreatment has been found to yield higher amount of sugars if subjected to microwave-assisted NaOH (0.1 g of alkali/g LCB) (Kumar et al. 2009). It has been found that NaOH loading within specified range (0.04–0.1 g per g LCB) in case of corn stover has bearing on the lignification rate. With the highest concentration of NaOH (say 0.1 g per g LCB), the highest delignification (79.9%) could be achieved (Chen et al. 2013). This increase in delignification has however been found to accompany with degradation of hemicelluloses and thus results in 20% more degradation of xylose. The result suggests that a balance is to be stricken between the extent of delignification and the degradation of carbohydrate while standardizing the alkaline pretreatment process (Chen et al. 2013). A comparative study on the affectivity of three alkalies, viz., NaOH, $Ca(OH)_2$, and liquid NH_3, on the pretreatment of populus wood indicated dilute NaOH to be the most effective one followed by NH_3 and $Ca(OH)_2$.

Liquid ammonia (NH_3) is another useful alkali for the pretreatment of LCB. Pretreatment of maize (stover and cobs mixture) and switch grass applying ammonia percolation process at 170 °C, NH_3 concentration of 2.5–20%, and residence time of 1 h has been studied; the result showed a 60–85% delignification for these LCBs (Iyer et al. 1996). In case of liquid NH_3 pretreatment, longer residence times rather than higher temperature were found to be more suitable (Bali et al. 2014). A strong alkali (say NaOH) in general requires low temperature and shorter residence time vis-à-vis weak alkali (say lime). Other disadvantages of strong alkali or higher concentration of alkali process are the possibility of inhibitory phenolic compounds formation and process cost escalation because of requirement for pH control in the next stage (saccharification). On the other hand, weak alkali is non-corrosive, easy to handle, and inexpensive (say lime).

The advantages of alkaline method are:

(a) Practicable at ambient conditions.
(b) Results in less sugar degradation.
(c) Salts mostly are regenerable or recoverable (Kumar et al. 2009).
(d) Cellulose mostly (95%) is preserved as found in case of corn stover (Lai 1991; Gupta and Lee 2010), resulting in higher sugar yield.

The disadvantages are:

(a) Suitable for low lignin containing LCB, e.g., agricultural residues, not for woody LCB (Sahoo et al. 2018).
(b) Application of NaOH leaves behind sodium in the effluent which is toxic and injurious to the fertility of soil. It is also difficult to be recovered (Zheng et al. 2009).
(c) KOH, the alternative alkali, is also toxic to the microbes and costly.

(d) Lime that tends to precipitate in the pretreatment mix, sodium (from NaOH) that tends to form toxic salts inhibitory to enzymes and fermenting microbes, and KOH that forms black liquor whose remediation/removal is costly are some of other constraints (Li et al. 2015).

7.4.4.2.3 Acid Methods

It has been one of the earliest methods pursued to get lignocellulosic ethanol. A two-stage dilute acid protocol was developed by NREL, USA, in 1985 (Hsu 1996) primarily because 80–90% hemicellulosic sugars can be obtained. Later on, focus was shifted to single-stage dilute acid pretreatment to get enzymatically hydrolyzable LCB as a step to reduce the use of corrosive acid. Potentially acid may disrupt the van der Waals forces, covalent and hydrogen bonds present in the LCB (Li et al. 2010). Thus the linkages between hemicellulose and cellulose and also those present within cellulose and hemicellulose are attacked (Lloyd and Wyman 2005). The hemicelluloses and amorphous regions of cellulose are dissolved quicker to soluble sugars (Foston and Ragauskas 2010) or under milder conditions to oligomers (Tao et al. 2011) because of breakage of xylosidic bonds and disruption of acetyl ester groups (Tang et al. 2011). Owing to preferential degradation of amorphous regions, the proportion of crystalline region in cellulose increases (crystallinity index enhancement) (Samuel et al. 2010). Under the influence of acids, lignin present in LCB is subjected to both disruption and condensation reactions giving rise to modified lignin molecules that tend to precipitate (Candido et al. 2012).

Mineral (sulfuric, hydrochloric, nitric, and phosphoric) as well as organic (formic, oxalic, and maleic) acids have been tested, and their effect has generally been found to correspond to their chemical strength. Therefore, acid with more negative pKa value exhibits more effect on the pretreatment process (Bobleter 1994). Because of the ease of availability, low cost, and affectivity, H_2SO_4 and phosphoric (H_3PO_4) acids are the preferred ones. Of the two, H_3PO_4 is milder and causes less damage to the environment. Hydrochloric (HCl) (Demirbas 2008) and nitric acids (HNO_3) are equally effective (Tutt et al. 2012a), but they are costlier as compared to sulfuric acid.

Broadly, two approaches for the application of acid pretreatment have been reported, viz., concentrated acid (30–70%) and dilute acid (0.1–10%), the former even at lower temperature (<100 °C) release sugars more efficiently, while in the diluted form (0.1–10%), it requires higher temperatures (100–250 °C) and more residence time. In either condition however, higher side of severity produces sugar and lignin degradation products inhibitory to enzymes and fermenting microbes. Other constraints in respect of concentrated acids are their toxic and corrosive nature that enhances maintenance cost. Organic acids are non-corrosive and free from the blemish of yielding degradation products (Qing et al. 2015), but they are costlier. The process is affected by size of feedstock, solid loading, reaction time, temperature, and acid concentration.

Acid pretreatment process generally involves drying of LCB to ~10% moisture content, milling to achieve smaller particle size (~1 mm × 0.5 cm × 0.5 cm for wood materials of 1 mm for straws), and impregnation in acid for an optimized period (for few hours) under normal temperature or subsequently subjected to higher temperature condition for an optimized period. The liquor containing soluble sugars is then filtered to separate solid matters (cellulose); the latter is then subjected to washing (to extract sugars and remove acids) and/or neutralization with lime prior to enzymatic saccharification. To demonstrate the impact of acid pretreatment on LCB, a modified equation (combined severity factor) based on the severity factor equation of Overend and Chornet incorporating the acid component was proposed as given below (Chum et al. 1990):

$$\text{Combined severity factor}\left(\text{CSF}\right) = \text{LogR}_{\text{o}} - \text{pH}, \text{where R}_{\text{o}} = t^{\exp\left[\frac{TR-TH}{14.75}\right]} \quad (7.2)$$

where pH represents the pH of reaction mix, "t" represents the residence time, "TR" represents reaction temperature, and "TH" represents the reference temperature (100 °C). The enzyme digestibility of pretreated LCB and release of glucose are said to increase with increase in the CSF. For instance, in case of acid pretreatment of corn straw biomass with CSF increase from 0.5 to 2.2, an increase of glucose yields from 32% to 57% (g glucose per g LCB) has been reported after enzymatic digestion of pretreated LCB (Lloyd and Wyman 2005).

Dilute Acid Method

Pretreatment with acid concentration in the range of 0.2–2.5% (w/w) is called dilute acid method which is generally carried out at a temperature range of 120–210 °C and usually under higher pressures to stop steam escape. At lower temperature, the pretreatment leads to less sugar yields (Chen et al. 2009). Earlier, a two-stage variant of dilute acid pretreatment was developed: first stage targeted hemicelluloses and the second stage aimed at solubilizing cellulose of course with the use of a higher acid concentration (Kazi et al. 2010). Currently, dilute acid pretreatment is a single-stage process in moderately harsh environment aiming at getting higher hemicellulosic sugars as well as enhancing enzyme digestibility of cellulose present in LCB to achieve favorable economics (Jain et al. 2013). At least two alternative methods have been given with respect to dilute acid pretreatment process:

1. Continuous flow at elevated temperature (T > 160 °C, 5–10 wt% substrate concentration) (Kaar and Holtzapple 2000)
2. Batch process at moderate temperature especially suitable for high solid loading process (T <= 160 °C, 10–40% substrate concentrate) (Kumar et al. 2009)

Among mineral acids, H_2SO_4 has been the most studied one, though reports on the applications of nitric acid (HNO_3) and hydrochloric acid (HCl) are also available. It has been found that dilute HNO_3 is better than H_2SO_4 as far as the glucose yield (after saccharification of pretreated LCB) is concerned as found in case of rye straw (Tutt et al. 2012b). The problem however is with the difficulty in the removal of by-products from pretreated LCB by washing (Tutt et al. 2012a). Some investigations have been made to test the utility of dilute phosphoric acid (H_3PO_4) in LCB pretreatment. Using potato peels as substrate, a sugar yield reaching 82.5% of the theoretical value was obtained with H_3PO_4 (Lenihan et al. 2010). Similarly, H_3PO_4 pretreatments of bamboo and corn cob at 170 °C for 45 min (Hong et al. 2012) and 140 °C for 10 min, respectively (Satimanont et al. 2012), have been reported to yield higher amount of sugar. Studies have also been made to test the effect of combinations of acids. For example, pretreatment with combination of HCl and H_2SO_4 of sweet sorghum bagasse has been studied, and the result revealed no significant improvement in sugar yield as compared to the effect of acids used individually (Heredia-Olea et al. 2012). On the other hand, combined H_2SO_4 and H_3PO_4 pretreatment of oil palm empty fruit bunch showed higher xylose yield as compared to single acid applications (Zhang et al. 2012).

Organic acids are other potential catalysts that have been studied for their effect on the pretreatment of LCB. Oxalic and maleic acids, for example, have been reported to be more effective than H_2SO_4 in solubilizing hemicelluloses (Lee and Jeffries 2011). Oxalic acid at a concentration of 200 mM and under conditions of 160 °C and residence time 10 min has been reported to be effective in pretreating corn stover with the optimum sugar yield (Mtui 2012). Organic acids are said to cause comparatively less degradation of sugars/lignin and thus less production of inhibitors (cf H_2SO_4) (Modenbach and Nokes 2012).

The advantages of the method are:

(a) Higher content of cellulose in pretreated substrates.
(b) Low requirement of enzymes.
(c) The lower volume of acid consumption (cf concentrated acid process), a feature that ensures lower cost and less process severity.
(d) Sulfur and phosphorus released from dilute acids (<1% w/v sulfuric and phosphoric acids, respectively) serve as nutrients for fermenting microbes.
(e) Furfural, a commercially important product, is produced from biomass applying dilute H_2SO_4. If it is targeted as co-product, it must help in economizing biomass biorefinery.
(f) Commercial process for LCB pretreatment applying dilute acid is available (O'Donovan et al. 2013).

The disadvantages are:

(a) Overall cost is higher (cost of acid and energy) as compared to many other processes especially physicochemical ones such as steam explosion.
(b) Need for expensive corrosion-resistant reactors.
(c) Desired particle size is required for higher efficiency, thus escalating cost of process.
(d) The process is less effective than alkaline method in lowering the lignin content.
(e) Hydrolysate needs neutralization before fermentation that creates considerable amount of solid waste (gypsum).
(f) Since a part of sugar is degraded to produce inhibitors (furfural, HMF, etc.), the overall yield of sugar is reduced.

(g) Both degradation products of lignin and modified (repolymerized) pseudolignin also inhibit cellulases and fermenting microbes.
(h) Enhancing severity to promote cellulose hydrolysis by enzyme does lead to increased degradation products and pseudolignin, resulting in reduced lignin recovery.

Concentrated Acid Method

Concentrated acid pretreatment method occurs at low temperature conditions. The process may be applied to get (a) hydrolyzable cellulose or (b) solubilizing biomass to get sugars (Harmsen et al. 2010). The acids used are H_2SO_4 (65–86% w/v), HCl (41%), or H_3PO_4 (85% w/w) to pretreat dried (5–10% moisture) LCB at moderate temperatures (30–60 °C) and pressures. Initially, this was a two-stage process involving in the first stage treatment of LCB with sulfuric acid (10%) at 100 °C for 2–6 h to get hemicellulosic sugars (mix of pentoses and hexoses) and then in the second stage the washed solid fraction obtained from the first stage treatment which was first dried and then impregnated in moderately concentrated H_2SO_4 (30–40%) for 1–4 h followed by drying of the acid to raise its concentration to 70% and keeping it for 1–4 h at 100 °C under this condition. The sugar thus released was separated, and acid is recovered to provide for the first stage hydrolysis (Shahbazi and Zhang 2010). The process has now largely been replaced with single-stage dilute acid pretreatment method. Since LCB is composed of heterogenous mass of sugars and the wild-type *Saccharomyces cerevisiae* cannot ferment pentoses, getting a heterogenous mixture of sugars via concentrated acid hydrolysis of LCB would only mean extra investment in separation of sugars or wasting or uneconomic uses of high proportion of pentoses. Thus the application of concentrated acid seems to be advantageous only to pretreat the LCB. After pretreatment, the solid fraction is washed and neutralized (Liu et al. 2012) and subjected to enzymatic hydrolysis. The factors affecting the efficiency of concentrated acid pretreatment process are concentration, reaction temperature, residence or reaction time, and acid/LCB ratio.

The example of successes of the concentrated acid (using 70% H_2SO_4) pretreatment is its commercialization by Arkenol Inc. as early as in 1999 (http://www.arkenol.com/Arkenol%20Inc/tech01.html). The process is based on the pretreatment of LCB with 70% H_2SO_4 at <50 °C and acid-to-solid ratio of 1.25 (van Groenestijn et al. 2007). Later on, many variations have been introduced to enhance the efficiency of this technique. One such variation is addition of some organic solvents, e.g., acetone, to pretreated LCB and subjecting the mix to agitation to terminate the reaction and wash out solid matters (Zhang et al. 2007).

Another potential acid for application under this category is H_3PO_4, and the advantages associated with this acid are least or no production of inhibitory substances and high sugar yield (Zhang et al. 2010). The higher concentration of H_3PO_4 above a threshold value dissolves cellulose rather to swell it (Mancini et al. 2016); such dissolved cellulose when regenerated has been found to be amorphous (cellulase hydrolyzable) (Sathitsuksanoh et al. 2012). One variation of H_3PO_4-based process is COSLIP (cellulose solvent- and organic solvent-based lignocellulose fractionation) involving pretreatment of LCB with concentrated H_3PO_4 at 50 °C for

60 min and then separating the solid fraction from reaction mix by adding an organic solvent, viz., ethanol (95% v/v), at ambient temperature for 10 min (Sathitsuksanoh et al. 2012). COSLIP process however suffers from shortcomings such as very slow rate of reaction and very high loads of solvent (Sathitsuksanoh et al. 2012). Another variation of H_3PO_4-based method is the use of PHP (containing 80% H_3PO_4 and 1.77% H_2O_2) for pretreatment. Applying 2.0 g of LCB (wheat straw) and 20.0 g PHP at 50 °C and for 1–3 h incubation leads to solubilization of all hemicelluloses and 70% of lignin and recovery of 90% of cellulose (Wang et al. 2018).

The advantages of the method are:

(a) Lower reaction temperature (e.g., 40 °C).
(b) Higher recovery of sugars (90%) from LCB (Badger 2002) as compared to dilute acid process.
(c) The less severity conditions used during process permit use of reactors or piping made up of relatively cheaper substance, e.g., fiberglass.

The disadvantages are:

(a) Very high concentration of acid (e.g., 30–70%) is used.
(b) Concentrated acid is corrosive and dangerous.
(c) Require costly acid-resistant reactor (expensive alloy made) for better performance.
(d) Expensive due to higher cost involved in neutralization process, remediation of large amounts of gypsum. and maintenance of the process.

A conclusive data as to the most feasible acid pretreatment technology is yet to be available, sulfuric acid has been an acid, and dilute acid method has been a technique of choice (Zhang et al. 2014a, b) possibly because of its low cost and environment friendliness. Exact pretreatment conditions (acid concentration, temperature, and residence time) however vary with habits (herbaceous or woody), parts (young stem/old stem, leaf, flower/cob), and species of plants yielding LCB. As to the acid pretreatment methods, reaction kinetic models have been reported (Fig. 7.3). These models exhibit first-order reaction with kinetic rate parameters depending upon severity factors indicated by SPs or severity

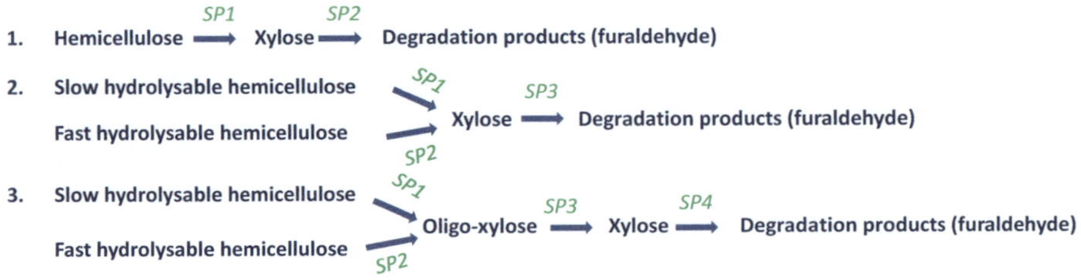

Fig. 7.3 Three kinetic models of dilute acid hydrolysis of hemicellulose fraction of LCB based on different severity parameters (Tanjore and Richard 2015)

parameters. In model 1, the cardinal point is temperature 180 °C, and below this, hemicelluloses are hydrolyzed to monomers, while above this, degradation of monomer (xylose) takes place (Lee et al. 1999). In model 2, two types of hemicelluloses requiring low and high severity parameters (SP1 and SP2) for hydrolysis are depicted. In model 3, both SP1 and SP2 are kept low to get oligo-xylan and to reduce the chance of monomer degradation. In the next stage, applying higher severity parameter (SP3) oligo-xylan can be hydrolyzed to its monomers (xylose) (Lloyd and Wyman 2005; Tanjore and Richard 2015).

The kinetic models given mainly highlight the importance of severity factors in monitoring hemicellulose hydrolysis. The constraint in their universal application is the variable chemical composition with respect to relative proportion of cellulose, hemicelluloses, and lignin in different LCB. Furthermore, increase of CSF and increased rate of sugars degradation go hand in hand. So it is a tricky problem how a balance between suitable sugar yield and inhibitor production is to be struck in case of single or multiple types of LCB.

7.4.4.2.4 Microwave-Assisted Method

The heating sources such as the heating elements, resistance rods, or burners radiate infrared waves which heat a very thin layer at the surface of the biomaterial; the heat then goes in the interior slowly through conduction as the thermal conductivity of the biomaterials is not high. Microwave (MW) radiation is electromagnetic wave that quickly enters inside, interacting with materials (of different properties, e.g., polar or non-polar) generating energy. The energy is deposited throughout the volume of the biomaterials; thus, the latter is heated faster and thoroughly (Yadav et al. 2016). Microwave mediated pretreatment has thus been said to be fast heating, easy to operate, energy-efficient, and less degradation by-product yielder (Tayyab et al. 2018) process. Moreover, MW heating is more regulated than conventional heating. It does not require energy-intensive size reduction processing of LCB.

The method however has scarcely been studied as independent pretreatment process (Irmak et al. 2018); its relevance has been found more in combination with alkali (Liu et al. 2018), acid (Gabhane et al. 2011), water, salt, and ionic liquid (Xu 2015). In a study with wheat straw as feedstock, MW was found to show more synergism with dilute acid pretreatment (0.5% H_2SO_4 w/v, 160 °C, and 10 min) as compared to hydrothermal (54.4% w/w, 200 °C, and 10 min) and alkali (10% w/w, 160 °C, and 10 min) in terms of sugar yield following hydrolysis of LCB (Saha et al. 2008). MW-assisted ionic liquid pretreatment has been reported to enhance cellulose hydrolysis rate (Zhu et al. 2006). As in case of *Crotalaria juncea* fibers, for example, this combination of methods has been found to yield 78.7% glucose at 160 °C and in 46 min (Paul and Dutta 2018). Moreover, a comparison of MW and autoclave as heating methods in dilute acid pretreatment of LCB has proved that both are almost at par in affectivity (Sahay and Rana 2017). Also, as to the production of inhibitors due to degradation of sugars and lignins, MW as source of heating has similar effect as compared to other heating source.

MW technology has been in use in pulp industries for delignifying the LCB (Wang et al. 2016). It may be applied to produce thermal effect (via fast heating) and non-thermal effect (at mild temperatures), and the two conditions lead to MW-assisted pyrolysis (thermal) and MW-assisted solvolysis (non-thermal) technologies, respectively. MW-assisted pyrolysis may be used to convert lignin into commercially important chemicals (such as guaiacol, phenol, etc.) efficiently, while MW-assisted solvolysis is useful in cleaving ether and Caryl-Cα bonds. Lignin is an underutilized fraction of LCB, and its utilization and valorization have tremendous potential in overall economics of the biomass biorefinery technology. It consists of phenylpropane units (*p*-coumaryl alcohol, syringyl alcohol, and guaiacyl alcohol) (Fan et al. 2017) in large amount that can be converted to liquid fuels and/or aromatic compounds (such as vanillin, guaiacol, syringaldehyde, etc.) (Wen et al. 2013). Under mild condition, MW-assisted solvolysis has been found as an effective strategy to achieve this (Zhou et al. 2018). For example, MW-assisted liquefaction of biomass in hydrogen donor solvents such as wheat straw in phenol and catalyst H_2SO_4 (Ouyang et al. 2015) and pine sawdust in methanol (Xu et al. 2012) have been reported to yield more monophenolic compounds as compared to conventional heating process. Despite many lab-scale studies have been made on the affectivity of MW in pretreatment of LCB, its industrial exploitation at commercial level has hardly been reported.

7.4.4.2.5 Ultrasonication-Mediated Method

Ultrasonic radiation generates shear forces that weakens/breaks linkages between hemicellulose, cellulose, and lignin in LCB (Ravindran and Jaiswal 2016). The treatment is suitable in combination with mineral and organic acids, alkalis, or ionic liquids (Koutsianitis et al. 2015). Among factors regulating the sonication treatment, sonication time and temperature have been found to be important (Liyakathali et al. 2016). However, this is energy-intensive process, and its commercial-level application is yet to be studied.

7.4.4.3 Chemical Pretreatment

7.4.4.3.1 Cold Alkali Method

This method is based on an existing process called cold alkaline extraction (CAE) applied for dissolving pulp (Arnoul-Jarriault et al. 2015). The process occurs at ambient temperature involving cheap chemicals (alkali); thus, the process is attractive. During this pretreatment, substantially non-degraded hemicelluloses and lignin mostly bound to hemicelluloses are leached out from LCB (Park and Kim 2012). The process however is not very efficient in terms of releasing lignin and hemicelluloses especially from woody biomass (Carvalho et al. 2016).

7.4.4.3.2 Ozonolysis

Ozone has been found to reduce lignin from LCB, affecting the structure of the hemicellulose to a little extent but not the cellulose at all. The process involving ozone thus is attractive that in addition does not produce inhibitors. This pretreatment method has been successfully applied in case of many types of LCBs to remove lignin and hemicelluloses. For example, Ben-Ghedalia and Miron (1981) applied successfully this method to pretreat wheat straw; Ben-Ghedalia and Shefet (1983) to cotton straw; Neely (1984) to pine, peanut, green hay, and bagasse; and Vidal and Molinier (1988) to poplar sawdust. The process however is costly, since ozone itself is a costly chemical and its required amount for the pretreatment is also huge magnifying the overall process cost (Kumar et al. 2009).

7.4.4.3.3 Organosolv Method

Application of organic solvents singly or in mixed form for the pretreatment of LCB at higher temperatures (i.e., 100–250 °C) aiming at degrading the lignin by cleaving of ether linkages (McDonough 1993) is called the organosolv pretreatment process. Many potential organic solvents have been applied successfully to pretreat a variety of LCBs naming a few are acetone, alcohols (ethanol and methanol), organic acids (usually acetic or formic acids), organic peracid, etc. (Zhang et al. 2016). These organic solvents may be used as pure or their aqueous solution (variously diluted). Some mineral acid, base, or salts may be used as catalyst to improve pretreatment efficiency (Borand and Karaosmanoglu 2018). The pretreatment of LCB with these combined organosolv and catalysts gives rise to cellulose in solid fraction, hemicellulosic sugars in aqueous phase, and lignin dissolved in organic phase. Generally acid especially H_2SO_4 has been found to be more effective as a catalyst, but its corrosive nature is a disadvantage for industrial application (Park et al. 2010). Applications of phenol, acetic acid (Villaverde et al. 2010), and formic acid (Zhao and Liu 2012) in combination with HCl and water have been used to pretreat LCB effectively at ambient conditions. Ethanosolv process applying ethanol as solvent however requires higher temperature and pressure that consequently results in reprecipitation of lignin. The process has been successfully applied to LCBs such as hybrid poplar (Pan et al. 2006) and Japanese cypress (Hideno et al. 2013). Based on ethanosolv process, Lignol process has been developed that applies aqueous ethanol (50%w/w) for the pretreatment of LCB at 200 °C and 400 psi (Pan et al. 2005). The process variables are temperature, retention time, solvent and acid type, and their concentrations. The recovery of the organic solvents is essential as they might inhibit enzyme and fermentative microbes. The advantages of the process are easy recovery (by distillation) and recycling of the solvents and recovery of pure lignin as co-product (Sun and Cheng 2002). The disadvantages are highly inflammable nature, higher cost of solvents, and higher energy requiring solvent recovery process (distillation) that makes the method a costlier one (Borand and Karaosmanoglu 2018).

7.4.4.3.4 N-Methylmorpholine-N-oxide (NMMO) Method

NMMO is solid with melting point 184 °C, but its aqueous solution has melting point 76 °C. Its aqueous solution is used to solubilize cellulose during Lyocell process that yields fiber (Tan et al. 2009). NMMO is a strong oxidant requiring an antioxidant (e.g., propyl gallate) to stabilize cellulose/NMMO mixture (Rosenau et al. 2002). Lignin itself serves as an antioxidant, lignocelluloses thus would not require any antioxidant, and a direct pretreatment of LCB with NMMO is possible. It is a non-toxic, biodegradable, and recyclable liquid that works at room temperature (Liebert 2008). It has high electron density on its oxygen atom, and its highly polar N–O bond shows strong dipole activity. These features make it water-soluble and enable it to disrupt hydrogen bonds present in cellulose and hemicelluloses, thus reducing its crystallinity (Kuo and Lee 2009). NMMO dissolves cellulose which can then be regenerated applying just water (serving as anti-solvent). The regenerated cellulose is of low crystallinity easily hydrolyzable by cellulases; thus, higher sugar recovery is attained. The process does not produce any toxic by-products and proceeds in mild environment, and up to 98% solvent itself can be recovered (Shafiei et al. 2010). Both fresh and recycled NMMO have been reported alike in effect as in case of sugarcane bagasse, the pretreatment led to reduction in cellulose crystallinity and enhancement of sugar yields (Kuo and lee 2009). LCB with higher concentration of lignin (up to 18.4%) was also found to be dissolved in NMMO/H_2O solution under conditions of constant stirring, a temperature of 85 °C, and residence time 5 h if preceded by glycerol swelling and mechanical extrusion (Zhang et al. 2017). Potential of NMMO to pretreat various LCBs such as sugarcane bagasse (Khodaverdi et al. 2012), spruce and oak (Kuo and Lee 2009), and paddy straw and poplar (Shafiei et al. 2010) has been demonstrated. An 85% (w/w) NMMO in the pretreatment of oak and spruce at 90–130° and ambient pressure for 1–3 h has been found to yield 64.6% and 83.5% sugars, respectively, after enzymatic hydrolysis (Shafiei et al. 2010). After pretreatment, as high as 99.5% (w/w) of solvent recovery has been reported in Lyocell fiber production process (Chou et al. 2010). The solvent is however highly degradable at higher temperature which is a constraint in its industrial application.

7.4.4.3.5 Ionic Liquid (IL) Method

IL is salt in liquid phase at room temperature. ILs are newer class of solvents, comprising organic cations (e.g., imidazolium, aliphatic ammonium, alkylated phosphonium, pyridinium, and sulfonium ions) linked with either organic or inorganic anions (Liu et al. 2012). Of these, imidazolium-based ILs are the most common ones (Zhang et al. 2017). The cation and anion fractions of ILs together take part in solubilizing the components of LCB during pretreatment. The advantages ILs offer as means to pretreat LCB are their reusability, non-volatility, non-toxicity, environment benignness, and thermal and chemical stability.

In the pioneering study, the IL [BMIM][Cl] was found to dissolve up to 25% (w/w) of cellulose with assistance of microwave heating (Swatloski et al. 2002). An anti-solvent such as water could regenerate the dissolved cellulose; thus, obtained cellulose was amorphous one that could subsequently be digested with cellulases or IL. With microwave assistance the same IL was found to hydrolyze both woody (e.g., oak, sumac wood, etc.) and herbaceous (e.g., grape stem, triticale straw, etc.) LCB in less than 2 min. Another IL [EMIM][CH3COO] was also reported to solubilize southern yellow pine (softwood) and a red oak (hardwood) (Sun et al. 2009). The same IL was found to dissolve sugarcane bagasse in 15–16 h at 100 °C, but only in 10 min at 175 °C and in 5 min at 185 °C indicating the role of severity in IL pretreatment (Li et al. 2011). Also it was concluded that IL could dissolve all types of LCBs and all components of LCB (Brandt et al. 2013).

There are at least four approaches to apply ILs in the LCB pretreatment:

1. Reduce crystallinity of native cellulose in LCB by treating with IL and regenerate applying antisolvent, e.g., water. To cite, cellulose from wheat straw if pretreated with water only, or steam explosion method only, IL (BMIM][Cl]) only or combination of IL (BMIM][Cl]) and steam explosion methods showed 42.78%, 68.78%, 70.37%, and 100% sugar yields, respectively, after enzyme hydrolysis. The result clearly indicates a positive and significant impact of IL ([BMIM][Cl]) on reducing crystallinity of cellulose or enhancing latter's proneness to cellulases (Liu and Chen 2006). To improve water and energy use efficiency during recycling of IL [EMIM]CH_3Coo], kosmotropic salt (K_3PO_4 or K_2HPO_4) was used to partition the pretreatment mix into IL-rich, salt-rich, and LCB-rich phases (Shill et al. 2011). The technique applied to LCB such as corn stover, etc. was found to show higher glucose yield after enzyme hydrolysis and efficient IL recovery (cf IL-pretreated and water-precipitated cellulose) (Shill et al. 2011).
2. Enzymatic hydrolysis in situ in ionic liquids is a highly potent approach toward reducing the process complexity and costs by abandoning cellulose regeneration step using antisolvent. However, this requires ILs to be non-toxic/non-inhibitory toward microbes and enzymes. Unfortunately, the most studied ILs such as 1,3-dialkylimidazolium containing chloride, dicyanamide, formate, and acetate anions may inhibit enzyme activity because these tend to denature enzymes by snatching hydrogen bonds from them (Xie and Zhao 2011).
3. Acid-catalyzed IL pretreatment is another approach that overcomes shortcoming of acid pretreatment such as harsh conditions, high capital costs, and the inefficiency of the process (Xie et al. 2012) and at the same time accelerates IL pretreatment rate. For example, study on acids (H_2SO_4, HCl, HNO_3, and H_3PO_4) hydrolysis of Avicel cellulose in [BMIM][Cl] medium (with acid/cellulose mass ratios as low as 0.11) showed faster reaction; the rate however depended on acid strength (e.g., H_2SO_4 exhibited higher rate as compared to H_3PO_4) (Li and Zhao 2007). Significant sugar yields up to 66%, 74%, 81%, and 68% from acid (7% HCl)-catalyzed IL (C_4mimCl) pretreatment of corn stalk, rice straw, pine wood, and bagasse, respectively, at 100 °C and residence time of 60 min have been achieved. Similar results were also obtained with various combinations of ionic liquids (C_6mimCl, C_4mimBr, AmimCl, C_4mimHSO$_4$, and SbmimHSO$_4$) and acids (H_2SO_4, HNO_3, H_3PO_4, and maleic acid) though in extended residence times (cf the combination of HCl and C_4mimCl) (Li et al. 2008).
4. Since ILs can dissolve diverse kinds of compounds, the resultant LCB hydrolysate thus contains diverse types of chemicals from which sugar separation is a difficult and costly affair. To overcome this, possibility of the use of solid catalysts instead of mineral acids has been studied (Rinaldi et al. 2008). The additional advantages of using solid acids are the ease of recycling of the acid catalyst and getting rid of requirement of special and costly reactor. But there are some disadvantages of using solid catalysts too, the reaction rate being very slow and hydrolysis of cellulose is incomplete only requiring additional step to get glucose

from partially hydrolyzed cellulose. Further study on the use of solid catalyst such as zeolite in cellulose hydrolysis in IL medium revealed that pore sizes and acid amount of zeolites play critical roles in cellulose hydrolysis (Cai et al. 2012), and thus mesoporous zeolites, e.g., ZSN-5 ($SiO_2/Al_2O_3 = 38$), has been prepared. ZSM-5 in IL has been reported to hydrolyze cellulose (76.2%) with a yield of higher total sugar (49.6%) (Chen et al. 2018).

The properties of ILs can be tailored by changing their cations and/or anions (Chen et al. 2017). ILs thus have been tailored to extract specifically lignin or cellulose from LCB. ILs [Bmim]Ace and [Emim]Ace, for example, have been shown to extract lignin selectively from wood biomass, including *Pinus radiate* and *Eucalyptus nitens* (Pinkert et al. 2011). These two ILs were synthesized from [Bmim]Cl and [Emim]Cl by substituting their anions Cl^- for acesulfamate (Ace^-). These bulky anions affect their cellulose solubilization ability negatively. The amine-sulfonate functionalized (ASF) IL especially ASF-IL$[Et_4N][Me_2NC_4SO_3]$ has been shown to remove 40% lignin from *Eucalyptus* bark without dissolving xylan and cellulose (Yan et al. 2015). Biomaterial-derived cholinium-based ILs have also been found to be useful in extracting lignin from rice straw (An et al. 2015). For example, cholinium argininate ([Ch][Arg]) was found to fractionate rice straw into 46% lignin-rich material (LRM) and rest as carbohydrate-rich material (CRM) (An et al. 2015), while [Cho][Ac] was shown to selectively solubilize hemicellulose and lignin in case of southern yellow pine wood (Cheng et al. 2014). Similarly, CO_2- and CS_2-based switchable ILs were found to extract hemicelluloses (64%) and lignin (70%) from *Betula pendula* wood (Eta and Mikkola 2016), and nitrile-based IL 1-propyronitrile-3-benzyllimidazolium chloride ([C2CNBzim]Cl) was shown to extract lignin (53%) from the bamboo (Muhammad et al. 2013).

ILs tailored to selectively extract cellulose from biomass such as 1-alkyl-3-methylimidazolium phosphonate and phosphinate-type ILs have also been developed (Abe et al. 2010), of which the latter has been applied to selectively hydrolyze polysaccharides from bran at medium temperature (25–50 °C). Another similar type of IL, tetra-n-butylphosphonium hydroxide, in aqueous solution was found to selectively dissolve polysaccharides (glucan and xylan) from wood (Abe et al. 2014). The IL, [Emim]Ac, along with water and ethanol or acetone was developed to fractionate hemicelluloses and cellulose from paper-grade kraft pulp treating at 60 °C for 3 h (Froschauer et al. 2013), and the same IL with water was found to fractionate cellulose and hemicelluloses in the extract derived from ozonized cotton linter (Stepan et al. 2016). The method was called IONCELL-P.

Toward reducing application cost of ILs, a relatively cheaper solvent (low-cost IL) triethylammonium hydrogen sulfate has been developed and successfully applied to extract lignin (75%) and hemicelluloses (100%) from *Miscanthus giganteus* grass. A yield of 77% glucose by enzymatic saccharification of the above pretreated *M. giganteus* grass could be obtained; at the same time, the solvent could also be recovered (with 90% recovery rate) and reused four times (Brandt-Talbot et al. 2017). Researches have also been carried out to device efficient separation techniques for sugars from heterogenous LCB hydrolysate in IL medium. Taking advantage of chemical affinity of boronates (e.g., phenylboronic acid and naphthalene-2-boronic acid) for sugars, boronate-based technique has been developed to isolate glucose, xylose, and cellobiose from LCB hydrolysate in aqueous 1-ethyl-3-methylimidazolium acetate medium. Applying this technique sugars up to 90% from aqueous IL solution and 100% from IL solution and corn hydrolysate in IL medium could be isolated, facilitating recovery of concentrated sugars and cleanup and recycle of IL (Brennan et al. 2010). ILs such as dimethylammonium cation and dimethylcarbamate anion whose parent chemicals are dimethylamine and CO_2 (in 2:1 ratio) show high amenability to distillation that may be distilled at a temperature as low as 45 °C (De Maria 2014). These types of ILs can reduce the recycling and thus overall cost (De Maria 2014). Another study on the reusability of [EMIM][CH3COO] showed that this IL loses lignin removing ability by 24% in the tenth batch but does not lose any percentage of ability to disrupt the crystallinity of the cellulose in case of corn stover feedstock; thus, the overall yield of sugars is not reduced even in ten times use of the same IL (Wu et al. 2011). In order to compensate high cost of ILs, possibility of high solid loading in case of switch grass pretreatment using [C_2mim][OAc] has been studied and concluded that despite inherent increase in viscosity in this case, there is a compensatory increase in sheer thinning of the slurry that reduces its complex viscosity (Cruz et al. 2013).

The advantages of the method are:

(a) Expensive corrosion-resistant reactor is not required (except in case of Cl^- based ILs).
(b) Very effective in decreasing LCB recalcitrance.
(c) Tailorable, with numerous cations/anions options.
(d) Indifferent to feedstock types (e.g., low or high lignin containing).
(e) Operates at milder conditions (lower temperature, etc.).
(f) Lignin obtainment and valorization are possible.
(g) Biocompatible IL may avoid the problems (toxicity to enzymes/microbes) associated with normal IL.

The disadvantages in their uses are:

(a) They are synthesized by non-green methods since most of them are produced from fossil resources (Zhao et al. 2007; Petkovic et al. 2012). Only, few cholinium-based ILs (e.g., choline acesulfamate, choline saccharinate, etc.) are produced from biodegradable and renewable choline chloride and artificial sweeteners such as acesulfamate and saccharinate (Hou et al. 2012). Cholinium amino acid (e.g., cholinium lysine) is another IL produced through green synthesis; it has also been found very efficient in pretreatment of sugarcane bagasse (Hou et al. 2013). But its cost and other features need to be assessed.
(b) Mostly ILs lose activity in presence of water (water intolerance), while the processing environment is generally an aqueous one. IL like [EMIM][CH3COO] is free from this deficiency, but it is very costly (USD20/kg) as per Simmons (2013) and highly viscous measuring 162 mPa/s at 20 °C (Zhang et al. 2006).

(c) Higher viscosities that may range from 4.8 to 1110 mPa/s (Zhang et al. 2006) are another problem that impedes stirring and pumping of slurry and also diffusion of enzyme in IL medium if in situ hydrolysis is opted. This enhances the operational cost and energy requirement. To overcome this, the IL [EMIM] [CH3COO] was mixed with 20% (w/w) water and ethylammonium hydrogen sulfate, and the mixture was found free from viscosity-related problems in the pretreatment of switch grass (George et al. 2015).
(d) Costly production – not only raw materials are costly; cost is also incurred on removal of impurities such as 1-methylimidazole (Chidambaram and Bell 2010) from the synthesis of imidazolium ILs.
(e) Costly recovery process, they are recovered by washing with excess of water and then drying.
(f) Generally toxic, especially chloride bearing imidazolium cations to enzyme/fermenting microbes. The highly studied IL [EMIM] [CH3COO] does cause acetylation of cellulose at elevated temperatures (Zhang 2013). The water requirement to remove and recover IL from slurry is very high, as also the energy requirement for dehydrating the recovered IL for recycling; the two disadvantages make the process unattractive altogether (Nguyen et al. 2010).

7.4.4.3.6 Deep Eutectic Solvent (DES) Method

DESs are eutectic mixture of salts and hydrogen bond donors (HBD), the resulting chemical (DES) has members bound with hydrogen bonds. One essential feature of DES is that it melts at lower temperature than its components, e.g., ChCl/urea (1:2) has melting point 12 °C, while ChCl and urea have melting point 302 °C and 133 °C, respectively. They are shown with general formula Cat^+X^-zY (Abbott et al. 2001), where Cat^+ is a cation (e.g., sulfonium, phosphonium or ammonium one), X^- represents a Lewis base (mostly a halide anion), and Y and z are Lewis or Brønsted acid and its number, respectively. Various X^- and Y in combinations constitute different anionic species (Smith et al. 2014) responsible for a variety of structures and thus properties suitable for various applications (Hammond et al. 2016). Choline chloride (ChCl) being non-toxic and cheaply available is the most commonly used salt, and urea, glycerol, carboxylic acids, and polyols are for the same reasons the most commonly used HBDs in making ILs (Zdanowick et al. 2018).

Studies on the pretreatment efficiency of DES containing ChCl-formic acid with respect to corn stover (Xu et al. 2016) and those containing monocarboxylic acid-ChCl, dicarboxylic acid-ChCl, and polyalcohol-ChCl with respect to corn cob (Zhang et al. 2016) have demonstrated their superiority over ILs. Further study showed that the strength and amount of hydroxyl groups in DESs are the main determinants of efficiency and that these hydroxyl groups interact with those present in lignin (Zhang et al. 2016). DESs can solubilize LCB by breaking intermolecular hydrogen bonds (Ren et al. 2016). DES, ChCl/imidazole pretreated corncob at 80 °C has been on hydrolysis reported to yield 92.3% glucose (Procentese et al. 2015). An interesting report on the application of combination of techniques such as microwave heating, acid (H_2SO_4, HCl, HNO_3, H_3PO_4, and p-toluenesulfonic acid) catalysis, and DES [(choline chloride-oxalic acid (ChCl/ox) and choline chloride-urea (ChCl/urea)] medium for solubilizing different kinds of LCBs (pine nit shell, carnauba leaves, and macauba shell) showed enhanced sugar yields (as high as 83.7% in ChCl/ox for carnauba leaves); the highest yield and reduced production of furfural and HMF were found under conditions of HNO_3 (10%), temperature 120 °C, retention time 30 min, and either of DES as medium (da Silva et al. 2016).

Major advantages of DESs lie in their structural and thus chemical flexibility as designed by changing HBD and HBA components, ease of production using non-toxic and readily available raw materials, and their non-inflammable, non-toxic, low-volatile, biocompatible, and biodegradable nature. As compared to ILs, DESs are thus cheaper and their production is greener. A typical method of DES formation involves continuous stirring of raw materials (HBA/HBD) in a fixed molar ratio say 1:2 at medium temperature. The raw materials used are natural ones, viz., glucose, succinic acid, malic acid, ChCl, etc. (Dai et al. 2013). IL preparation however is accomplished typically in two steps: alkylation of an amine/phosphine/sulfide to form an intermediate salt and then an anion exchange step (Hallett and Welton 2011). Though comparatively more benign than ILs, DESs do have some toxicity (Xu et al. 2017), and like ILs they also suffer from higher viscosity which varies with temperature inversely (Zhao et al. 2015).

The solvent-based methods are inherently mild (low energy requiring), comparatively greener, and able to pretreat biomass at higher loadings. Of the four solvent types, ionic liquids and DESs seem to be potential ones.

7.4.4.4 Biological Pretreatment

Use of lignin-degrading microbes (usually white-rot, brown-rot, or soft-rot fungi) or enzymes (e.g., laccases, lignin peroxidase, manganese peroxidase, etc.) in the pretreatment of LCB is called biological pretreatment.

7.4.4.4.1 Whole Cell Method

White-rot fungi are capable of utilizing lignin and hemicelluloses selectively without consuming cellulose in a LCB; this forms the basis for applying them in LCB pretreatment. To cite, *Trametes hirsute* when grown on paddy straw for 10 days showed increase in sugar recover after saccharification by 11.1% (Saritha et al. 2012). The fungus exhibited ligninase activity but low cellulase activity. Another white-rot fungus *Irpex lacteus* when used to biologically pretreat corn stalks for 28 days, the latter on subsequent enzymatic hydrolysis yielded 82% sugars (Du et al. 2011). Among all the fungus, *Phanerochaete chrysosporium* has been the most studied and highly potent one for pretreating various LCBs.

7.4.4.4.2 Enzyme Method

Application of enzymes to pretreat LCB seems to be more promising. Lignin, which is main target of pretreatment, may be degraded under the influence of phenol oxidase (laccase) and peroxidases [lignin peroxidase (LiP), versatile peroxidase (VP), and manganese peroxidase (MnP)] (Zamocky et al. 2014). Laccases (Lac) (EC 1.10.3.2) are multicopper containing glycoprotein that catalyzes oxidation of aromatic amines and phenolic compounds with the production of water (Sahay et al. 2019). They can oxidize non-phenolic component of lignin as well with the help of some low molecular mass compounds (e.g., hydroxybenzotriazole). Laccase production has been reported from a number of fungi (e.g., *Trametes versicolor*), bacteria (*Bacillus subtilis*), actinomycetes (*Streptomyces coelicolor*), etc. (Strong and Claus 2011). The enzyme catalyzes degradation and modification of lignin; these features make it potentially useful for pretreatment of LCB. The application *Myceliophthora thermophila* laccase thus in the pretreatment of wood of *Eucalyptus globules* showed removal of up to 50% of lignin (Rico et al. 2014).

LiP (EC 1.11.1.4) and MnP (EC 1.11.1.13) are iron containing and hydrogen peroxide dependent (Hammel and Cullen 2008). MnP requires Mn to be able to catalyze reactions. While LiP commonly oxidizes non-phenolic lignin, MnP can oxidize both phenolic and non-phenolic lignins. MnP-catalyzed reaction occurs through lipid peroxidation reactions resulting into the formation of various phenolic compounds (3-ethylthiazoline-6-sulfonate, 2, 6-dimethyloxyphenol syringol, guaiacol) and non-phenol compound (e.g., alcohol) (Brown and Chang 2014). During MnP catalysis, Mn^{2+} is oxidized to Mn^{3+}, the latter takes up the task of oxidizing the phenol to phenoxy radicals resulting into degradation of the compounds. VP (EC 1.11.1.16) on the other hand exhibits catalytic activities of both MnP (transform Mn^{2+} to Mn^{3+}) and LiP (acts on non-phenolic lignin) (Zavarzina et al. 2018). It is thus versatile as its name suggests in oxidizing the varieties of phenolic and non-phenolic compounds including lignin directly, without redox mediators.

The advantages of biological pretreatment are:

(a) Less costly, lesser energy-consuming, and eco-friendly (Sindhu et al. 2016)

The major disadvantages of this method are:

(a) Low efficiency and time-consuming (Tian et al. 2012)
(b) Requirement of very precise growth conditions and large space
(c) Possibility of consumption by microbes a part of carbon nutrient

7.4.4.5 Nanoscale Pretreatment

These methods are currently investigated at nanoscale. The methods may be based on the ability of nanoscale actors (nanoparticles) variously equipped to penetrate the wall of LCB and interact there with the major constituents under low severity conditions to release target molecule(s) say lignin and hemicellulose or method that creates very high shear in tiny reactor to amplify the effect of a chemical catalyst on disruption of LCB recalcitrance.

7.4.4.5.1 Acid-Functionalized Magnetic Nanoparticle (AFMAN) Method

These are nanoscale magnetic particles blessed with the features of acid and so also called solid acid nanocatalysts. They behave like acid (as in case of acid pretreatment) in hydrolyzing LCB during pretreatment, but their magnetic nature enables their recovery with the use of magnet to apply them again and again (Fig. 7.4)

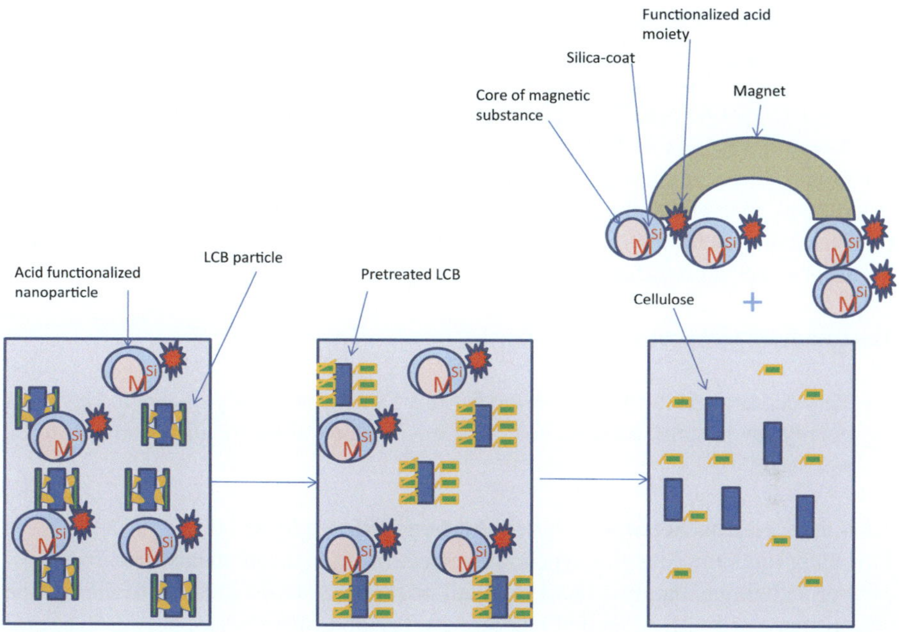

Fig. 7.4 Schematic representation of application of acid-functionalized nanoparticle in LCB pretreatment

(Pena et al. 2012, 2014). These nanoparticles contain a nano-size core of composite materials consisting of magnetic particle coated with mesoporous silica; this composite is then functionalized with an acid. For example, nanoparticles containing iron oxide in the core as magnetic material and functionalized with sulfonic acid (Lai et al. 2011), Fe/Fe3O4 core/shell in the core and functionalized with same acid (Wang et al. 2015), and cobalt spinel ferrite in the core and functionalized with perfluoroalkylsulfonic (PFS) and alkylsulfonic (AS) acids (Pena et al. 2012) have been reported. These nanoparticles have been found effective in the pretreatment of LCB, though with variable efficiencies. For example, PFS- and AS-functionalized and cobalt spinel ferrite core containing nanoparticles were evaluated to pretreat wheat straw under two conditions (1) 80 °C for 24 h and (2)160 °C for 2 h. Under condition (1), PFS-functionalized nanoparticles exhibited more hemicelluloses hydrolysis (about 24%) than AS-functionalized one (about 9.1%), while under condition (2), both PFS- and AS-functionalized nanoparticles showed higher hemicelluloses hydrolysis to the extent of about 46% and 45%, respectively, as compared to control (about 35%) (Pena et al. 2012). Another nanocatalyst containing silica-coated cobalt iron oxide functionalized with propylsulfonic acid (PS) has been tested for the effect of various catalyst loadings (0.1, 0.2 and 0.3 g/g corn stover) at different temperatures (160 °C, 180 °C, and 200 °C) for 1 h. It was found that catalyst loading had hardly any effect on glucose yield; of course severity (as temperature) was found to have a significant effect so that hydrolysis achieved were 59%,

90%, and about 100% at 160, 180, and 200 °C, respectively, with the same catalyst loading (0.2 g/g LCB). The advantages of this method are:

(a) Similar to solid acid, it is corrosion-free, requires no neutralization of hydrolysate, and releases no waste.
(b) Owing to having magnetic property, they can be easily recovered and reused several times; thus, the process seems to be green and economic.
(c) Flexibility as to the use of type of magnetic particle and acids has the potential to provide impetus for further advancement of the process.

The disadvantages are:

(a) The technique is in infancy, tested for few LCBs and at only microtube level; its energy investment, process cost, and environmental implications are yet to be assessed.

7.4.4.5.2 Nanoscale Shear Hybrid Alkaline (NSHA) Method

This technique employs the synergistic effect of high-speed shear force and presence of a chemical catalyst (acid or alkali or even a volatile solvent) to disrupt the recalcitrance of LCB. A special reactor (say Taylor-Couette reactor) is used to provide a shear of the order of about 10,000 s^{-1} to LCB for a shorter period (few minutes) in presence of chemical catalyst, the result in the removal of lignin and hemicelluloses and recovery of cellulose in higher quantity (say 90%) and in an easily digestable form. A 1:1 mixture of sodium hydroxide and corn stover, for example, that was subjected to a shear of 12,500 s^{-1} just for 2 min eliminated major portion of lignin and hemicelluloses leaving behind easily digestable cellulose by the enzymes (Wang et al. 2013). Microscopic observation confirmed that NSHA treatment caused major disruption of cell wall and disintegration of cellulose fibers in corn stover (Wang et al. 2013). In another study, effect of a cationic polyelectrolyte additive poly(diallyldimethylammonium chloride) on NSHA treatment of corn stover revealed that lignin instead of being extracted got redistributed on both surfaces of the cell wall; cellulose however was found to become highly prone to enzymatic hydrolysis (Ji and Lee 2013). For more discussion, readers may refer to Ingle et al. (2019).

The advantages of the technique are:

(a) Impressive recovery of purer cellulose and also lignin and hemicelluloses would help economization of the process on biorefinery concept.

The disadvantages are:

(a) This technique is also in infancy, tested for few LCBs and at only microtube level; its energy investment, process cost, and environmental and human health implications (Rai et al. 2016) are yet to be assessed.

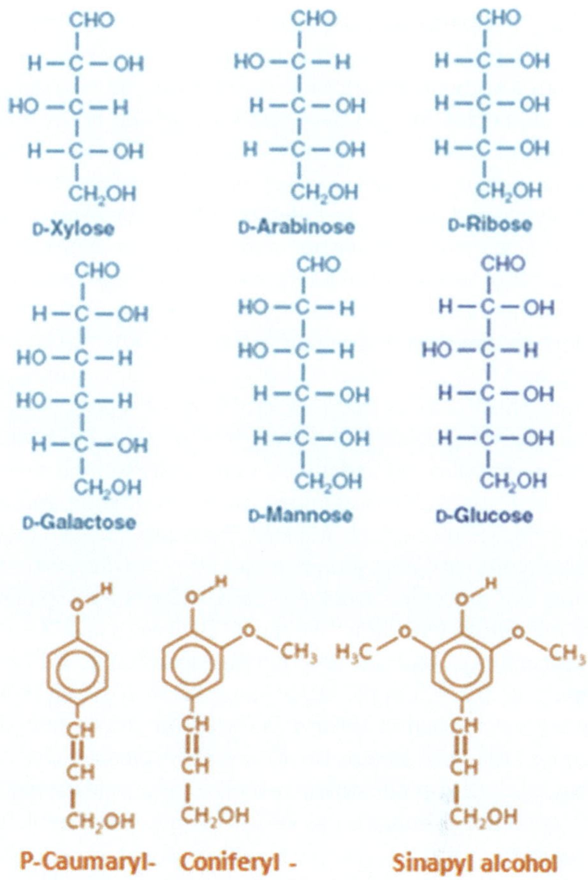

Fig. 7.5 Target monomers (primary platform chemicals) from LCB

7.5 Biorefinery Approach and Pretreatment

Basically, biorefinery approach implies use of every fraction of LCB to produce a useful product with commercial value. The main reason to think in terms of this is the cost of pretreatment, which is very high nearly 30–35% of total processing cost. This can be offset only by selling many value-added products derived from LCB. Some of the primary platform chemicals that can be obtained from LCB are given in Fig. 7.5. Depending upon the severity of the process, many sugar/lignin degradation but valuable products (say furfural and HMF) may also be obtained to make the process commercially attractive. Additionally, this approach has potential to reduce the sewage load and consequently reduce the cost of remediation as well. So any pretreatment strategy must incorporate this dimension as well.

7.6 Conclusions

From the above description, it can easily be conceived that in spite of impressive developments in the science and arts of the pretreatment processes, the real challenges, i.e., cost and energy economy and environmental sustainability, are still alive. Among the most studied pretreatment methods, dilute acid, alkaline autohydrolysis, and ionic liquid methods do suffer from serious constraints. While dilute acid is corrosive to reactor and known for generating inhibitors, alkalies hardly hydrolyze hemicelluloses; thus, the solid fraction in treated LCB contains both cellulose and hemicellulases requiring enzymes for both in subsequent hydrolysis step only to generate mixture of sugars, autohydrolysis at higher severity conditions (required for generating hydrolyzable cellulose) does generate inhibitors, and ionic liquid itself suffers from the problems of higher cost and toxicity toward enzymes and microbes. Nanotechnology in this wake has brought some relief as this technique is applicable under mild conditions so high severity-related problems should not be there. But the technique itself is at infancy, and it will take time to cause positive impact on biofuel industry. Probably because they are predecessors, acid and autohydrolysis pretreatment technology are currently at the stage of commercialization. For example, "Abengoa" and "Blue Sugar" (USA) and "Sekab" (Sweden) all using dilute acid while "Weyland" (Norway) and "Izumi Biorefinery" (Japan) using concentrated acid are now having operational 2G bioethanol plants (Bensah and Mensah 2013). On the other hand, "Cometha" an industrial-scale pre-commercial plant established in Europe is based on autohydrolysis pretreatment. And now in India DBT-ICT Centre for Energy Biosciences (India) has developed technology based on acid pretreatment which is now transferred to BPCL (Bharat Petroleum Corporation Limited) and HPCL (Hindustan Petroleum Corporation Limited) for commercialization. Current emphasis is on biorefinery approach to employ LCB for getting as many products as possible for the sake of overall economy of the process. A citable example of BIOCORE technology claims to produce itaconic acid and value-added lignin products besides ethanol from plant biomass (Project ID:241566 funded under FP7-KBBE, EU Commission-final report). A pretreatment process going ahead with this approach is expected to win the race in the future. Moreover, more research is yet required to design an economic, non-viscous, and non-toxic ionic liquid and of course to answer how nanoscale method is scaled up to commercial level replacing its predecessors, viz., acid and autohydrolysis methods.

References

Abbott AP, Capper G, Davies DL, Munro HL, Rasheed RK, Tambyrajah V (2001) Preparation of novel, moisture-stable, Lewis-acidic ionic liquids containing quaternary ammonium salts with functional side chains. Chem Commun 19:2010–2011

Abe M, Fukaya Y, Ohno H (2010) Extraction of polysaccharides from bran with phosphonate or phosphinate-derived ionic liquids under short mixing time and low temperature. Green Chem 12:1274–1280

Abe M, Yamada T, Ohno H (2014) Dissolution of wet wood biomass without heating. RSC Adv 4:17136–17140

An YX, Zong MH, Wu H, Li N (2015) Pretreatment of lignocellulosic biomass with renewable cholinium ionic liquids: biomass fractionation, enzymatic digestion and ionic liquid reuse. Bioresour Technol 192:165–171

Arnoul-Jarriault B, Lachenal D, Chirat C, Heux L (2015) Upgrading softwood bleached kraft pulp to dissolving pulp by cold caustic treatment and acid-hot caustic treatment. Ind Crop Prod 65:565–571

Badger PC (2002) Ethanol from cellulose: a general review. In: Janick J, Whipkey A (eds) Trends in new crop and new uses. ASHS press, Alexandria, pp 17–21

Bali G, Meng X, Deneff JI, Sun Q, Ragauskas AJ (2014) The effect of alkaline pretreatment methods on cellulose structure and accessibility. ChemSusChem 00:1–5

Barakat A, Mayer C, Solhy A, Arancon RAD, De Vries H, Luque R, Barakat A, Mayer C, Solhy Arancon RAD, De Vries H (2014) Mechanical pretreatments of lignocellulosic biomass: towards facile and environmentally sound technologies for biofuels production. RSC Adv 4:48109–48127. https://doi.org/10.1039/C4RA07568D

Baruah J, Nath BK, Sharma R, Kumar S, Deka RC, Baruah DC, Kalita E (2018) Recent trends in the Pretreatment of Lignocellulosic biomass for value-added products. Front Energy Res 6:141. https://doi.org/10.3389/fenrg.2018.00141

Benazzi T, Calgaroto S, Dalla Rosa C, Vladimir Oliveira J, Mazutti MA (2013) Hydrolysis of sugarcane bagasse using supercritical carbon dioxide to obtain fermentable sugars. J Chem Technol Biotechnol 88:1766–1768

Ben-Ghedalia D, Miron J (1981) The effect of combined chemical and enzyme treatment on the saccharification and in vitro digestion rate of wheat straw. Biotechnol Bioeng 23:823–831

Ben-Ghedalia D, Shefet G (1983) Chemical treatments for increasing the digestibility of cotton straw. J Agric Sci 100:393–400

Bensah EC, Mensah M (2013) Chemical pretreatment methods for the production of cellulosic ethanol: technologies and innovations. Int J Chem Eng 2013, Article ID 719607, 21 pages. https://doi.org/10.1155/2013/719607

Bobleter O (1994) Hydrothermal degradation of polymers derived from plants. Prog Polym Sci 19(5):797–841

Borand MN, Karaosmanoglu F (2018) Effects of organosolv pretreatment conditions for lignocellulosic biomass in biorefinery applications: a review. J Renew Sust Energy 10:033104

Brandt A, Gräsvik J, Hallett JP, Welton T (2013) Deconstruction of lignocellulosic biomass with ionic liquids. Green Chem 15(3):550–583. https://doi.org/10.1039/c2gc36364j

Brandt-Talbot A, Gschwend FJ, Fennell PS, Lammens TM, Tan B, Weale J et al (2017) An economically viable ionic liquid for the fractionation of lignocellulosic biomass. Green Chem 19:3078–3102

Brennan TCR, Datta S, Blanch HW, Simmons BA, Holmes BM (2010) Recovery of sugars from ionic liquid biomass liquor by solvent extraction. Bioenerg Res 3:123–133

Brown ME, Chang MC (2014) Exploring bacterial lignin degradation. Curr Opin Chem Biol 19:1–7

Cai H, Li C, Wang A, Xu G, Zhang T (2012) Zeolite-promoted hydrolysis of cellulose in ionic liquid, insight into the mutual behavior of zeolite, cellulose and ionic liquid. Appl Catal B 123:333–338

Candido RG, Godoy GG, Gonçalves AR (2012) Study of Sugarcane bagasse pretreatment with sulfuric acid as a step of cellulose obtaining. World Acad Sci Eng Technol 61:101–105

Carvalho DM, Olena Sevastyanova O, Queiroz JH, Colodette JL (2016) Cold alkaline extraction as a pretreatment for bioethanol production from eucalyptus, sugarcane bagasse and sugarcane straw. Energy Convers Manag 124:315–324

Chang VS, Burr B, Holtzapple MT (1997) Lime pretreatment of switchgrass. Appl Biochem Biotechnol 63–65:3–19

Chang VS, Nagwani M, Holtzapple MT (1998) Lime pretreatment of crop residues bagasse and wheat straw. Appl Biochem Biotechnol 74:135–159

Chang VS, Nagwani M, Kim CH, Holtzapple MT (2001) Oxidative lime pretreatment of high-lignin biomass. Appl Biomass Biotechnol 94:1–28

Chen SX, Yong Q, Xu Y, Zhu JJ, Yu SY (2009) Study on dilute-acid pretreatment of corn stalk. Chem Ind For Prod 29:27–32

Chen Y, Stevens MA, Zhu Y, Holmes J, Xu H (2013) Understanding of alkaline pretreatment parameters for corn stover enzymatic saccharification. Biotechnol Biofuels 6:8

Chen H, Liu J, Chang X, Chen D, Xue Y, Liu P et al (2017) A review on the pretreatment of lignocellulose for high-value chemicals. Fuel Process Technol 160:196–206

Chen T, Xiong C, Tao Y (2018) Enhanced hydrolysis of cellulose in ionic liquid using Mesoporous ZSM-5. Molecules 23:529. https://doi.org/10.3390/molecules23030529

Cheng F, Wang H, Chatel G, Gurau G, Rogers RD (2014) Facile pulping of lignocellulosic biomass using choline acetate. Bioresour Technol 164:394–401

Chidambaram M, Bell AT (2010) A two-step approach for the catalytic conversion of glucose to 2,5-dimethylfuran in ionic liquids. Green Chem 12(7):1253–1262

Chou WT, Lai MY, Huang KS (2010) Method of recovering aqueous N-Methylmopholine-NOxide solution used in production of Lyocell fiber. US Patent no. 201102264271A

Chum HL, Johnson DK, Black SK, Overend RP (1990) Pretreatment-catalyst effects and the combined severity parameter. Appl Biochem Biotechnol 24(25):1–14

Cruz AG, Scullin C, Mu C, Cheng G, Stavila V, Varanasi P, Xu D, Mentel J, Chuang YD, Simmons AB, Singh S (2013) Impact of high biomass loading on ionic liquid pretreatment. Biotechnol Biofuels 6:52

da Silva LV, JBL S, Correa-guimarães A, Martín-Ramos P, Hernándeznavarro S, Sánchez-Bascones M, Navas-Gracia LM, Pérez-Lebeña E, Martín-Gil J (2016) Efficient microwave-assisted acid hydrolysis of lignocellulosic materials into total reducing sugars in ionic liquids. Cellul Chem Technol 50(7–8):761–770

Dai Y, van Spronsen J, Witkamp G-J, Verpoorte R, Choi YH (2013) Natural deep eutectic solvents as new potential media for green technology. Anal Chim Acta 766:61–68

De María PD (2014) Recent trends in (ligno)cellulose dissolution using neoteric solvents: switchable, distillable and bio-based ionic liquids. J Chem Technol Biotechnol 89(1):11–18. https://doi.org/10.1002/jctb.4201

Demirbas A (2008) Products from lignocellulosic materials via degradation processes. Energ Sources A 30(1):27–37

Doner LW, Hicks KB (1997) Isolation of hemicellulose from corn fiber by alkaline hydrogen peroxide extraction. Cereal Chem 74:176–181

Du W, Yu H, Song L, Zhang J, Weng C, Ma F et al (2011) The promising effects of by-products from Irpex lacteus on subsequent enzymatic hydrolysis of bio-pretreated corn stalks. Biotechnol Biofuels 4:37

Duff SJB, Murrayh WD (1996) Bioconversion of forest products industry waste cellulosic to fuel ethanol: a review. Bioresour Technol 55:1–33

Ebringerová A, Hromadkova Z, Heinze TH (2005) Hemicellulose. Adv Polym Sci 128:1–68

Eda S, Ohnishi A, Kato K (1976) Xvlan isolated from the stalk of *NicOliana labacul1l*. Agric Biol Chem 40:359–364

El-Naggar NE, Deraz S, Khalil A (2014) Bioethanol production from lignocellulosic feedstocks based on enzymatic hydrolysis: current status and recent developments. Biotechnology 13:1–21. https://doi.org/10.3923/biotech.2014.1.21

Eta V, Mikkola JP (2016) Deconstruction of nordic hardwood in switchable ionic liquids and acylation of the dissolved cellulose. Carbohydr Polym 136:459–465

Fan LL, Zhang YN, Liu SY, Zhou N, Chen P, Cheng YL, Addy M, Lu Q, Omar MM, Liu YH et al (2017) Bio-oil from fast pyrolysis of lignin: effects of process and upgrading parameters. Bioresour Technol 241:1118–1126

Farhat W, Venditti RA, Hubbe M, Taha M, Becquart F, Ayoub A (2017) A review of water-resistant hemicellulose-based materials: processing and applications. ChemSusChem 10:305–323

Faulon J, Carlson GA et al (1994) A three-dimensional model for lignocellulose from gymnospermous wood. Org Geochem 21:1169–1179

Foston M, Ragauskas AJ (2010) Changes in lignocellulosic supramolecular and ultrastructure during dilute acid pretreatment of populus and switchgrass. Biomass Bioenergy 34(12):1885–1895

Froschauer C, Hummel M, Iakovlev M, Roselli A, Schottenberger H, Sixta H (2013) Separation of hemicellulose and cellulose from wood pulp by means of ionic liquid/cosolvent systems. Biomacromolecules 14:1741–1750

Gabhane J, Prince William SPM, Vaidya AN, Mahapatra K, Chakrabarti T (2011) Influence of heating source on the efficacy of lignocellulosic pretreatment—a cellulosic ethanol perspective. Biomass Bioenergy 35:96–102

Galia A, Schiavo B, Antonetti C, Galletti AMR, Interrante L, Lessi M, Scialdone O, Valenti MG (2015) Autohydrolysis pretreatment of Arundo donax: a comparison between microwave-assisted batch and fast heating rate flow-through reaction systems. Biotechnol Biofuels 8:218. https://doi.org/10.1186/s13068-015-0398-5

George A, Brandt A, Tran K, Zahari SMSNS, Klein-Marcuschamer D, Sun N, Sathitsuksanoh N, Shi J, Stavila V, Parthasarathi R et al (2015) Design of low-cost ionic liquids for lignocellulosic biomass pretreatment. Green Chem 17:1728–1734

Gruppen H, Hamer RJ, Voragen AGJ (1992) Water-unextractable cell wall material from wheat flour. 2. Fractionation of alkali-extracted polymers and comparison with water extractable arabinoxylans. J Cereal Sci 16:53–67

Gupta R, Lee YY (2010) Investigation of biomass degradation mechanism in pretreatment of switchgrass by aqueous ammonia and sodium hydroxide. Bioresour Technol 101:8185–8191

Hallett JP, Welton T (2011) Room-temperature ionic liquids: solvents for synthesis and catalysis. Chem Rev 111(5):3508–3576

Hammel KE, Cullen D (2008) Role of fungal peroxidases in biological ligninolysis. Curr Opin Plant Biol 11:349–355

Hammond OS, Bowron DT, Edler KJ (2016) Liquid structure of the choline chloride-urea deep eutectic solvent (reline) from neutron diffraction and atomistic modelling. Green Chem 18(9):2736–2744

Harmsen PFH, Huijgen WJJ, Lopez Bermudez LM, Bakkar RRC (2010) Literature review of physical and chemical pretreatment processes for Lignocellulosic Biomass. ECN BioSynergy, Food and bio-based research Wageningien.

He L, Huang H, Zhang Z, Lei Z (2016) A review of hydrothermal pretreatment of lignocellulosic biomass for enhanced biogas production. Curr Org Chem 19:437–446

Heredia-Olea H, P'erez-Carrillo E, Serna-Saldívar SO (2012) Effects of different acid hydrolyses on the conversion of sweet sorghum bagasse into C5 and C6 sugars and yeast inhibitors using response surface methodology. Bioresour Technol 119:216–223

Heredia-Olea E, Pérez-Carrillo E, Montoya-Chiw M, Serna-Saldívar SO (2015) Effects of extrusion pretreatment parameters on sweet sorghum bagasse enzymatic hydrolysis and its subsequent conversion into bioethanol. Biomed Res Int: 325905. https://doi.org/10.1155/2015/325905

Hideno A, Kawashima A, Endo T, Honda K, Morita M (2013) Ethanol-based organosolv treatment with trace hydrochloric acid improves the enzymatic digestibility of Japanese cypress (*Chamaecyparis obtusa*) by exposing nanofibers onthe surface. Bioresour Technol 132:64–70

Hong H, Xue G, Weng L, Guo X (2012) Pretreatment of moso bamboo with dilute phosphoric acid. Bioresources 7(4):4902–4913

Hongdan Z, Shaohua X, Shubin W (2013) Enhancement of enzymatic saccharification of sugarcane bagasse by liquid hot water pretreatment. Bioresour Technol 143:391–396

Hou XD, Smith TJ, Li N, Zong M-H (2012) Novel renewable ionic liquids as highly effective solvents for pretreatment of rice straw biomass by selective removal of lignin. Biotechnol Bioeng 109(10):2484–2493

Hou XD, Li N, Zong MH (2013) Facile and simple pretreatment of sugar cane bagasse without size reduction using renewable ionic liquids-water mixtures. ACS Sustain Chem Eng 1(5):519–526

Hsu T-A (1996) Pretreatment of biomass. In: Wyman CE (ed) Handbook on bioethanol, production and utilization. Taylor & Francis, Washington, DC, pp 179–212

Imman S, Laosiripojana N, Champreda V (2018) Effects of liquid hot water pretreatment on enzymatic hydrolysis and physicochemical changes of corncobs. Appl Biochem Biotechnol 184:432–443

Ingle AP, Chandel AK, Antunes FAF, Rai M, da Silva SS (2019) New trends in application of nanotechnology for the pretreatment of lignocellulosic biomass. Biofuels Bioprod Biorefin 13:776–788

Irmak S, Meryemoglu B, Sandıp A, Subbiah J, Mitchell RB, Sarath G (2018) Microwave pretreatment effects on switchgrass and miscanthus solubilization in subcritical water and hydrolysate utilization for hydrogen production. Biomass Bioenergy 108:48–54

Isikgor FH, Becer CR (2015) Lignocellulosic biomass: a sustainable platform for the production of bio-based chemicals and polymers. Polym Chem 2015(6):4497–4559

Iyer V, Wu ZW, Kim SB, Lee YY (1996) Ammonia recycled percolation process for pretreatment of herbaceous biomass. Appl Biochem Biotechnol 57/58:121–132

Jain P, Rana RS, Sahay S (2013) Evaluation of a minimal energy-dependent technique for preparation of *Ailanthus excelsa* hemicellulose acid hydrolysate fermentable by *Schefferomyces stipitis*. J Biofuels 3:103–111

Ji S, Lee I (2013) Impact of cationic polyelectrolyte on the nanoshear hybrid alkaline pretreatment of corn stover: morphology and saccharification study. Bioresour Technol 133:45–50

Kaar WE, Holtzapple MT (2000) Using lime pretreatment to facilitate the enzymic hydrolysis of corn stover. Biomass Bioenergy 18(3):189–199

Kazi FK, Fortman J, Anex R et al (2010) Techno-economic analysis of biochemical scenarios for production of cellulosic ethanol. Tech. Rep. NREL/TP-6A2–46588, NREL

Khodaverdi M, Jeihanipour A, Karimi K, Taherzadeh MJ (2012) Kinetic modeling of rapid enzymatic hydrolysis of crystalline cellulose after pretreatment by NMMO. J Ind Microbiol Biotechnol 39:429–438

Kim S, Holtzapple MT (2006) Effect of structural features on enzyme digestibility of corn stover. Bioresour Technol 97:583–591

Kim HJ, Lee S, Kim J, Mitchell RJ, Lee JH (2013) Environmentally friendly pretreatment of plant biomass by planetary and attrition milling. Bioresour Technol 144:50–56

Kirk-Othmer (2001) In: Kirk-Othmer (ed) Encyclopedia of chemical technology, 4th edn. Wiley Publishing, New York

Kormelink FJM, Voragen AG (1993) Degradation of different [(glucurono)arabin';]xylans by a combination of purified xylan-degrading enzymes. Appl Microbiol Biotechnol 38:688695

Koutsianitis D, Mitani C, Giagli K, Tsalagkas D, Halász K, Kolonics O et al (2015) Properties of ultrasound extracted bicomponent lignocelluloses thin films. Ultrason Sonochem 23:148–155

Krassig H, Schurz J (2002) Ullmann's Encyclopedia of industrial chemistry, Sixth edn. Wiley-VCH, Weinheim

Kumar AK, Sharma S (2017) Recent updates on different methods of pretreatment of lignocellulosic feedstocks: a review. Bioresour Bioprocess 4:7

Kumar R, Mago G, Balan V, Wyman CE (2009) Physical and chemical characterizations of corn stover and poplar solids resulting from leading pretreatment technologies. Bioresour Technol 100(17):3948–3962

Kuo CH, Lee CK (2009) Enhanced enzymatic hydrolysis of sugarcane bagasse by Nmethylmorpholine-N-oxide pretreatment. Bioresour Technol 100(2):866–871

Lai YZ (1991) In wood and cellulose chemistry. In: DNS H, Shiraishi N (eds) Chemical degradation, 2nd edn. Marcel Dekker Inc, New York, pp 455–473

Lai DM, Deng L, Guo QX, Fu Y (2011) Hydrolysis of biomass by magnetic solid acid. Energy Environ Sci 4:3552. https://doi.org/10.1039/c1ee01526e

Lee JW, Jeffries TW (2011) Efficiencies of acid catalysts in the hydrolysis of lignocellulosic biomass over a range of combined severity factors. Bioresour Technol 102(10):5884–5890

Lee YY, Iyer P, Torget RW (1999) Dilute-acid hydrolysis of lignocellulosic biomass. Adv Biochem Eng Biotechnol 65:94–115

Lenihan L, Orozco A, O'Neill E, Ahmad MNM, Rooney DW, Walker GM (2010) Dilute acid hydrolysis of lignocellulosic biomass. Chem Eng J 156(2):395–403

Li C, Zhao ZK (2007) Efficient acid-catalyzed hydrolysis of cellulose in ionic liquid. Adv Synth Catal 349(11–12):1847–1850. https://doi.org/10.1002/adsc.200700259

Li C, Wang Q, Zhao ZK (2008) Acid in ionic liquid: an efficient system for hydrolysis of lignocellulose. Green Chem 10(2):177–182. https://doi.org/10.1039/B711512A

Li C, Knierim B, Manisseri C, Arora R, Scheller HV, Auer M, Vogel KP, Simmons BA, Singh S (2010) Comparison of dilute acid and ionic liquid pretreatment of switchgrass: biomass recalcitrance, delignification and enzymatic saccharification. Bioresour Technol 101:4900–4906. https://doi.org/10.1016/j.biortech.2009.10.066

Li W, Sun N, Stoner B, Jiang X, Lu X, Rogers RD (2011) Rapid dissolution of lignocellulosic biomass in ionic liquids using temperatures above the glass transition of lignin. Green Chem 13(8):2038–2047. https://doi.org/10.1039/c1gc15522a

Li HQ, Jiang W, Jia JX, Xu J (2014) pH pre-corrected liquid hot water pretreatment on corn stover with high hemicelluloses recovery and low inhibitors formation. Bioresour Technol 153:292–299

Li L, Chen C, Zhang R, He Y, Wang W, Liu G (2015) Pretreatment of corn stover for methane production with the combination of potassium hydroxideand calcium hydroxide. Energy Fuel 29:5841–5846. https://doi.org/10.1021/acs.energyfuels.5b01170

Liebert T (2008) CELL 139-Cellulose solvents: remarkable history and bright future. Abstracts of papers of the American Chemical Society, 235

Liu L, Chen H (2006) Enzymatic hydrolysis of cellulose materials treated with ionic liquid [BMIM]Cl. Chinese Sci Bull 51(20):2432–2436. https://doi.org/10.1007/s11434-006-2134-9

Liu ZS, Wu XL, Kida K, Tang YQ (2012) Corn stover saccharification with concentrated sulfuric acid: effects of saccharification conditions on sugar recovery and by-product generation. Bioresour Technol 119:224–233

Liu Y, Sun B, Zheng X, Yu L, Li J (2018) Integrated microwave and alkaline treatment for the separation between hemicelluloses and cellulose from cellulosic fibers. Bioresour Technol 247:859–863

Liyakathali NAM, Muley PD, Aita G, Boldor D (2016) Effect of frequency and reaction time in focused ultrasonic pretreatment of energy cane bagasse for bioethanol production. Bioresour Technol 200:262–271

Lloyd TA, Wyman CE (2005) Combined sugar yields for dilute sulfuric acid pretreatment of corn stover followed by enzymatic hydrolysis of the remaining solids. Bioresour Technol 96:1967–1977. https://doi.org/10.1016/j.biortech.2005.01.011

Mancini G, Papirio S, Lens PNL, Esposito G (2016) Solvent pretreatments of lignocellulosic materials to enhance biogas production: a review. Energy Fuel 30:1892–1903

Maurya DP, Singla A, Negi S (2015) An overview of key pretreatment processes for biological conversion of lignocellulosic biomass to bioethanol. 3 Biotech 5:597–609

McDonough TJ (1993) The chemistry of organosolv delignification. TAPPI J 76:186–193

McMillan JD (1993) Pretreatment of lignocellulosic biomass. In: Himmel ME, Baker JO, Overend RP (eds) Enzymatic conversion of biomass for fuel production. American Chemical Society, Washington, DC, pp 292–323

McMillan JD (1994) Pretreatment of lignocellulosic biomass. In: Himmel ME, Baker JO, Overend RP (eds) *Enzymatic Conversion of Biomass for Fuels Production*. ACS, Washington, DC, pp 292–324

Millet MA, Baker AJ, Scatter LD (1976) Physical and chemical pretreatment for enhancing cellulose saccharification. Biotech Bioeng Symp 6:125–153

Mirahmadi K, Mohseni Kabir M, Jeihanipour A, Karimi K, Taherzadeh M (2010) Alkaline pretreatment of spruce and birch to improve bioethanol and biogas production. Bioresources 5:928–938

Modenbach AA, Nokes SE (2012) The use of high-solids loadings in biomass pretreatment-a review. Biotechnol Bioeng 109(6):1430–1442

Mosier NS, Wyman C, Dale B, Elander R, Lee YY, Holtzapple M, Ladisch MR (2005) Features of promising technologies for pretreatment of lignocellulosic biomass. Bioresour Technol 96:673–686

Mtui GYS (2012) Oxalic acid pretreatment, fungal enzymatic saccharification and fermentation of maize residues to ethanol. Afr J Biotechnol 11(4):843–851

Mueller-Hartlev I, Hartlev RD, Harris PJ, Curzon EH (1986) Linkage of p-coumaroyi and feruloyl groups to cell-wall polysaccharides of barley straw. Carbohydr Res 148:71–85

Muhammad N, Man Z, Bustam MA, Mutalib MIA, Rafiq S (2013) Investigations of novel nitrile-based ionic liquids as pre-treatment solvent for extraction of lignin from bamboo biomass. J Ind Eng Chem 19:207–214

Narayanaswamy N, Faik A, Goetz DJ, Gu T (2011) Supercritical carbon dioxide pretreatment of corn stover and switchgrass for lignocellulosic ethanol production. Bioresour Technol 102:6995–7000

Neely WC (1984) Factors affecting the pretreatment of biomass with gaseous ozone. Biotechnol Bioeng 26:59–65

Nguyen TA, Kim KR, Han SJ, Cho HY, Kim JW, Park SM et al (2010) Pretreatment of rice straw with ammonia and ionic liquid for lignocellulose conversion to fermentable sugars. Bioresour Technol 101:7432–7438

O'Connor RP, Woodley R, Kolstad JJ, Kean R, Glassner DA, Mastel B, Ritzenthaler JM, John H, Warwick J, Hettenhaus JR, Brooks RK (2007) Process for fractionating ligno cellulosic biomass into liquid and solid products. assignee U. S. A. Nature-works LLC, patent number WO 2007120210

O'Donovan A, Gupta VK, Tuohy MG (2013) Recent updates in acid pretreatments and SEM analysis of acid pretreated grass biomass. In: Gupta VK, Tuohi MG (eds) Biofuel technologies: recent development. Springer Science Publishers, Berlin, pp 97–118

Ouyang X, Zhu G, Huang X, Qiu X (2015) Microwave assisted liquefaction of wheat straw alkali lignin for the production of monophenolic compounds. J Energy Chem 24:72–76

Overend RP, Chornet E (1987) Fractionation of lignocellulosics by steam-aqueous pretreatments. Phil Trans R Soc A 321(1561):523–536

Pan X, Arato C, Gilkes N et al (2005) Biorefining of softwoods using ethanol organosolv pulping: preliminary evaluation of process streams for manufacture of fuel-grade ethanol and coproducts. Biotechnol Bioeng 90(4):473–481

Pan X, Gilkes N, Kadla J et al (2006) Bioconversion of hybrid poplar to ethanol and co-products using an organosolv fractionation process: optimization of process yields. Biotechnol Bioeng 94(5):851–861

Park YC, Kim JS (2012) Comparison of various alkaline pretreatment methods of lignocellulosic biomass. Energy 47:31–35

Park N, Kim H-Y, Koo B-W, Yeo H, Choi I-G (2010) Organosolv pretreatment with various catalysts for enhancing enzymatic hydrolysis of pitch pine (Pinus rigida). Bioresour Technol 101(18):7046–7053

Patil JH, AntonyRaj M, Gavimath CC (2011) Study on effect of pretreatment methods on biomethanation of water hyacinth. Int J Adv Biotechnol Res 2:143–147

Paul S, Dutta A (2018) Challenges and opportunities of lignocellulosic biomass for anaerobic digestion. Resour Conserv Recycl 130:164–174

Pena L, Ikenberry M, Hohn KL, Wang D (2012) Acid-functionalized nanoparticles for pretreatment of wheat straw. J Biomater Nanobiotechnol 3:342–352

Pena L, Xu F, Hohn KL, Li J, Wang D (2014) Propyl-sulfonic acid functionalized nanoparticles as catalyst for pretreatment of corn stover. J Biomater Nanobiotechnol 5:8–16

Petkovic M, Hartmann DO, Adamova G, Seddon KR, Rebelo LPN, Pereira CS (2012) Unravelling the mechanism of toxicity of alkyltributylphosphonium chlorides in *Aspergillus nidulans* conidia. New J Chem 36(1):56–63

Pielhop T, Amgarten J, Rohr PR, Studer MH (2016) Steam explosion pretreatment of softwood: the effect of the explosive decompression on enzymatic digestibility. Biotechnol Biofuels 9:152

Pinkert A, Goeke DF, Marsh KN, Pang S (2011) Extracting wood lignin without dissolving or degrading cellulose: investigations on the use of food additive-derived ionic liquids. Green Chem 13:3124–3136

Procentese A, Johnson E, Orr V, Campanile AG, Wood JA, Marzocchella A, Rehmann L (2015) Deep eutectic solvent pretreatment and subsequent saccharification of corncob. Bioresour Technol 192:31–36

Qing Q, Huang M, He Y, Wang L, Zhang Y (2015) Dilute oxalic acid pretreatment for high total sugar recovery in pretreatment and subsequent enzymatic hydrolysis. Appl Biochem Biotechnol 177:1493–1507

Rabemanolontsoa H, Saka S (2016) Various pretreatments of lignocellulosics. Bioresour Technol 199:83–91

Rai M, dos Santos JC, Soler MF, Marcelino PRF, Brumano LP, Ingle AP, Gaikwad S, Gade A, da Silva SS (2016) Strategic role of nanotechnology for production of bioethanol and biodiesel. Nanotechnol Rev 5(2):231–250

Rajendran K, Drielak E, Varma VS, Muthusamy S, Kumar G (2017) Updates on the pretreatment of lignocellulosic feedstocks for bioenergy production–a review. Biomass Conver Biorefin 8:471–483

Ravindran R, Jaiswal AK (2016) A comprehensive review on pre-treatment strategy for lignocellulosic food industry waste: challenges and opportunities. Bioresour Technol 199:92–102

Redding AP, Wang Z, Keshwani RD, Cheng J (2011) High temperature dilute acid pretreatment of coastal Bermuda grass for enzymatic hydrolysis. Bioresour Technol 102:1415–1424

Ren HW, Chen CM, Wang QH, Zhao DS, Guo SH (2016) The properties of choline chloride-based deep eutectic solvents and their performance in the dissolution of cellulose. Bioresources 11(2):5435–5451

Rico A, Rencoret J, Río JCD, Martínez A, Gutiérrez A (2014) Pretreatment with laccase and a phenolic mediator degrades lignin and enhances saccharification of Eucalyptus feedstock. Biotechnol Biofuels 7(1):6

Rinaldi R, Palkovits R, Schüth F (2008) Depolymerization of cellulose using solid catalysts in ionic liquids. Angew Chem Int Edit 47(42):8047–8050. https://doi.org/10.1002/anie.200802879

Rosenau T, Potthast A, Adorjan I, Hofinger A, Sixta H, Firgo H, Kosma P (2002) Cellulose solutions in N-methylmorpholine-Noxide (NMMO)–degradation processes and stabilizers. Cellulose 9:283–291

Saha BC, Biswas A, Cotta MA (2008) Microwave pretreatment, enzymatic saccharification and fermentation of wheat straw to ethanol. J Biobaased Mater Bioenergy 2:210–217. https://doi.org/10.1166/jbmb.2008.412

Sahay S, Rana RS (2017) Hemicellulose hydrolysate from *Ailanthes excelsa* wood potentially fermentable to ethanol. J Trop For Res 29(2):172–178

Sahay S, Chauhan D, Chourse V (2019) Pb activated laccase from thermotolerant *Aspergillus nidulans* TTF6 showing Pb activation for smaller substrates and dyes remediation in all climates. Proc Natl Acad Sci Biol Sci. https://doi.org/10.1007/s40011-019-01092-y

Sahoo D, Ummalyma SB, Okram AK, Pandey A, Sankar M, Sukumaran RK (2018) Effect of dilute acid pretreatment of wild rice grass (Zizania latifolia) from Loktak Lake for enzymatic hydrolysis. Bioresour Technol 253:252–255

Samuel R, Pu Y, Foston M, Ragauskas AJ (2010) Solid-state NMR characterization of switchgrass cellulose after dilute acid pretreatment. Biofuels 1(1):85–90

Sánchez OJ, Cardona CA (2008) Trends in biotechnological production of fuel ethanol from different feedstocks. Bioresour Technol 99:5270–5295

Saritha M, Arora A, Nain L (2012) Pretreatment of paddy straw with Trametes hirsuta for improved enzymatic saccharification. Bioresour Technol 104:459–465

Sathitsuksanoh N, Zhu Z, Zhang YHP (2012) Cellulose solvent-based pretreatment for corn stover and avicel: concentrated phosphoric acid versus ionic liquid [BMIM]Cl. Cellulose 19(4):1161–1172

Satimanont S, Luengnaruemitchai A, Wongkasemjit S (2012) Effect of temperature and time on dilute acid pretreatment of corn cobs. Int J Chem Biol Eng 6:333–337

Schell D, Harwood C (1994) Milling of lignocellulosic biomass. Appl Biochem Biotechnol 45–46(1):159–168

Serna LD, Alzate CO, Alzate CC (2016) Supercritical fluids as a green technology for the pretreatment of lignocellulosic biomass. Bioresour Technol 199:113–120

Shafiei M, Karimi K, Taherzadeh MJ (2010) Pretreatment of spruce and oak ny N-methylmorpholine-N-oxide (NMMO) for efficient conversion of their cellulose to ethanol. Bioresour Technol 101:4914–4918

Shahbazi A, Zhang B (2010) Dilute and concentrate acid hydrolysis of lignocellulosic biomass. In: Waldron K (ed) Bioalcohol production, Woodhead publishing series in energy. Woodhead Publishing, Boca Raton, pp 143–158

Shibuya N, Iwasaki T (1985) Structural features of rice bran hemicellulose. Phytochemistry 24:285–289

Shill K, Padmanabhan S, Xin Q, Prausnitz JM, Clark DS, Blanch HW (2011) Ionic liquid pretreatment of cellulosic biomass: enzymatic hydrolysis and ionic liquid recycle. Biotechnol Bioeng 108(3):511–520

Shirkavand E, Baroutian S, Gapes DJ, Young BR (2016) Combination of fungal and physicochemical processes for lignocellulosic biomass pretreatment–a review. Renew Sust Energ Rev 54:217–234

Sindhu R, Binod P, Pandey A (2016) Biological pretreatment of lignocellulosic biomass – an overview. Bioresour Technol 199:76–82

Sjöström E (1993) Wood Chemistry – fundamentals and applications, 2nd edn. Academic Press, New York

Smith EL, Abbott AP, Ryder KS (2014) Deep eutectic solvents (DESs) and their applications. Chem Rev 114(21):11060–11082

Stepan AM, Monshizadeh A, Hummel M, Roselli A, Sixta H (2016) Cellulose fractionation with IONCELL-P. Carbohydr Polym 150:99–106

Strong PJ, Calus H (2011) Laccase: a review of its past and its future in bioremediation. Crit Rev Environ Sci Technol 41:373–434

Sun Y, Cheng J (2002) Hydrolysis of lignocellulosic materials for ethanol production: a review. Bioresour Technol 83:1–11

Sun N, Rahman M, Qin Y, Maxim ML, Rodríguez H, Rogers RD (2009) Complete dissolution and partial delignification of wood in the ionic liquid 1-ethyl-3- methylimidazolium acetate. Green Chem 11(5):646–655. https://doi.org/10.1039/b822702

Swatloski RP, Spear SK, Holbrey JD, Rogers RD (2002) Dissolution of cellulose with ionic liquids. J Am Chem Soc 124(18):4974–4975. https://doi.org/10.1021/ja025790m

Tan SSY, MacFarlane DR, Upfal J, Edye LA, Doherty WOS, Patti AF, Pringle JM, Scott JL (2009) Extraction of lignin from lignocellulose at atmospheric pressure using alkylbenzenesulfonate ionic liquid. Green Chem 11:339–345

Tang J, Chen K, Xu J, Li J, Zhao C (2011) Effects of dilute acid hydrolysis on composition and structure of cellulose in eulaliopsis binata. Bioresources 6(2):1069–1078

Tanjore D, Richard TL (2015) A systems view of lignocellulose hydrolysis. In: Ravindra P (ed) Advances in bioprocess technology. Springer International Publishing AG, Cham, pp 36–50

Tao L, Aden A, Elander RT et al (2011) Process and technoeconomic analysis of leading pretreatment technologies for lignocellulosic ethanol production using switchgrass. Bioresour Technol 102(24):11105–11114

Tarkow H, Feist WC (1969) A mechanism for improving the digestibility of lignocellulosic materials with dilute alkali and liquid ammonia. Adv Chem 95:197–218. https://doi.org/10.1021/ba-1969-0095.ch012

Tayyab M, Noman A, Islam W, Waheed S, Arafat Y, Ali F et al (2018) Bioethanol production from lignocellulosic biomass by environment friendly pretreatment methods: a review. Appl Ecol Environ Res 16:225–249

Thermowoodhandbook (2003) Helsinki, Finnish Thermowood Association.

Tian L, Branford-White C, Wang W, Nie H, Zhu L (2012) Laccase-mediated system pretreatment to enhance the effect of hydrogen peroxide bleaching of cotton fabric. Int J Biol Macromol 50:782–787

Timell TE (1967) Recent progress in the chemistry of wood hemicelluloses. Wood Sci Technol 1:45–70

Tutt M, Kikas T, Olt J (2012) Influence of different pretreatment methods on bioethanol production from wheat straw. Agron Res Biosyst Eng 1:269–276

Tutt M, Kikas T, Olt J (2012a) Comparison of different pretreatment methods on degradation of Rye straw, Engineering for Rural Development, Jelgava, Latvia

Tutt M, Kikas T, Olt J (2012b) Influence of different pretreatment methods on bioethanol production from wheat straw. Agron Res Biosys Eng 1:269–276

van Groenestijn JW, Hazewinkel JHO, Bakker RR (2007) Pre-treatment of ligno-cellulose with biological acid recycling (The biosulfurol process). Netherlands Organisation for Applied Scientific Research (TNO), Zeist

Vidal PF, Molinier J (1988) Ozonolysis of lignin-improvement of in vitro digestibility of poplar saw dust. Biomass 16:1–17

Villaverde JJ, Ligero P, de Vega A (2010) Miscanthus x giganteus as a source of biobased products through organosolv fractionation: a mini review. Open Agric J 4:102–110

Wang W, Ji S, Lee I (2013) Fast and efficient nanoshear hybrid alkaline pretreatment of corn stover for biofuel and materials production. Biomass Bioenergy 51:35–42

Wang H, Covarrubias J, Prock H, Wu X, Wang D, Bossmann SH (2015) Acid-functionalized magnetic nanoparticle as heterogeneous catalysts for biodiesel synthesis. J Phys Chem C 119:26020–26028

Wang Y, Fan L, Shan S, Liu Y, Ruan R (2016) Review of microwave assisted lignin conversion for renewable fuels and chemicals. J Anal Appl Pyrolysis 119:104–113

Wang Q, Tian D, Hu J, Shen F, Yang G, Zhang Y, Deng S, Zhang J, Zeng Y, Hu Y (2018) Fates of hemicellulose, lignin and cellulose in concentrated phosphoric acid with hydrogen peroxide (PHP) pretreatment. RSC Adv 2018(8):12714–12723

Wen JL, Xue BL, Xu F, Sun RC, Pinkert A (2013) Unmasking the structural features and property of lignin from bamboo. Ind Crop Prod 42:332–343

Wu H, Mora-Pale M, Miao J, Doherty TV, Linhardt RJ, Dordick JS (2011) Facile pretreatment of lignocellulosic biomass at high loadings in room temperature ionic liquids. Biotechnol Bioeng 108(12):2865–2875. https://doi.org/10.1002/bit.23266

Xie H, Zhao ZK (2011) Selective breakdown of (ligno)cellulose in ionic liquids. In: Kokorin A (ed) Ionic liquids: applications and perspectives. InTech, Rijeka, pp 61–80. https://doi.org/10.5772/14681

Xie H, Liu W, Beadham I, Gathergood N (2012) Biorefinery with ionic liquids. In: Xie H, Gathergood N (eds) The role of green chemistry in biomass processing and conversion. Wiley, Hoboken, pp 75–133. https://doi.org/10.1002/9781118449400.ch3

Xu J (2015) Microwave pretreatment. In: Pandey A, Negi S, Binod P, Larroch C (eds) Pretreatment of biomass processes and technologies, 1st edn. Elsevier, Waltham, pp 157–172

Xu J, Zhang X, Cheng JJ (2012) Pretreatment of corn stover for sugar production with switchgrass-derived black liquor. Bioresour Technol 111:255–260

Xu P, Zheng GW, Du PX, Zong MH, Lou WY (2016) Whole-cell biocatalytic processes with ionic liquids. ACS Sustain Chem Eng 4(2):371–386

Xu P, Zheng GW, Zong MH, Li N, Lou WY (2017) Recent progress on deep eutectic solvents in biocatalysis. Bioresour Bioprocess 4:34

Yadav P, Mundada G, Biswas B, Srivastava V, Singh R, Krishna BB, Kumar J, Bhaskar T (2016) Microwave assisted pyrolysis of biomass for liquid biofuels production: In Biofuels: production and future prospectives (Ram Sarup Singh, Ashok Pandey, Edgard Gnansounou) Boca Raton: CRC Press pp 164-182

Yan P, Xu Z, Zhang C, Liu X, Xu W, Zhang ZC (2015) Fractionation of lignin from eucalyptus bark using amine-sulfonate functionalized ionic liquids. Green Chem 17:4913–4920

Yang B, Tao L, Wyman CE (2018) Strengths, challenges, and opportunities for hydrothermal pretreatment in lignocellulosic biorefineries. Biofpr Biofuels Bioprod Biorefin 12(1):125–138
Zamocky M, Gasselhuber B, Furtmüller PG, Obinger C (2014) Turning points in the evolution of peroxidase-catalase superfamily: molecular phylogeny of hybrid heme peroxidises. Cell Mol Life Sci 71:4681–4696
Zavarzina AG, Lisov AV, Leontievsky AA (2018) The role of ligninolytic enzymes laccase and a versatile peroxidase of the white-rot fungus Lentinus tigrinus in biotransformation of soil humic matter: comparative in vivo study. J Geophys Res Biogeosci 123:2727–2742
Zdanowick M, Wilpiszewska K, Spychaj T (2018) Deep eutectic solvents for polysaccharides processing. A review. Carbohydr Polym 200:361–380
Zhang ZC (2013) Catalytic transformation of carbohydrates and lignin in ionic liquids. WIREs Energy Environ 2(6):655–672. https://doi.org/10.1002/wene.67
Zhang S, Sun N, He X, Lu X, Zhang X (2006) Physical properties of ionic liquids: database and evaluation. J Phys Chem Ref Data 35(4):1475–1517
Zhang YHP, Ding S-Y, Mielenz JR et al (2007) Fractionating recalcitrant lignocellulose at modest reaction conditions. Biotechnol Bioeng 97(2):214–223
Zhang YHP, Zhu Z, Rollin R, Sathitsuksanoh N (2010) Advances in cellulose solvent- and organic solvent-based lignocelluloses fractionation (COSLIF) in cellulose solvents. In: Liebert T, Heinze T, Edgar K (eds) Cellulose solvents: for analysis, shaping and chemical modification, American Chemical Society symposium series, vol 1033. American Chemical Society, Washington, DC, pp 365–379
Zhang D, Ong YL, Li Z, Wu JC (2012) Optimization of dilute acid-catalyzed hydrolysis of oil palm empty fruit bunch for high yield production of xylose. Chem Eng J 181–182:636–642
Zhang J, Wang Y, Zhang L, Zhang R, Liu G, Cheng G (2014a) Understanding changes in cellulose crystalline structure of lignocellulosic biomass during ionic liquid pretreatment by XRD. Bioresour Technol 151:402–405
Zhang Y, Zhao J, Xu F, Li Y (2014b) Pretreatment of lignocellulosic biomass for enhanced biogas production. Prog Energy Combust Sci 42:35–53. https://doi.org/10.1016/j.pecs.2014.01.001
Zhang CW, Xia SQ, Ma PS (2016) Facile pretreatment of lignocellulosic biomass using deep eutectic solvents. Bioresour Technol 219:1–5
Zhang L, Lu H, Yu J, Wang Z (2017) Dissolution of Lignocelluloses with a high lignin content in a *N*-Methylmorpholine-*N*-oxide Monohydrate solvent system via simple glycerol-swelling and mechanical pretreatments. J Agric Food Chem 65(44):9587–9594
Zhao Z, Liu D (2012) Fractionating pretreatment of sugarcane bagasse by aqueous formic acid with direct recycle of spent liquor to increase cellulose digestibility-the Formiline process. Bioresour Technol 117:25–32
Zhao DB, Liao YC, Zhang ZD (2007) Toxicity of ionic liquids. Clean Soil Air Water 35(1):42–48
Zhao C, Ding W, Chen F, Cheng C, Shao Q (2014) Effects of compositional changes of AFEX-treated and H-AFEX-treated corn stover on enzymatic digestibility. Bioresour Technol 155:34–40
Zhao BY, Xu P, Yang FX, Wu H, Zong MH, Lou WY (2015) Biocompatible deep eutectic solvents based on choline chloride: characterization and application to the extraction of rutin from Sophora japonica. ACS Sustain Chem Eng 3(11):2746–2755
Zheng M, Li X, Li L, Yang X, He Y (2009) Enhancing anaerobic biogasification of corn stover through wet state NaOH. Bioresour Technol 100: 5140-5145
Zhou M, Xu J, Jiang J, Sharma BK (2018) A review of microwave assisted liquefaction of lignin in hydrogen donor solvents: effect of solvents and catalysts. Energies 11(11):2877
Zhu S, Wu Y, Chen Q, Yu Z, Wang C, Jin S (2006) Dissolution of cellulose with ionic liquids and its application: a mini-review. Green Chem 8:325–327. https://doi.org/10.1039/b601395c
Zhuang X, Wang W, Yu Q, Qi W, Wang Q, Tan X et al (2016) Liquid hot water pretreatment of lignocellulosic biomass for bioethanol production accompanying with high valuable products. Bioresour Technol 199:68–75

Impact of Pretreatment Technology on Cellulosic Availability for Fuel Production

8

Nesrine BenYahmed, Mohamed Amine Jmel, and Issam Smaali

Abstract

Cellulosic substrates represent an attractive source for the production of renewable energies such as biofuels. However, the rigid structure of most of these substrates characterized by the presence of lignin presents an obstacle for their bioconversion. Therefore, a pretreatment step aimed at the degradation of these cellulosic feedstocks making the cellulose available to hydrolysis is required for biofuels production. This crucial step influences the whole biofuels production process as well as co-products. Currently, various pretreatment techniques have been developed. They affect the chemical composition and the physical structure of the biofuels substrates and increase hydrolysis rates. In this chapter, numerous pretreatment process methods for treatment of different lignocellulosic and algal biomasses are presented. Their impacts on these cellulosic substrates are demonstrated and their advantages and limitations are discussed.

Keywords

Cellulosic substrates · Biofuels · Chemical composition · Structure · Pretreatment techniques

8.1 Introduction

Nowadays, according to the International Energy Agency (IEA), meeting the increasing demand for fossil fuels will require the replacement of existing deposits, which will need significant investments. The realization of these investments will have a direct effect on the increase in the price of a barrel. Given that resources are

N. BenYahmed (✉) · M. A. Jmel · I. Smaali
Laboratoire d'ingénierie des protéines et des molécules bioactives, Université de Carthage, INSAT-BP 676, Centre urbain nord, Carthage Cedex, Tunisia

N. Srivastava et al. (eds.), *Substrate Analysis for Effective Biofuels Production*, Clean Energy Production Technologies, https://doi.org/10.1007/978-981-32-9607-7_8

concentrated in a number of producing countries, wide consuming countries will therefore have problems of security of supply and will be very dependent on imports (AIE 2016). On the other hand, according to the Intergovernmental Panel on Climate Change, global warming and the emission of greenhouse gases (GHG) are caused by human activities and mainly the activities of the energy sector (Percebois and Mandil 2012).

Thus, in view of the world's population growth and global warming problem associated with the decrease of fossil sources, the production of renewable energies such as biofuels represents an increased interest. Lignocellulosic and algal substrates are promising renewable biomasses for the production of second- and third-generation biofuels. They represent an abundant source of unutilized biomass and their availability does not necessarily affect arable areas use. However, the rigid and highly organized crystalline structure of cellulose represents an obstacle to its hydrolysis (Ben Yahmed et al. 2017). Besides, most potential cellulosic substrates for biofuels production are heavily lignified (Pandey 2008). Thus, physical or chemical pretreatment aimed at the destructuration of the lignocellulosic matrix consisting of cellulose, hemicelluloses, and lignin to render the cellulose accessible to the hydrolytic enzymes seems to be essential prior to their energetic bioconversion. Biomass conversion process includes five major steps: choice of suitable biomass, effective pretreatment, production of hydrolytic enzymes, fermentation of hexoses and pentoses, and downstream processing (Menon and Rao 2012). Pretreatment of biomass is considered one of the most expensive steps in the whole process. With the high progress in developing lignocellulose biomass- and algal biomass-based biorefineries, global interest has been on developing pretreatment methods and technologies that are technico-economically feasible (Binod and Pandey 2015).

The search for effective pretreatment methods has been recognized as one of the main axes for the processing of biomass to biofuels. The choice of pretreatment process depends on the nature of cellulosic material, namely, its biochemical composition and structure, the cost of this process, and its environmental impact. Pretreatments can be divided into physical, chemical, physicochemical, and biological methods, but there is a strong interdependence between these processes (Bajpai 2016; Menon and Rao 2012; Monlau et al. 2013).

Various pretreatment techniques are developed such as mechanical treatment, steam explosion, ammonia fiber explosion (AFEX), acid or alkaline pretreatment, organosolv pretreatment, biological treatment using fungi or enzymes, etc. (Ben Yahmed et al. 2016, 2017; Chen et al. 2015,2016; Cheng and Timilsina 2011; Jmel et al. 2017a; Kumar et al. 2009; Menon and Rao 2012; Rodriguez et al. 2015; Zhang et al. 2007).

Nevertheless, some technical and scientific issues related to pretreatment and hydrolysis remain to be focused on.

8.2 Cellulosic Biomass for Biofuels Production: Structure and Composition

Understanding cellulosic biomass feedstock composition and structure is mandatory for developing effective pretreatment technologies and therefore ensuring its best conversion to fuels. Different biomasses are used for the production of the three biofuels types (first, second, and third generations).

8.2.1 Biomass for First-Generation Biofuels

First-generation biofuels are produced from sugars or starchy residues such as sugarcane in Brazil and maize seeds and sugar beet in the United States and Europe for bioethanol and by transesterification of vegetable oils for biodiesel (Fatih Demirbas 2009). This makes the countries of Europe, the United States, and Brazil the strongest producers in the world of biofuels (Subhadra and Edwards 2010).

The structural simplicity of this biomass makes it easily fermentable with good production yields, but the problem with its use is the competition with the food sector and the need for large areas of arable land (Ruane et al. 2010).

8.2.2 Biomass for Second-Generation Biofuels

Second-generation biofuels are produced from lignocellulosic biomass. Lignocellulosics are the most abundant source of misused biomass and their availability does not affect land use (Monlau et al. 2013).

Lignocellulose is the primary building block of plant cell walls consisting mainly of three polymeric components: cellulose, hemicelluloses, and lignin.

Cellulose, the main structural constituent in plant cell walls, is found in an organized fibrous structure (Kumar et al. 2009). These fibrils are attached to each other by hemicelluloses and amorphous polymers and covered by lignin. The cellulose microfibrils are devised in crystalline and amorphous regions. Crystalline cellulose comprises the major proportion of cellulose, while a lesser percentage of unorganized cellulose chains form amorphous cellulose. The crystalline cellulose is more resistant to biodegradation than the amorphous parts (Menon and Rao 2012). Cellulose $[C_6H_{10}O_5]_n$ is a linear homopolymer formed by the sequence of cellobiose units (Fig. 8.1), a unit composed of two D-glucopyranose linked by a β- (1 → 4) glucosidic bond.

The degree of polymerization (DP) which represents the number of glucosidic units per cellulose chain is between 100 and 20,000 depending on its origin and its location within the cell wall (Kolpak and Blackwell 1976). Cellulose with low degree of polymerization (DP) will be more susceptible to cellulolytic enzymes (Menon and Rao 2012).

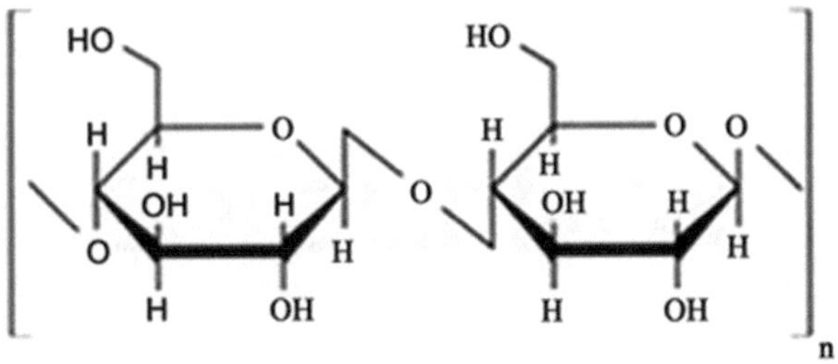

Fig. 8.1 Illustration of a cellulose chain (Muktham et al. 2016)

Hemicelluloses or polyoses represent, after cellulose, the most abundant polysaccharide in nature. Unlike cellulose, they cannot be described succinctly because of their high diversity. In addition, their structure depends on their varietal origin, the tissue or cell type, the age of the cells, and their location in the plant wall (Hoch 2007; Sun et al. 1996). They differ from cellulose due to the heterogeneity of their monosaccharide composition. These are heteropolysaccharides consisting of shorter molecular chains with a degree of polymerization often less than 200 (Yang 2007). They have branches with short lateral chains that consist of different sugar monomers such as xylose, mannose, galactose, rhamnose, and arabinose (Fig.8.2) which can be easily hydrolyzed by several hemicellulase enzymes as well as by chemicals (dilute acid or base). While cellulose is crystalline and resistant to hydrolysis, hemicelluloses have a random, amorphous structure with little strength. Hemicelluloses have a lower molecular weight than cellulose (Ren and Sun 2010; Sun et al. 1996).

Lignin, a major component of plant walls, represents the third most abundant polymer after polysaccharides (cellulose and hemicelluloses). Lignin is an amorphous phenolic polymer with high molecular weight, closely associated with cellulose in lignocellulosic biomass. It provides plants with structural support, ensures impermeability, and provides resistance against microbial attack (Perez et al. 2003). This three-dimensional polymer comes from the enzymatic radical copolymerization of three phenylpropanoic alcohols: coniferyl alcohol, sinapyl alcohol, and coumaryl alcohol (or p-hydroxycinnamyl alcohol) (Fig. 8.3). These precursors are produced by plants from L-tyrosine and L-phenylalanine (Monlau et al. 2013).

Lignocellulosic biomass generally consists of 40–50% cellulose, 25–30% hemicelluloses, and 15–20% lignin and other extractable components (Knauf and Moniruzzaman 2004). However, this composition varies depending on the type of the feedstock. The compositions of various lignocellulosic substrates are presented in Table 8.1.

8.2.3 Biomass for Third-Generation Biofuels

In recent years, several studies are interested in finding an alternative to the use of lignocellulosic biomass. In fact, algal biomasses, namely, macro- and microalgae, have received increased attention as source of third-generation biofuels (Ben Yahmed et al. 2016; Jard et al. 2013; Jazzar et al. 2015; Jmel et al. 2017a; Montingelli et al. 2016;

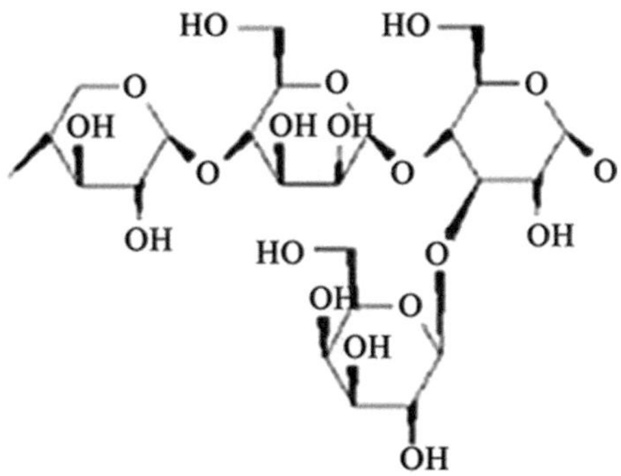

Fig. 8.2 Hemicellulose structure (Muktham et al. 2016)

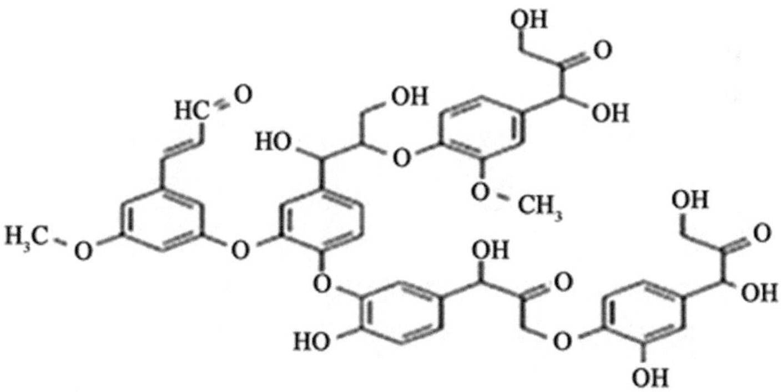

Fig. 8.3 Lignin polymer structure (Muktham et al. 2016)

Trivedi et al. 2013). Algal substrates possess several advantages including characteristic composition rich on carbohydrates with low lignin content making them suitable substrates for biofuels production (Chen et al. 2015; Jones and Mayfield 2012), no competition with land use and water consumption necessary for food crops (Singh et al. 2011), easy harvest with low cost of collection and null environmental damage (Ruiz et al. 2013), and having higher photosynthetic activity (6–8%) than terrestrial biomass (1.8–2.2%) (Ruiz et al. 2013).

The cell wall of algae differs significantly in its organization and biochemical composition from those of other plant organisms. It generally has a bi-phasic structure composed of a crystalline fibrous skeletal phase and a matrix phase whose structure and composition are variable according to the species (Robic et al. 2009). The fibrous phase is ordered and embedded in a mucilaginous amorphous matrix. Microfibrils are generally composed of β-1,4-glucopyranoses which form cellulosic chains arranged in layers and stabilized by hydrogen bonds. Other polymers such as xylans participate, just like cellulose, in the structuring of the cell wall and protect the cells from the hydrodynamics to which they are subjected. The hydrosoluble matrix phase which is sometimes associated with proteins is very complex in nature. The amorphous *Chlorophyceae* matrix consists of ulvan, a high-molecular-weight

Table 8.1 Composition of common lignocellulosic substrates

Feedstocks	Carbohydrates composition (% dry wt)			References
	Cellulose	Hemicelluloses	Lignin	
Barley hull	34	36	19	Kim et al. (2008)
Barley straw	36–43	24–33	6.3–9.8	García-Aparicio et al. (2006)
Banana waste	13	15	14	Monsalve et al. (2006)
Corn stover	35.1–39.5	20.7–24.6	11–19	Mosier et al. (2005)
Corn stalk	31	11	30	Rubio et al. (1998)
Hardwood stems	40–55	24–40	18–25	Howard et al. (2003)
Rice straw	29.2–34.7	23–25.9	17–19	Brylev et al. (2001)
Rice husk	28.7–35.6	11.9–29.3	15.4–20	Abbas and Ansumali (2010)
Wheat straw	35–39	22–30	12–16	Prasad et al. (2007)
Wheat bran	10.5–14.8	35.5–39.2	8.3–12.5	Miron et al. (2001)
Grasses	25–40	25–50	10–30	Howard et al. (2003)
Sugarcane bagasse	25–45	28–32	15–25	Alves et al. (2010)
Poplar wood	45–51	25–28	10–21	Torget and Teh-An (1994)
Switchgrass	35–40	25–30	15–20	Howard et al. (2003)

Adapted from Menon and Rao (2012)

sulfated polyanionic polysaccharide, composed mainly of neutral oses (xylose, glucose, galactose, and rhamnose) and a small amount of uronic acids (galacturonic and glucuronic acids) (Ray 2006). Figure 8.4 shows the disposition of the different polysaccharide compounds of the cell wall of *Ulva* sp.

The biochemical composition varies depending on the species of the algae. The main composition of different algal feedstocks is shown in Table 8.2.

Macroalgae contain generally only 10–15% dry matter including approximately 60% carbohydrates (Chen et al. 2015) which makes them suitable substrates for biofuels production.

However, the biochemical composition of macroalgae depends greatly on the growth conditions, the harvest period, and the pollution of water bodies (Lamare and Wing 2001). Approximately, according to Ben Yahmed et al., *Ulva* sp. collected in March consists of 33.2 ± 0.8% carbohydrate containing 12.4 ± 0.2% of glucose based on total solids and high protein content (11.4 ± 0.5%TS) compared to a low content of lipids (1.8 ± 0.05%TS) and an absence of lignin (Ben Yahmed et al. 2017). Similarly, it was demonstrated by Briand and Morand that the green macroalga *Ulva* sp. possesses very low content of lignin associated with a high fraction of hemicelluloses and ulvans which give a characteristic structure, advancing accessibility to saccharolytic enzymes with a good hydrolysis output (Briand and Morand 1997).

About the algae cellulose crystallinity, which is an important factor in biomass bioconversion, green macroalgae species can be divided into three major classes taking into account their cell wall constituents (Nicolai and Preston 1952). The first class includes green algae characterized by highly crystalline native cellulose representing the main constituent of their cell walls (*Cladophora*, *Chaetomorpha*,

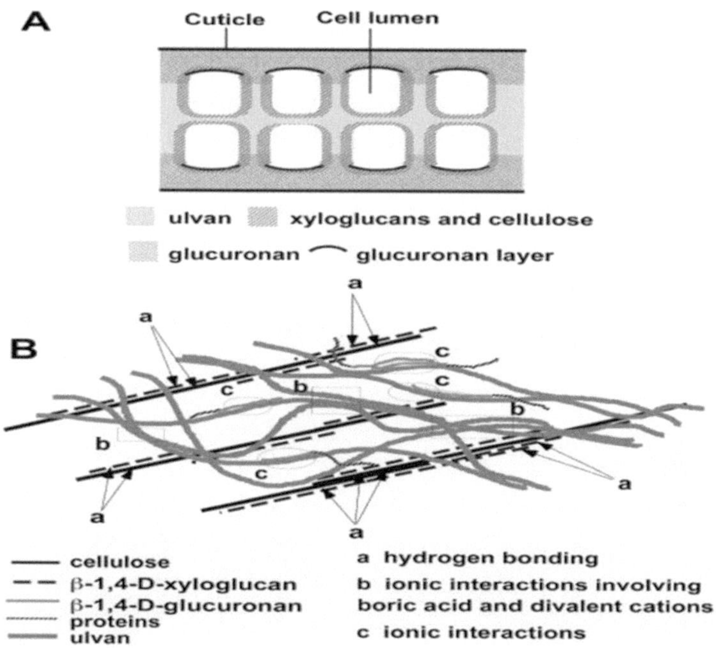

Fig. 8.4 Distribution of polysaccharides in the cell wall of *Ulva* sp. in a schematic cross section of a thallus (**a**) and proposed associations between the different cell wall polysaccharides (**b**). Based on Lahaye and Robic (2007)

Rhizoclonium, and *Microdyction*). The second class is formed by green algae with cell walls containing an important amount of mercerized-like cellulose (cellulose II). This cellulose has a low degree of crystallinity and the chains are randomly oriented. Most of the algae species like *Enteromorpha* sp. (crystallinity index of 36%) belong to this second class (Jmel et al. 2016). The third class includes heterogeneous algae where cellulose does not represent a major fraction of the cell walls (*Vaucheria* and *Spirogyra*).

8.3 From Cellulosic Substrates to Biofuels: Pretreatment as Crucial Step

The valorization of cellulosic biomass involves transforming it into high-value derivatives such as biofuels. The main pathways of transformation are biological pathway for biogas and bioethanol production and chemical pathway for biodiesel and thermal transformation (Naik et al. 2010; Ndiaye 2008; Ruane et al. 2010).

In order to transform the cellulosic substrates to biofuels, different stages are required:

Table 8.2 Chemical composition of various macro- and microalgae species (w/w% dry biomass)

Composition	Macroalgae			Microalgae
	Green algae	Red algae	Brown algae	
Polysaccharides	Cellulose (38–52)	Agar (up to 52)	Alginate (up to 40)	Starch
	Mannan	Carrageenan	Laminarin (up to 35)	**Total carbohydrates**
	Ulvan	Cellulose	Mannitol	
	Starch	Lignin	Glucan	
			Cellulose	Glucose
				Galactose
				Fucose
				Arabinose
				Rhamnose
				Xylose
Monosaccharides	Glucose	Glucose	Glucose	Ribose
	Mannose	Galactose	Galactose	
	Rhamnose	Agarose	Uronic acid	
	Uronic acid			
Algae example	***Ulva lactuca***	***Gelidium amansii***	***Laminaria japonica***	
Carbohydrates	54.3	83.6	59.5	
Protein	20.6	12.2	30.9	
Lipid	6.2	0.9	1.5	
Ash	18.9	3.3	8.1	

Based on Chen et al. (2015)

– A pretreatment step to alter the structure, the size, and the chemical composition of the cellulosic material facilitating therefore its subsequent hydrolysis by enzymes.
– A saccharification step transforming polysaccharides to monomers which can be fermented by yeast or bacteria.
– A bioconversion stage such as fermentation for the production of bioethanol or anaerobic digestion for the production of biogas.

Figure 8.5 illustrates the different stages of the biofuels production.

Pretreatment plays an important role in cellulosic biomass conversion process. It is required to alter the structure of cellulosic feedstock and to make it accessible for enzymatic hydrolysis. This crucial process step allows the physical separation of the lignocellulosic matrix into lignin, hemicelluloses, and cellulose (Fig. 8.6) and reduces the crystallinity of cellulose (Binod and Pandey 2015; Figoli et al. 2016). Several studies have been interested in this step given its importance in biofuels production (Chen et al. 2015,2016; Cheng and Timilsina 2011; Kumar et al. 2009; Menon and Rao 2012; Rodriguez et al. 2015). Indeed, this step influences the entire

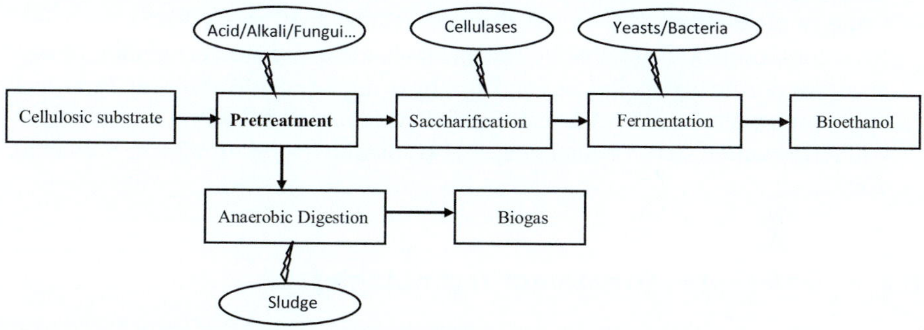

Fig. 8.5 Schematic illustration of biological conversion of cellulosic substrate

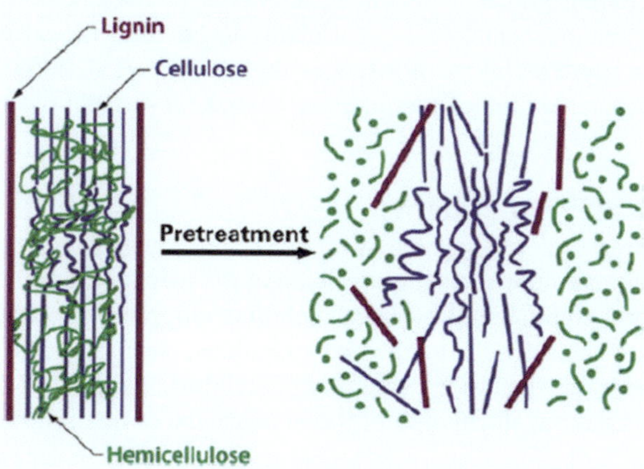

Fig. 8.6 Schematic illustration of pretreatment effect on lignocellulosic biomass. Based on Mosier et al. (2005)

biofuels production process as well as co-products. Moreover, it is considered as the most expensive step in the whole biofuels production approach. In fact, Mosier et al. (2005) revealed that pretreatment accounts for about 30 US cents/gallon of cellulosic ethanol produced (Mosier et al. 2005).

The choice of an appropriate pretreatment technology is very significant. The selection of the suitable pretreatment method must take into account different factors, namely:

- The type and composition of the substrate: each biofuels substrate has its own structure and chemical composition and therefore requires a specific pretreatment method (Kumar et al. 2019; Zhang 2008).
- The cost-effectiveness of the process: pretreatment reactors must be inexpensive through minimizing their volume and keeping mild operating conditions. The chemicals used in pretreatment and subsequent neutralization should be minimal and low cost. Moreover, the energy required for pretreatment like power and heat should be low and/or be compatible for being thermally integrated with the rest of the process (Badiei et al. 2014; Binod and Pandey 2015).
- The concentration of released sugars should be high enough to ensure a perfect bioconversion. Cellulose from pretreatment should be highly digestible, with yields of greater than 90% in less than five (Binod and Pandey 2015).
- Inhibitors formation: suitable pretreatment should avoid the degradation of sugars derived from hemicelluloses and minimize the formation of by-products such as furfural, phenols, and acids that inhibit to the enzymatic hydrolysis and microbial fermentation steps (Badiei et al. 2014; Jönsson and Martín 2016; Weil et al. 2002).

8.4 Different Pretreatment Technologies

In the bioconversion of cellulosic substrates to fuel, the substrate needs to be treated so that the cellulose in the plant fibers becomes exposed. Pretreatment uses several techniques, including physical treatment, ammonia fiber explosion, steam explosion, chemical treatment, and biological treatment, to alter the structure of cellulosic biomass to make cellulose more accessible (Kumar et al. 2009). Each kind of substrate requires some kind of pretreatment to make it suitable for bioprocess.

8.4.1 Physical Pretreatment

Most of the cellulosic substrates require some of the mechanical treatments for size reduction. Several pretreatment methods including chipping, grinding, milling, and irradiation (using gamma rays, microwave radiations, electron beam, etc.) are commonly used to enhance the biodegradability of cellulosic materials. Mechanical pretreatment allows the transformation of the biomass into a fine powder. It decreases therefore the cellulose crystallinity while increasing the surface area. Moreover, it reduces the degree of polymerization of cellulose and hemicelluloses (Sun and Cheng 2002). Palmowski and Müller (2003) have studied the effect of comminution on various organic feedstocks such as apples, rice, sunflower seeds, hay, and maple leaves. They found that, after pretreatment, cells were destroyed and/or the organic components were dissolved through newly generated accessible areas which permit the release of soluble sugars (Palmowski and Müller 2003). Nevertheless, Bridgeman et al. (2007) showed that mechanical pretreatment has a negative effect on switchgrass as it causes significant carbohydrate losses that ultimately result in a small amount of reducing sugars (Bridgeman et al. 2007). Furthermore, this process is not cost-effective since it requires too much energy and likely will not be used in a full-scale process.

8.4.2 Physicochemical Pretreatment

Physicochemical pretreatments combine both physical and chemical processes. The most important processes include steam explosion, ammonia fiber explosion (AFEX) and ammonia recycle percolation (ARP), microwave-chemical pretreatment, and liquid hot water pretreatment.

8.4.2.1 Steam Explosion

During steam explosion pretreatment, the cellulosic substrate is exposed to a high pressure generating its destructuration. Steam explosion is the most studied and applied physicochemical method of biomass pretreatment (Agbor et al. 2011). It is generally initiated at a temperature of 160–260 °C (corresponding pressure to 0.69–4.83 MPa) for several seconds to a few minutes before the material is exposed to atmospheric pressure. This pretreatment causes hemicelluloses degradation and lignin transformation due to high temperature, thus increasing the potential of cellulose hydrolysis (Cheng and Timilsina 2011; Toussaint et al. 1991). Previous study reported that steam explosion improved the efficiency of enzymatic hydrolysis of poplar chips (from 15% for untreated material to 90% for pretreated ones) (Grous et al. 1986). The steam explosion technique possesses a lot of advantages such as the low capital investment, the high sugar recovery, and the low environmental impacts (Focher et al. 1991). However, this well-known technique is affected by numerous factors, namely, moisture content, temperature, residence time, and the size of substrates (Duff and Murray 1996).

An innovative technique usually employed is the use of the steam explosion in association with a catalyst such as carbon or sulfur dioxide or sulfuric acid. The addition of a catalyst increases the recovery of hemicelluloses sugars while decreasing the production of inhibitory by-products and improving the enzymatic hydrolysis (Mosier et al. 2005; Sun and Cheng 2002). Nevertheless, it is necessary to take into account its environmental and economic effect.

8.4.2.2 Instant Controlled Pressure Drop (DIC)

Recently, an innovative pretreatment technology "détente instantanée controlée" (DIC) or "instant controlled pressure drop" has been utilized for several biomasses.

As thermo-mechanical pretreatment, the DIC technique is comparable to steam explosion. It represents a sudden transition from high steam pressure to a vacuum that causes a rapid cooling of the treated material and an auto-evaporation of the water it contains leading to the ending of sugars' thermal degradation (Sarip et al. 2016). DIC technology destructurates the cellulosic material by destroying cell walls and creating porous areas. It increases then the accessibility of its organic constituents within short treatment time (Mounir et al. 2012).

Nonetheless, this technique requires first a heating step of substrates at high temperature (180–240 °C,) which decreases its cost-effectiveness (Mounir et al. 2014).

8.4.2.3 Ammonia Fiber Explosion (AFEX) and Ammonia Recycle Percolation (ARP)

Ammonia fiber explosion (AFEX) and ammonia recycle percolation (ARP) are two physicochemical pretreatment methods that use liquid ammonia to treat the biomass (Kumar et al. 2009). AFEX is a pretreatment process in which the substrate is exposed to liquid ammonia at high temperature and pressure for a period of time, and then the pressure is quickly reduced. However, in ARP process, aqueous ammonia (10–15 wt %) passes through biomass at high temperature, after which ammonia is recovered.

The AFEX technology is very similar to steam explosion although ARP is carried out at higher temperatures (Bajpai 2016). The AFEX process has been used for the pretreatment of many lignocellulosic materials including alfalfa, wheat straw, and wheat chaff (Holtzapple et al. 1991; Menon and Rao 2012). AFEX pretreatment leads to the decrystallization of cellulose, the partial depolymerization of hemicelluloses, and the delignification of the materials. Thus, it increases accessible surface area due to structural disruption and therefore wetting ability of the treated biomass (Gollapalli et al. 2002; Kumar et al. 2009; Monlau et al. 2013). Indeed, Gupta and Lee (2010) reported that AFEX pretreatment applied to switchgrass allowed the solubilization of 68%, 45%, and 1% of lignin, hemicelluloses, and cellulose, respectively (Gupta and Lee 2010). Besides, Holtzapple et al. (1991) showed that over 90% hydrolysis of cellulose and hemicelluloses was obtained after AFEX pretreatment of bermudagrass (approximately 5% lignin) and bagasse (15% lignin) (Holtzapple et al. 1991). However, AFEX is not a very efficient technology for lignocellulosic biomass with relatively high lignin content such as wood and nut shells (McMillan 1994; Taherzadeh and Karimi 2008).

The AFEX method possesses many advantages: it is more cost-effective compared to various pretreatment techniques for bioethanol production (Teymouri et al. 2004); the ammonia used during the process can be recovered and reused, and the subsequent processing is less complex than other pretreatment processes. Nevertheless, a recycling step for ammonia is needed after the pretreatment in order to reduce the cost and protect the environment (Holtzapple et al. 1992).

The second method of alkaline process utilizing ammonia is the ammonia recycle percolation (ARP) technique. In this technique, the ammonia is separated and recycled. Under these conditions, aqueous ammonia reacts primarily with lignin and engenders depolymerization of lignin and cleavage of lignin-carbohydrate linkages. The ammonia pretreatment does not generate inhibitors for the downstream biological processes, so a water wash step is not necessary (Menon and Rao 2012).

8.4.2.4 Microwave-Chemical Pretreatment

The microwave-chemical pretreatment represents a more effective pretreatment than the conventional heating chemical pretreatment. Indeed, microwave irradiation heats the whole volume of a sample, whereas conventional heating techniques heat the sample in contact with the reaction vessel accelerating therefore reactions during the pretreatment process (Zhu et al. 2005; Zhu et al. 2006). Normally, microwave irradiation disorders the ultrastructure of cellulose, degrades hemicelluloses, and increases the substrate accessibility (Zhu et al. 2005). Nonetheless, the microwave-chemical pretreatment has some drawbacks, including complicated operation procedures, high energy consumption, and strict monitoring of equipment (Monlau et al. 2013).

8.4.2.5 Liquid Hot Water Pretreatment

The liquid hot water (LHW) process is a physicochemical pretreatment considered as the most environment-friendly among the chemical and physicochemical ones (Zhuang et al. 2016). In fact, the absence of catalysts makes this pretreatment nontoxic and more economical since the secondary pollution caused by the high consumption of water to wash the pretreated material is avoided. During this pretreatment, water is generally kept at high temperatures between 160 and 240 °C using pressure. This high pressure usually generates some organic acids such as acetic acid. These acids are beneficial for the pretreatment process as they are able to catalyze the hemicelluloses degradation to monosugars (Michelin and Teixeira 2016).

On the other hand, the absence of catalysts during the LHW pretreatment eliminates the risks of materials corrosion and makes the pretreatment process more sustainable and viable (Imman et al. 2018).

Some previous works reported that the LHW pretreatment is not quite effective in eliminating lignin from the pretreated biomass or partially removes it. The lignin remaining after the LHW process aggregates and inhibits the consequent enzymatic hydrolysis by preventing the cellulolytic enzymes from accessing to cellulose or by blocking their binding mechanism to the substrate (Li et al. 2017). This inhibitory effect of lignin was not observed for substrates such as green macroalgae since they do not contain lignin and the enzymatic hydrolysis will not be affected by the abovementioned phenomenon. The LHW pretreatment constitutes then the perfect procedure for enhancing biomass digestibility as reported by Jmel et al. (2017b) where the liquid hot water process was demonstrated to be the best pretreatment method to enhance the enzymatic digestibility of the green macroalga *Ulva lactuca*. The LWH pretreatment showed also interesting results related to the improvement of the enzymatic digestibility of sugarcane bagasse (Yu et al. 2013), rice straw (Imman et al. 2014), *Miscanthus* (Li et al. 2013), and hardwood (Kim et al. 2014).

8.4.3 Chemical Pretreatment

Chemical pretreatment is the most studied pretreatment technique among pretreatment categories. Its principal objective is improving the biodegradability of cellulose by removing lignin and/or hemicelluloses and decreasing the degree of polymerization and crystallinity of the cellulose component. The most used chemical pretreatments include acid, alkaline, organic acid, and ionic liquid.

8.4.3.1 Acid Pretreatment

The acid biomass pretreatment is technically the most studied and industrially applied chemical pretreatment process (Deshavath et al. 2017). Acid pretreatment involves the use of concentrated and dilute acids to break the rigid structure of the lignocellulosic material. Dilute sulfuric acid (H_2SO_4) is the most commonly employed. It has been commercially used for a wide range of biomass types such as switchgrass (Li et al. 2010), spruce (Shuai et al. 2010), and poplar (Wyman et al. 2009). Dilute acid pretreatment (with 0.6% H_2SO_4) was also applied by Ben Yahmed et al. (2016) for the green algal biomass, but it was accompanied by a significant loss of solid matter and led to low saccharification yield (Ben Yahmed et al. 2016).

Like any other pretreatment technology, the acid pretreatment presents numerous advantages related to lignocellulosic biomass disruption and cellulose hydrolysis but also some drawbacks.

The pretreatment based on the acid catalyst has been proven to be extremely effective on disrupting the glycosidic bonds and consequently producing monomers and oligomers from the pretreated polysaccharides. The most affected polysaccharide during acid hydrolysis is hemicellulose as it is generally hydrolyzed by low acid concentrations. The cellulose and lignin fractions are also affected by the acid pretreatment, and this is why it is highly studied and researched since it affects all the lignocellulosic compartments which makes it the most commercialized hydrolysis process. Nevertheless, some biomasses need combined pretreatments for their complete valorization and associate other pretreatment types with acid for a higher conversion efficiency (Liu and Bao 2019).

Many discussions are dealing with the acid pretreatment drawbacks. The most common difficulties faced while applying the acid process are related to the important production of inhibitory molecules from the degradation of the pretreated sugars and lignin (Deshavath et al. 2017). Another problem is the significant amount of wastewater produced while biomass washing at the end of the pretreatment step. This wastewater possesses generally a high aquatic and human toxicity potential and also a high acidification one.

8.4.3.2 Alkaline Pretreatment

Alkaline pretreatment implies the use of bases such as sodium, calcium, potassium, and ammonium hydroxide for the treatment of biofuels substrates. This pretreatment allows the destruction of lignin, the partial decrystallization of the cellulose, and the degradation of hemicelluloses. Sodium hydroxide solution is the most employed pretreatment catalyst (Menon and Rao 2012). Alkaline method was used for various lignocellulosic substrates such as corn stover, switchgrass, bagasse, wheat, and rice straw (Hu et al. 2007; Sun and Cheng 2002). The effectiveness of sodium hydroxide pretreatment was demonstrated for hardwood, switchgrass, wheat straw, and softwood containing less than 26% of lignin (Zhao et al. 2008). Alkaline pretreatment was also employed for algal biomass. The NaOH pretreatment assayed on the green macroalga *Chaetomorpha linum* at low concentration gave interesting results. It allowed a significant destructuration of the algal substrates increasing therefore the enzymes accessibility with a recovery of an important residual pretreated biomass (Ben Yahmed et al. 2016). Besides, compared to liquid hot water, ionic liquid, and organosolv pretreatments, Jmel et al. found that alkaline pretreatment (with 6% NaOH) was the most effective method in hemicellulose elimination from *Ulva* sp. by breaking the ester bonds between cell wall polysaccharides (Jmel et al. 2017a).

Alkaline pretreatment permits also to swell the fibers and increase the pore size, facilitating therefore the diffusion of the hydrolytic enzymes (Monlau et al. 2013). Although alkaline pretreatment can cleave the bonds between lignin and cellulose or hemicelluloses, a significant fraction of the lignin could still be mixed with the cellulose after pretreatment which can inhibit cellulase enzymes during subsequent enzymatic hydrolysis (Cheng and Timilsina 2011). A neutralizing step to remove lignin and inhibitors (salts, furfural, aldehydes, and phenolic acids) is consequently needed before enzymatic hydrolysis.

8.4.3.3 Green Solvents (Ionic Liquids)

An increasing interest is given in the recent years to ionic liquids and their important effects regarding biomass pretreatment, fractionation, and conversion instead of the conventional use of organic solvent catalysts (Shill et al. 2011). Ionic liquids are defined as solutions fully made of ions and that remain liquid at low temperatures (lower than 100 °C). The electrochemical stability of the ionic liquid solutions is mainly due to the strong electrostatic interactions between ions (Schutt et al. 2017).

Ionic liquids are applied in several fields and especially in catalysis and analytical chemistry applications. Recently, the application spectra of ionic liquids reached the pretreatment, characterization, and bioconversion of lignocellulosic biomass (Hou et al. 2017).

One of the main challenges for an effective pretreatment is the dissolution or the disruption of the rigid lignocellulosic biomass. The main target of any biomass pretreatment is the cellulose which represents generally the most abundant fraction in lignocellulosic biomass (Moniruzzaman and Goto 2018). Several works reported the ionic liquids' ability to dissolve cellulose by neutralizing its intra- and inter-molecular hydrogen bonds (Labbe et al. 2012). The dissolved cellulose can be regenerated and its physicochemical characteristics can be maintained (Viell et al. 2013).

On the other hand, lignin represents an important barrier and maintains the high recalcitrance of lignocellulosic biomass. Several works dealt with lignin dissolution in ionic liquids; the results showed that ionic liquids are not only effective for lignin neutralization but also for its recovery as they maintain its original structure and it can be further processed for different applications (Passos et al. 2014).

Ionic liquids were applied for the pretreatment of numerous biomasses such as oil palm biomass (Mohtar et al. 2015), soybean (da Cunha-Pereira et al. 2016), bamboo (Si et al. 2017), and also green macroalgae (Jmel et al. 2017b), and they showed quite promising results concerning biomass fractionation and cellulose digestibility except for green algae where the high viscosity of the employed ionic liquid represented a major problem for algal biomass fractionation. Despite their impressive abilities toward biomass valorization, numerous issues are related to the use of ionic liquids. The main issues that need to be solved concern the enzyme activity inhibition, the high toxicity of many ionic liquids toward living organisms, the high amounts of water used to regenerate the dissolved molecules, and most importantly, the high purchase costs of ionic liquids.

8.4.3.4 Organosolv

The organic solvent pretreatment generally known as organosolv pretreatment gained an increased interest in lignocellulosic biomass pretreatment. In fact, the organosolv pretreatment separates cellulose from hemicelluloses and lignin while conserving its structure and avoiding further degradation (Zhang et al. 2016).

Regarding cellulose, the organosolv pretreatment allows its recuperation by dissolving hemicelluloses and lignin in the employed organic solvent (Gandolfi et al. 2014). The lignin neutralization allows the decrease in cellulose recalcitrance and an important increase of its surface area allowing consequently an increase in cellulolytic enzymes accessibility (Zhang et al. 2016). The organosolv pretreatment allowed the achievement of important cellulose hydrolysis yields for several biomasses such as wood (Alio et al. 2019), sorghum bagasse (Teramura et al. 2018), rice straw (Raita et al. 2017), and green algae (Jmel et al. 2017b).

It was demonstrated that the organosolv pretreatment allows the recovery of important yields of cellulose while minimizing its degradation and making its physicochemical properties favorable for its saccharification by hydrolytic enzymes with low yields of fermentation inhibitors (Chen et al. 2018; Jmel et al. 2017b; Zhang et al. 2016).

On the other hand, the highly valuable lignin obtained after the organosolv pretreatment showed several interesting properties such as its hydrophobicity and the low amount of sulfur opening a wide range of applications such as adhesives or as a precursor for the production of high-added-value chemicals (Matsakas et al. 2018; Zhu et al. 2015).

Another important fraction is the hemicellulosic one. The dissolved hemicellulose fraction is generally hydrolyzed to hemicellulosic sugars and then valorized mainly by fermentation to bioethanol or for the production of xylitol (Amiri and Karimi 2015; Gandolfi et al. 2014).

8.4.4 Biological Pretreatment

Biological pretreatment uses microorganisms such as fungi and bacteria in the treatment of biomass. These microorganisms allow the modification of the biochemical structure of the treated biomass. In general, brown-rot fungi primarily attack cellulose with minor modifications to lignin, and white-rot fungi degrade lignin (Sun and Cheng 2002). Currently, research is focused on finding organisms that can degrade lignin more specifically. White-rot fungi such as basidiomycetes are the most commonly used microorganisms for pretreatment of biomass (Sanchez 2009). Hwang et al. (2008) studied the effect of biological pretreatment on wood chips using four different white-rot fungi for 30 days. They found that the glucose yield of pretreated wood by *Trametes versicolor* MrP 1 reached 45% by enzymatic hydrolysis, whereas 35% of solid substrate was converted to glucose during fungal incubation (Hwang et al. 2008). Fungal pretreatment using solid-state fermentation (SSF) of filamentous fungus *Aspergillus fumigatus* was also investigated by Ben Yahmed et al. for the degradation of algal substrate. Results showed that the proposed SSF-based pretreatment enhanced the biogas production by 21% and permitted an eco-friendly valorization of large amounts of abundant macroalga *Ulva* sp. (Ben Yahmed et al. 2017).

Biological pretreatment represents a promising process because it reduces the use of chemicals and does not require a large amount of energy which reduces the cost of the process while having a positive impact on the environment. However, biological pretreatment is very slow and requires careful control of growth conditions of microorganisms (Ben Yahmed et al. 2017; Rouches et al. 2016). Moreover, most lignolytic microorganisms solubilize not only lignin but also hemicelluloses and cellulose (Menon and Rao 2012).

In addition to the biological pretreatment using filamentous fungi, there also exists an enzymatic pretreatment using the commercial enzymatic cocktails produced by fungi mainly the ascomycete *Trichoderma* and *Aspergillus* genus (Singhvi et al. 2014). These cocktails primarily consist of cellulases and xylanases. Compared to fungal pretreatments, enzymatic pretreatments are generally more complex and expensive since they require additional steps for enzyme production and extraction (Ben Yahmed et al. 2017).

The choice of the suitable treatment process depends on the substrate nature, the cost of the process, and its environmental impact. Table 8.3 presents the main pretreatment processes for biofuels production as well as their advantages and disadvantages.

Table 8.3 Main methods of biomass pretreatment: characteristics, advantages, and limitations

Pretreatment method	Sugars production	Inhibitors generation	Reuse of chemicals	Cost	Advantages	Limitations
Mechanical	Low	Nil	No	High	Cellulose crystallinity reduction	High power consumption than inherent biomass energy
Steam explosion	High	High	–	High	Hemicelluloses removal and alteration in lignin structure	Incomplete destruction of lignin-carbohydrate matrix
AFEX	High	Low	–	High	Removal of lignin and hemicelluloses	Not efficient for biomass with high lignin content
Liquid hot water	High	High	No	–	Removal of hemicelluloses making enzymes accessible to cellulose	Long residence time, less lignin removal
Acid	High	High	Yes	High	Hydrolysis of cellulose and hemicelluloses and alteration of lignin structure	Risk of corrosion and inhibition. Negative impact on the environment
Alkaline	High	Low	Yes	–	Removal of lignin and hemicelluloses, increases accessible surface area	Long residence time, irrecoverable salts formed
Ionic liquids	High/low	Low	Yes	–	Dissolution of cellulose, increased accessibility to cellulases	Still in initial stages
Organosolv	High	High	Yes	High	Hydrolysis of lignin and hemicelluloses	Solvents need to be drained, evaporated, condensed, and reused
Biological	High	Low	–	Low	Hydrolysis of lignin and cellulose, increased accessibility to surface areas, eco-friendly	Very slow and requires careful control of growth conditions of microorganisms

Adapted from Menon and Rao (2012)

8.5 Pretreatment Major Challenges

Several pretreatment technologies have been developed to enhance the biofuels production. Many of them have given promising results for industrial applications such as chemical and thermo-chemical pretreatments. However, there is no pretreatment method that leads to a complete conversion of cellulosic substrates into fermentable sugars. There is always a loss of matter that affects the final conversion rate and increases the cost of production (Maurya et al. 2015). Therefore, many challenges remain in order to ensure the maximum conversion yield, the scale-up, and the large commercialization of these biofuels. In fact, to increase the economic feasibility of the process, carbohydrates and value-added by-products should be recovered and the chemicals and water use must be reduced (Zheng et al. 2009). Characteristics of different cellulosic substrates must be also considered for optimizing pretreatment (Paudel et al. 2017). Besides, the reduction of inhibitory compounds formation should be taken into account (Yang and Wyman 2008). Also the combination of two or more treatments could improve the performance of the bioconversion process (Rabemanolontsoa and Saka 2016).

8.6 Conclusion

The cellulosic substrates conversion includes different steps. The pretreatment stage represents an essential step which influences the whole biofuels production process. Various pretreatment technologies have been developed. These technologies are based on physical, chemical, and biological methods. Each method has its advantages and limitations. Chemical pretreatments are actually the most used at industrial scale. However, no treatment leads to a total conversion of the biomass. The combination of two or more treatment techniques is a good way to improve the efficiency of the pretreatment step. Nevertheless, considerable research is still needed to enhance the upgrade of the pretreatment process and to improve its eco-feasibility. Among the future perspectives to be considered is the development of the models allowing the selection of the appropriate pretreatment technology for each substrate (according to its composition and characteristics) as well as the design and the optimization of this technology.

References

Abbas A, Ansumali S (2010) Global potential of rice husk as a renewable feedstock for ethanol biofuel production. Bioenergy Res 3:328–334

Agbor VB, Cicek N, Sparling R, Berlin A, Levin DB (2011) Biomass pretreatment: fundamentals toward application. Biotechnol Adv 29:675–685

AIE (2016) Key world energy statistics 2016. Paris, OECD Publishing

Alio MA, Tugui OC, Vial C, Pons A (2019) Microwave-assisted Organosolv pretreatment of a sawmill mixed feedstock for bioethanol production in a wood biorefinery. Bioresour Technol 276:170–176. https://doi.org/10.1016/j.biortech.2018.12.078

Alves EF, Bose SK, Francis RC, Colodette JL, Iakovlev M, Van Heiningen A (2010) Carbohydrate composition of eucalyptus, bagasse and bamboo by a combination of methods. Carbohydr Polym 82:1097–1101

Amiri H, Karimi K (2015) Improvement of acetone, butanol, and ethanol production from woody biomass using organosolv pretreatment. Bioprocess Biosyst Eng 38:1959–1972. https://doi.org/10.1007/s00449-015-1437-0

Badiei M, Asim N, Jahim JM, Sopian K (2014) Comparison of chemical pretreatment methods for cellulosic. Biomass APCBEE Procedia 9:170–174. https://doi.org/10.1016/j.apcbee.2014.01.030

Bajpai P (2016) Pretreatment of lignocellulosic biomass for biofuel production. Springer, Singapore

Ben Yahmed N, Jmel MA, Ben Alaya M, Bouallagui H, Marzouki MN, Smaali I (2016) A biorefinery concept using the green macroalgae Chaetomorpha linum for the coproduction of bioethanol and biogas. Energy Convers Manag 119:257–265. https://doi.org/10.1016/j.enconman.2016.04.046

Ben Yahmed N, Carrere H, Marzouki MN, Smaali I (2017) Enhancement of biogas production from Ulva sp. by using solid-state fermentation as biological pretreatment. Algal Res 27:206–214. https://doi.org/10.1016/j.algal.2017.09.005

Binod P, Pandey A (2015) Chapter 1: Introduction. In: Pandey A, Negi S, Binod P, Larroche C (eds) Pretreatment of biomass. Elsevier, Amsterdam, pp 3–6. https://doi.org/10.1016/B978-0-12-800080-9.00001-3

Briand X, Morand P (1997) Anaerobic digestion of Ulva sp. 1. Relationship between Ulva composition and methanisation. J Appl Phycol 9:511–524

Bridgeman T et al (2007) Influence of particle size on the analytical and chemical properties of two energy crops. Fuel 86:60–72

Brylev A, Adylov D, Tukhtaeva G, Kamal'dinova N, Abidova L, Rakhimov D (2001) Polysaccharides of rice straw. Chem Nat Compd 37:569–570

Chen H, Zhou D, Luo G, Zhang S, Chen J (2015) Macroalgae for biofuels production: Progress and perspectives. Renew Sust Energ Rev 47:427–437. https://doi.org/10.1016/j.rser.2015.03.086

Chen L, Li J, Lu M, Guo X, Zhang H, Han L (2016) Integrated chemical and multi-scale structural analyses for the processes of acid pretreatment and enzymatic hydrolysis of corn Stover. Carbohydr Polym 141:1–9

Chen T, Li Z, Zhang X, Min D, Wu Y, Wen J, Yuan T (2018) Effects of hydrothermal pretreatment on the structural characteristics of Organosolv lignin from Triarrhena lutarioriparia. Polymers 10. https://doi.org/10.3390/polym10101157

Cheng JJ, Timilsina GR (2011) Status and barriers of advanced biofuel technologies: a review. Renew Energy 36:3541–3549. https://doi.org/10.1016/j.renene.2011.04.031

da Cunha-Pereira F, Rech R, Zachia Ayub MA, Pinheiro Dillon A, Dupont J (2016) Liberation of fermentable sugars from soybean hull biomass using ionic liquid 1-butyl-3-methylimidazolium acetate and their bioconversion to ethanol. Biotechnol Prog 32:312–320. https://doi.org/10.1002/btpr.2207

Deshavath NN, Mohan M, Veeranki VD, Goud VV, Pinnamaneni SR, Benarjee T (2017) Dilute acid pretreatment of sorghum biomass to maximize the hemicellulose hydrolysis with minimized levels of fermentative inhibitors for bioethanol production. 3 Biotech 7:139. https://doi.org/10.1007/s13205-017-0752-3

Duff SJ, Murray WD (1996) Bioconversion of forest products industry waste cellulosics to fuel ethanol: a review. Bioresour Technol 55:1–33

Fatih Demirbas M (2009) Biorefineries for biofuel upgrading: a critical review. Appl Energy 86:S151–S161. https://doi.org/10.1016/j.apenergy.2009.04.043

Figoli A, Marino T, Galiano F (2016) Polymeric membranes in biorefinery. In: Membrane technologies for biorefining. Woodhead Publishing/Elsevier, pp 29–59

FocherB, MarzettiA, CrescenziV (1991) Steam explosion techniques: fundamentals and industrial applications. In: Proceedings of the international workshop on steam explosion techniques: fundamentals and industrial applications, Milan, Italy, 20–21 October 1988, CRC Press

Gandolfi S, Ottolina G, Consonni R, Riva S, Patel I (2014) Fractionation of hemp hurds by organosolv pretreatment and its effect on production of lignin and sugars. ChemSusChem 7:1991–1999. https://doi.org/10.1002/cssc.201301396

García-Aparicio MP, Ballesteros I, González A, Oliva JM, Ballesteros M, Negro MJ (2006) Effect of inhibitors released during steam-explosion pretreatment of barley straw on enzymatic hydrolysis. Appl Biochem Biotechnol 129:278–288

Gollapalli LE, Dale BE, Rivers DM (2002) Predicting digestibility of ammonia fiber explosion (AFEX)-treated rice straw. Appl Biochem Biotechnol 98–100:23–35

Grous WR, Converse AO, Grethlein HE (1986) Effect of steam explosion pretreatment on pore size and enzymatic hydrolysis of poplar. Enzym Microb Technol 8:274–280. https://doi.org/10.1016/0141-0229(86)90021-9

Gupta R, Lee YY (2010) Investigation of biomass degradation mechanism in pretreatment of switchgrass by aqueous ammonia and sodium hydroxide. Bioresour Technol 101:8185–8191. https://doi.org/10.1016/j.biortech.2010.05.039

Hoch G (2007) Cell wall hemicelluloses as mobile carbon stores in non-reproductive plant tissues. Funct Ecol 21:823–834

Holtzapple MT, Jun J-H, Ashok G, Patibandla SL, Dale BE (1991) The ammonia freeze explosion (AFEX) process. Appl Biochem Biotechnol 28:59–74

Holtzapple MT, Lundeen JE, Sturgis R, Lewis JE, Dale BE (1992) Pretreatment of lignocellulosic municipal solid waste by ammonia fiber explosion (AFEX). Appl Biochem Biotechnol 34:5

Hou Q, Ju M, Li W, Liu L, Chen Y, Yang Q (2017) Pretreatment of Lignocellulosic biomass with ionic liquids and ionic liquid-based solvent systems. Molecules 22. https://doi.org/10.3390/molecules22030490

Howard R, Abotsi E, Van Rensburg EJ, Howard S (2003) Lignocellulose biotechnology: issues of bioconversion and enzyme production. Afr J Biotechnol 2:602–619

Hu Z, Wang Y, Wen Z (2007) Alkali (NaOH) pretreatment of switchgrass by radio frequency-based dielectric heating. In: Biotechnology for fuels and chemicals. Humana Press/Springer, Totowa, pp 589–599

Hwang S-S, Lee S-J, Kim HK, Ka J-O, Kim K-J, Song H-G (2008) Biodegradation and saccharification of wood chips of Pinus strobus and Liriodendron tulipifera by white rot fungi. J Microbiol Biotechnol 18:1819–1825

Imman S, Arnthong J, Burapatana V, Champreda V, Laosiripojana N (2014) Effects of acid and alkali promoters on compressed liquid hot water pretreatment of rice straw. Bioresour Technol 171:29–36. https://doi.org/10.1016/j.biortech.2014.08.022

Imman S, Laosiripojana N, Champreda V (2018) Effects of liquid hot water pretreatment on enzymatic hydrolysis and physicochemical changes of corncobs. Appl Biochem Biotechnol 184:432–443. https://doi.org/10.1007/s12010-017-2541-1

Jard G, Marfaing H, Carrere H, Delgenes JP, Steyer JP, Dumas C (2013) French Brittany macroalgae screening: composition and methane potential for potential alternative sources of energy and products. Bioresour Technol 144:492–498. https://doi.org/10.1016/j.biortech.2013.06.114

Jazzar S et al (2015) A whole biodiesel conversion process combining isolation, cultivation and in situ supercritical methanol transesterification of native microalgae. Bioresour Technol 190:281–288

Jmel MA, Ben Messaoud G, Marzouki MN, Mathlouthi M, Smaali I (2016) Physico-chemical characterization and enzymatic functionalization of Enteromorpha sp. cellulose. Carbohydr Polym 135:274–279. https://doi.org/10.1016/j.carbpol.2015.08.048

Jmel MA, Anders N, Ben Yahmed N, Schmitz C, Marzouki MN, Spiess A, Smaali I (2017a) Variations in physicochemical properties and bioconversion efficiency of Ulva lactuca polysaccharides after different biomass pretreatment techniques. Appl Biochem Biotechnol. https://doi.org/10.1007/s12010-017-2588-z

Jmel MA, Anders N, Yahmed NB, Schmitz C, Marzouki MN, Spiess A, Smaali I (2017b) Variations in physicochemical properties and bioconversion efficiency of Ulva lactuca polysaccharides after different biomass pretreatment techniques. Appl Biochem Biotechnol. https://doi.org/10.1007/s12010-017-2588-z

Jones CS, Mayfield SP (2012) Algae biofuels: versatility for the future of bioenergy. Curr Opin Biotechnol 23:346–351. https://doi.org/10.1016/j.copbio.2011.10.013

Jönsson LJ, Martín C (2016) Pretreatment of lignocellulose: formation of inhibitory by-products and strategies for minimizing their effects. Bioresour Technol 199:103–112. https://doi.org/10.1016/j.biortech.2015.10.009

Kim TH, Taylor F, Hicks KB (2008) Bioethanol production from barley hull using SAA (soaking in aqueous ammonia) pretreatment. Bioresour Technol 99:5694–5702

Kim Y, Kreke T, Mosier NS, Ladisch MR (2014) Severity factor coefficients for subcritical liquid hot water pretreatment of hardwood chips. Biotechnol Bioeng 111:254–263. https://doi.org/10.1002/bit.25009

Knauf M, Moniruzzaman M (2004) Lignocellulosic biomass processing: a perspective. Int Sugar J 106:147–150

Kolpak FJ, Blackwell J (1976) Determination of the structure of cellulose II. Macromolecules 9:273–278. https://doi.org/10.1021/ma60050a019

Kumar P, Barrett DM, Delwiche MJ, Stroeve P (2009) Methods for pretreatment of lignocellulosic biomass for efficient hydrolysis and biofuel production. Ind Eng Chem Res 48:3713–3729

Kumar MN, Ravikumar R, Thenmozhi S, Kumar MR, Shankar MK (2019) Choice of pretreatment technology for sustainable production of bioethanol from lignocellulosic biomass: bottle necks and recommendations. Waste Biomass Valoriz 10:1693–1709

Labbe N, Kline LM, Moens L, Kim K, Kim PC, Hayes DG (2012) Activation of lignocellulosic biomass by ionic liquid for biorefinery fractionation. Bioresour Technol 104:701–707. https://doi.org/10.1016/j.biortech.2011.10.062

Lahaye M, Robic A (2007) Structure and functional properties of ulvan, a polysaccharide from green seaweeds. Biomacromolecules 8:1765–1774

Lamare MD, Wing SR (2001) Calorific content of New Zealand marine macrophytes. N Z J Mar Freshw Res 35:335–341

Li C et al (2010) Comparison of dilute acid and ionic liquid pretreatment of switchgrass: biomass recalcitrance, delignification and enzymatic saccharification. Bioresour Technol 101:4900–4906

Li HQ, Li CL, Sang T, Xu J (2013) Pretreatment on Miscanthus lutarioriparius by liquid hot water for efficient ethanol production. Biotechnol Biofuels 6:76. https://doi.org/10.1186/1754-6834-6-76

Li M, Cao S, Meng X, Studer M, Wyman CE, Ragauskas AJ, Pu Y (2017) The effect of liquid hot water pretreatment on the chemical-structural alteration and the reduced recalcitrance in poplar. Biotechnol Biofuels 10:237. https://doi.org/10.1186/s13068-017-0926-6

Liu G, Bao J (2019) Constructing super large scale cellulosic ethanol plant by decentralizing dry acid pretreatment technology into biomass collection depots. Bioresour Technol 275:338–344. https://doi.org/10.1016/j.biortech.2018.12.061

Matsakas L, Karnaouri A, Cwirzen A, Rova U, Christakopoulos P (2018) Formation of lignin nanoparticles by combining Organosolv pretreatment of birch biomass and homogenization processes. Molecules 23. https://doi.org/10.3390/molecules23071822

Maurya DP, Singla A, Negi S (2015) An overview of key pretreatment processes for biological conversion of lignocellulosic biomass to bioethanol. 3. Biotech 5:597–609. https://doi.org/10.1007/s13205-015-0279-4

McMillan JD (1994) Pretreatment of lignocellulosic biomass. In: Enzymatic conversion of biomass for fuels production, vol 566. American Chemical Society, Washington, DC, pp 292–324

Menon V, Rao M (2012) Trends in bioconversion of lignocellulose: biofuels, platform chemicals & biorefinery concept. Prog Energy Combust Sci 38:522–550. https://doi.org/10.1016/j.pecs.2012.02.002

Michelin M, Teixeira JA (2016) Liquid hot water pretreatment of multi feedstocks and enzymatic hydrolysis of solids obtained thereof. Bioresour Technol 216:862–869. https://doi.org/10.1016/j.biortech.2016.06.018

Miron J, Yosef E, Ben-Ghedalia D (2001) Composition and in vitro digestibility of monosaccharide constituents of selected byproduct feeds. J Agric Food Chem 49:2322–2326

Mohtar SS et al (2015) Extraction and characterization of lignin from oil palm biomass via ionic liquid dissolution and non-toxic aluminium potassium sulfate dodecahydrate precipitation processes. Bioresour Technol 192:212–218. https://doi.org/10.1016/j.biortech.2015.05.029

Moniruzzaman M, Goto M (2018) Ionic liquid pretreatment of Lignocellulosic biomass for enhanced enzymatic delignification. Adv Biochem Eng Biotechnol. https://doi.org/10.1007/10_2018_64

Monlau F, Barakat A, Trably E, Dumas C, Steyer J-P, Carrère H (2013) Lignocellulosic materials into biohydrogen and biomethane: impact of structural features and pretreatment. Crit Rev Environ Sci Technol 43:260–322. https://doi.org/10.1080/10643389.2011.604258

Monsalve GJF, Medina de Perez VI, Ruiz Colorado AA (2006) Ethanol production of banana shell and cassava starch. Dyna (Medellin) 73:21–27

Montingelli ME, Benyounis KY, Stokes J, Olabi AG (2016) Pretreatment of macroalgal biomass for biogas production. Energy Convers Manag 108:202–209. https://doi.org/10.1016/j.enconman.2015.11.008

Mosier N, Wyman C, Dale B, Elander R, Lee YY, Holtzapple M, Ladisch M (2005) Features of promising technologies for pretreatment of lignocellulosic biomass. Bioresour Technol 96:673–686. https://doi.org/10.1016/j.biortech.2004.06.025

Mounir S, Allaf T, Mujumdar AS, Allaf K (2012) Swell drying: coupling instant controlled pressure drop DIC to standard convection drying processes to intensify transfer phenomena and improve quality—an overview. Dry Technol 30:1508–1531. https://doi.org/10.1080/07373937.2012.693145

Mounir S, Allaf T, Berka B, Hassani A, Allaf K (2014) Instant controlled pressure drop technology: from a new fundamental approach of instantaneous transitory thermodynamics to large industrial applications on high performance–high controlled quality unit operations. C R Chim 17:261–267. https://doi.org/10.1016/j.crci.2013.10.019

Muktham RK, Bhargava S, Bankupalli SS, Ball A (2016) A review on 1st and 2nd generation bioethanol production-recent progress. J Sustain Bioenergy Syst 06:72–92. https://doi.org/10.4236/jsbs.2016.63008

Naik SN, Goud VV, Rout PK, Dalai AK (2010) Production of first and second generation biofuels: a comprehensive review. Renew Sust Energ Rev 14:578–597. https://doi.org/10.1016/j.rser.2009.10.003

Ndiaye FT (2008) Pyrolyse de la biomasse en réacteur cyclone-recherche des conditions optimales de fonctionnement. Vandoeuvre-les-Nancy, INPL

Nicolai E, Preston RD (1952) Cell-wall studies in the Chlorophyceae. I. A general survey of submicroscopic structure in filamentous species. Proc R Soc Lond Ser B Biol Sci 140:244–274

Palmowski L, Müller J (2003) Anaerobic degradation of organic materials-significance of the substrate surface area. Water Sci Technol 47(12):231–238

Pandey A (2008) Handbook of plant-based biofuels. CRC press, Boca Raton

Passos H, Freire MG, Coutinho JA (2014) Ionic liquid solutions as extractive solvents for value-added compounds from biomass. Green Chem 16:4786–4815. https://doi.org/10.1039/C4GC00236A

Paudel SR, Banjara SP, Choi OK, Park KY, Kim YM, Lee JW (2017) Pretreatment of agricultural biomass for anaerobic digestion: current state and challenges. Bioresour Technol 245:1194–1205. https://doi.org/10.1016/j.biortech.2017.08.182

Percebois J, Mandil C (2012) Les conclusions du rapport Energies 2050 sur les perspectives du nucléaire en France. In: Annales des Mines-Réalités industrielles, vol 3. Eska, pp 53–58

Perez S, Rodriguez-Carvajal MA, Doco T (2003) A complex plant cell wall polysaccharide: rhamnogalacturonan II. A structure in quest of a function. Biochimie 85:109–121

Prasad S, Singh A, Joshi H (2007) Ethanol as an alternative fuel from agricultural, industrial and urban residues. Resour Conserv Recycl 50:1–39

Rabemanolontsoa H, Saka S (2016) Various pretreatments of lignocellulosics. Bioresour Technol 199:83–91. https://doi.org/10.1016/j.biortech.2015.08.029

Raita M, Denchokepraguy N, Champreda V, Laosiripojana N (2017) Effects of alkaline catalysts on acetone-based organosolv pretreatment of rice straw. 3 Biotech 7:340. https://doi.org/10.1007/s13205-017-0969-1

Ray B (2006) Polysaccharides from Enteromorpha compressa: isolation, purification and structural features. Carbohydr Polym 66:408–416

Ren J-L, Sun R-C (2010) Chapter 4: Hemicelluloses. In: Sun R-C (ed) Cereal straw as a resource for sustainable biomaterials and biofuels. Elsevier, Amsterdam, pp 73–130. https://doi.org/10.1016/B978-0-444-53234-3.00004-3

Robic A, Gaillard C, Sassi JF, Lerat Y, Lahaye M (2009) Ultrastructure of ulvan: a polysaccharide from green seaweeds. Biopolymers 91:652–664

Rodriguez C, Alaswad A, Mooney J, Prescott T, Olabi AG (2015) Pre-treatment techniques used for anaerobic digestion of algae. Fuel Process Technol 138:765–779. https://doi.org/10.1016/j.fuproc.2015.06.027

Rouches E, Herpoël-Gimbert I, Steyer JP, Carrere H (2016) Improvement of anaerobic degradation by white-rot fungi pretreatment of lignocellulosic biomass: a review. Renew Sust Energ Rev 59:179–198. https://doi.org/10.1016/j.rser.2015.12.317

Ruane J, Sonnino A, Agostini A (2010) Bioenergy and the potential contribution of agricultural biotechnologies in developing countries. Biomass Bioenergy 34:1427–1439. https://doi.org/10.1016/j.biombioe.2010.04.011

Rubio M, Tortosa JF, Quesada J, Gómez D (1998) Fractionation of lignocellulosics. Solubilization of corn stalk hemicelluloses by autohydrolysis in aqueous medium. Biomass Bioenergy 15:483–491

Ruiz HA, Rodríguez-Jasso RM, Fernandes BD, Vicente AA, Teixeira JA (2013) Hydrothermal processing, as an alternative for upgrading agriculture residues and marine biomass according to the biorefinery concept: a review. Renew Sust Energ Rev 21:35–51. https://doi.org/10.1016/j.rser.2012.11.069

Sanchez C (2009) Lignocellulosic residues: biodegradation and bioconversion by fungi. Biotechnol Adv 27:185–194. https://doi.org/10.1016/j.biotechadv.2008.11.001

Sarip H, Hossain MS, Azemi MNM, Allaf K (2016) A review of the thermal pretreatment of lignocellulosic. Biomass towards glucose production: autohydrolysis with DIC technology. Bioresources 11(4):29

Schutt TC, Hegde GA, Bharadwaj VS, Johns AJ, Maupin CM (2017) Impact of water-dilution on the solvation properties of the ionic liquid 1-Methyltriethoxy-3-ethylimidazolium acetate for model biomass molecules. J Phys Chem B 121:843–853. https://doi.org/10.1021/acs.jpcb.6b09101

Shill K, Padmanabhan S, Xin Q, Prausnitz JM, Clark DS, Blanch HW (2011) Ionic liquid pretreatment of cellulosic biomass: enzymatic hydrolysis and ionic liquid recycle. Biotechnol Bioeng 108:511–520. https://doi.org/10.1002/bit.23014

Shuai L, Yang Q, Zhu J, Lu F, Weimer P, Ralph J, Pan X (2010) Comparative study of SPORL and dilute-acid pretreatments of spruce for cellulosic ethanol production. Bioresour Technol 101:3106–3114

Si W, Li Y, Zheng J, Wei S, Wang D (2017) Enhanced hydrolysis of bamboo biomass by chitosan based solid acid catalyst with surfactant addition in ionic liquid. Carbohydr Polym 174:154–159. https://doi.org/10.1016/j.carbpol.2017.05.082

Singh A, Nigam PS, Murphy JD (2011) Renewable fuels from algae: an answer to debatable land based fuels. Bioresour Technol 102:10–16. https://doi.org/10.1016/j.biortech.2010.06.032

Singhvi MS, Chaudhari S, Gokhale DV (2014) Lignocellulose processing: a current challenge. RSC Adv 4:8271–8277. https://doi.org/10.1039/C3RA46112B

Subhadra B, Edwards M (2010) An integrated renewable energy park approach for algal biofuel production in United States. Energy Policy 38:4897–4902

Sun Y, Cheng J (2002) Hydrolysis of lignocellulosic materials for ethanol production: a review. Bioresour Technol 83:1–11

Sun R, Lawther JM, Banks WB (1996) Fractional and structural characterization of wheat straw hemicelluloses. Carbohydr Polym 29:325–331. https://doi.org/10.1016/S0144-8617(96)00018-5

Taherzadeh MJ, Karimi K (2008) Pretreatment of lignocellulosic wastes to improve ethanol and biogas production: a review. Int J Mol Sci 9:1621–1651

Teramura H, Sasaki K, Oshima T, Kawaguchi H, Ogino C, Sazuka T, Kondo A (2018) Effective usage of sorghum bagasse: optimization of organosolv pretreatment using 25% 1-butanol and subsequent nanofiltration membrane separation. Bioresour Technol 252:157–164. https://doi.org/10.1016/j.biortech.2017.12.100

TeymouriF, Laureano-PérezL, AlizadehH, DaleBE (2004) Ammonia fiber explosion treatment of corn stover. In: Proceedings of the twenty-fifth symposium on biotechnology for fuels and chemicals held May 4–7, 2003, Springer, Breckenridge, CO, pp 951–963

Torget R, Teh-An H (1994) Two-temperature dilute-acid prehydrolysis of hardwood xylan using a percolation process. Appl Biochem Biotechnol 45:5–22

Toussaint B, Excoffier G, Vignon MR (1991) Effect of steam explosion treatment on the physico-chemical characteristics and enzymic hydrolysis of poplar cell wall components. Anim Feed Sci Technol 32:235–242. https://doi.org/10.1016/0377-8401(91)90028-Q

Trivedi N, Gupta V, Reddy CR, Jha B (2013) Enzymatic hydrolysis and production of bioethanol from common macrophytic green alga Ulva fasciata Delile. Bioresour Technol 150:106–112. https://doi.org/10.1016/j.biortech.2013.09.103

Viell J, Wulfhorst H, Schmidt T, Commandeur U, Fischer R, Spiess A, Marquardt W (2013) An efficient process for the saccharification of wood chips by combined ionic liquid pretreatment and enzymatic hydrolysis. Bioresour Technol 146:144–151. https://doi.org/10.1016/j.biortech.2013.07.059

Weil JR, Dien B, Bothast R, Hendrickson R, Mosier NS, Ladisch MR (2002) Removal of fermentation inhibitors formed during pretreatment of biomass by polymeric adsorbents. Ind Eng Chem Res 41:6132–6138. https://doi.org/10.1021/ie0201056

Wyman CE et al (2009) Comparative sugar recovery and fermentation data following pretreatment of poplar wood by leading technologies. Biotechnol Prog 25:333–339

Yang S-T (2007) Chapter 1: Bioprocessing – from biotechnology to biorefinery. In: Yang S-T (ed) Bioprocessing for value-added products from renewable resources. Elsevier, Amsterdam, pp 1–24. https://doi.org/10.1016/B978-044452114-9/50002-5

Yang B, Wyman CE (2008) Pretreatment: the key to unlocking low-cost cellulosic ethanol. Biofuels Bioprod Biorefin 2:26–40

Yu Q et al (2013) Liquid hot water pretreatment of sugarcane bagasse and its comparison with chemical pretreatment methods for the sugar recovery and structural changes. Bioresour Technol 129:592–598. https://doi.org/10.1016/j.biortech.2012.11.099

Zhang YP (2008) Reviving the carbohydrate economy via multi-product lignocellulose biorefineries. J Ind Microbiol Biotechnol 35:367–375. https://doi.org/10.1007/s10295-007-0293-6

Zhang W, Liang M, Lu C (2007) Morphological and structural development of hardwood cellulose during mechanochemical pretreatment in solid state through pan-milling. Cellulose 14:447–456

Zhang K, Pei Z, Wang D (2016) Organic solvent pretreatment of lignocellulosic biomass for biofuels and biochemicals: a review. Bioresour Technol 199:21–33. https://doi.org/10.1016/j.biortech.2015.08.102

Zhao Y, Wang Y, Zhu J, Ragauskas A, Deng Y (2008) Enhanced enzymatic hydrolysis of spruce by alkaline pretreatment at low temperature. Biotechnol Bioeng 99:1320–1328

Zheng Y, Pan Z, Zhang R (2009) Overview of biomass pretreatment for cellulosic ethanol production. Int J Agric Biol Eng 2:51–68

Zhu S, Wu Y, Yu Z, Liao J, Zhang Y (2005) Pretreatment by microwave/alkali of rice straw and its enzymic hydrolysis. Process Biochem 40:3082–3086

Zhu S et al (2006) Comparison of three microwave/chemical pretreatment processes for enzymatic hydrolysis of rice straw. Biosyst Eng 93:279–283

Zhu MQ, Wen JL, Su YQ, Wei Q, Sun RC (2015) Effect of structural changes of lignin during the autohydrolysis and organosolv pretreatment on Eucommia ulmoides Oliver for an effective enzymatic hydrolysis. Bioresour Technol 185:378–385. https://doi.org/10.1016/j.biortech.2015.02.061

Zhuang X et al (2016) Liquid hot water pretreatment of lignocellulosic biomass for bioethanol production accompanying with high valuable products. Bioresour Technol 199:68–75. https://doi.org/10.1016/j.biortech.2015.08.051

Application of Metabolic Engineering for Biofuel Production in Microorganisms

9

Amirhossein Nazhand

Abstract

There are some intrinsic drawbacks with fossil fuels including non-renewability, susceptibility to geopolitical instabilities, and negative influences on universal environment and human health. Renewable microbial biomass can be an attractive substitute for fossil fuels in biofuel production, but it is associated with enormous technological difficulties. A large body of investigations concerns attempts to surmount such technical complications by assistance of metabolic engineering and synthetic biology techniques. These techniques design and construct effective biosynthetic pathways in suitable microorganisms and optimize them for high titers, production rate, and yields of biofuels. This chapter reviews recent developments in engineering cellular metabolism for the promotion of biofuel molecules (e.g., higher alcohols, isoprenoids, and fatty acids) and addresses the status quo and potential of biofuels in the future.

Keywords

Metabolic engineering · Biofuels · Higher alcohols · Fatty acids · Isoprenoids · Commercialization

9.1 Introduction

Exploitation of renewable resources has arisen from climate change-related concerns and energy security, diverting public attention toward the generation of low-cost biofuels (Peralta-Yahya and Keasling 2010). As most microorganisms are capable of generating biofuels, they account for sustainable resources for

A. Nazhand (✉)
Biotechnology Department, Sari University of Agricultural Sciences and Natural Resources, Mazandaran, Sari, Iran

N. Srivastava et al. (eds.), *Substrate Analysis for Effective Biofuels Production*, Clean Energy Production Technologies, https://doi.org/10.1007/978-981-32-9607-7_9

biofuel production. However, microorganisms are not exploited at commercial scale due to the lack of sufficient genetic equipment and physiological information. Among various suggested methods, metabolic engineering has been proven to be superior because of both time- and cost-effective for manufacturing eco-friendly products. A variety of fuels, chemicals, food ingredients, and pharmaceuticals have been produced from renewable resources through making cellular metabolism by engineering metabolic network (Stephanopoulos and Gill 2001; Nielsen and Keasling 2016).

In metabolic engineering, various methods and instruments are employed for determination, designing, and rewiring of cellular metabolism (Tai and Stephanopoulos 2015; Ekas et al. 2019). Accordingly, metabolic engineering of desired products consists of some steps: (1) using microorganisms with high capacity, (2) understanding biosynthetic network, (3) cofactor equilibrium, (4) analysis of stoichiometry and thermodynamics, (5) transformation technology accessibility, and (6) being able to reduce toxic intermediates and products (Stephanopoulos and Vallino 1991). Tai and Stephanopoulos (2015) have also proposed that the above technology is relevant in the improvement of fuel transportation systems in the future.

Production of biofuels has seen improvements in recent years with groundbreaking developments achieved through metabolic engineering techniques (Fig. 9.1). This chapter presents a survey of investigations on typical biosynthetic pathways of biofuels (e.g., higher alcohols, fatty acids, isoprenoids), followed by emphasizing on achievements in this field.

9.2 Biofuels Derived from Fatty Acids

The researcher had represented various biosynthetic pathways for fatty acids biofuel in (Fig. 9.2). There are also various strategies for tunning the oleochemical products, namely, (1) increasing metabolic precursors of fatty acids (e.g., fatty acyl-CoAs, acetyl-CoA, and malonyl-CoA) through metabolic engineering of microorganisms, (2) redox metabolism engineering, (3) controlling products' chain length by pathway compartmentalization and fatty acid synthase (FAS) engineering such as FAS I (in eukaryotic cells) and FAS II (in most bacteria) and reverse β-oxidation, and (4) raising fatty acid intermediates into products (Jiang et al. 2018; López-Lara and Soto 2018; Marella et al. 2018).

As an oleaginous industrial organism, *Yarrowia lipolytica* can synthesize high amounts of intracellular lipids (Pfleger et al. 2015; Lazar et al. 2018; Shi and Zhao 2017; Abdel-Mawgoud et al. 2018; Markham and Alper 2018b). Yuzbasheva et al. (2017b) 41% enhanced lipid accumulation in *Y. lipolytica* via preparation of a more pool of co-factors as well as over-expression of *ZWF1* and *ACBP* genes. Reverse engineering of oxidative stress defense pathways is a potential route for the development of affordable lipid production from renewable resources in the *Y. lipolytica*. Accordingly, creation of microbial lipid cell factories through detoxification of reactive aldehydes and up-regulation of oxidative stress can yield a titer (72.7 $g{\bullet}L^{-1}$) in the bioreactors (Xu et al. 2017). A challenge for increasing substrate rate in

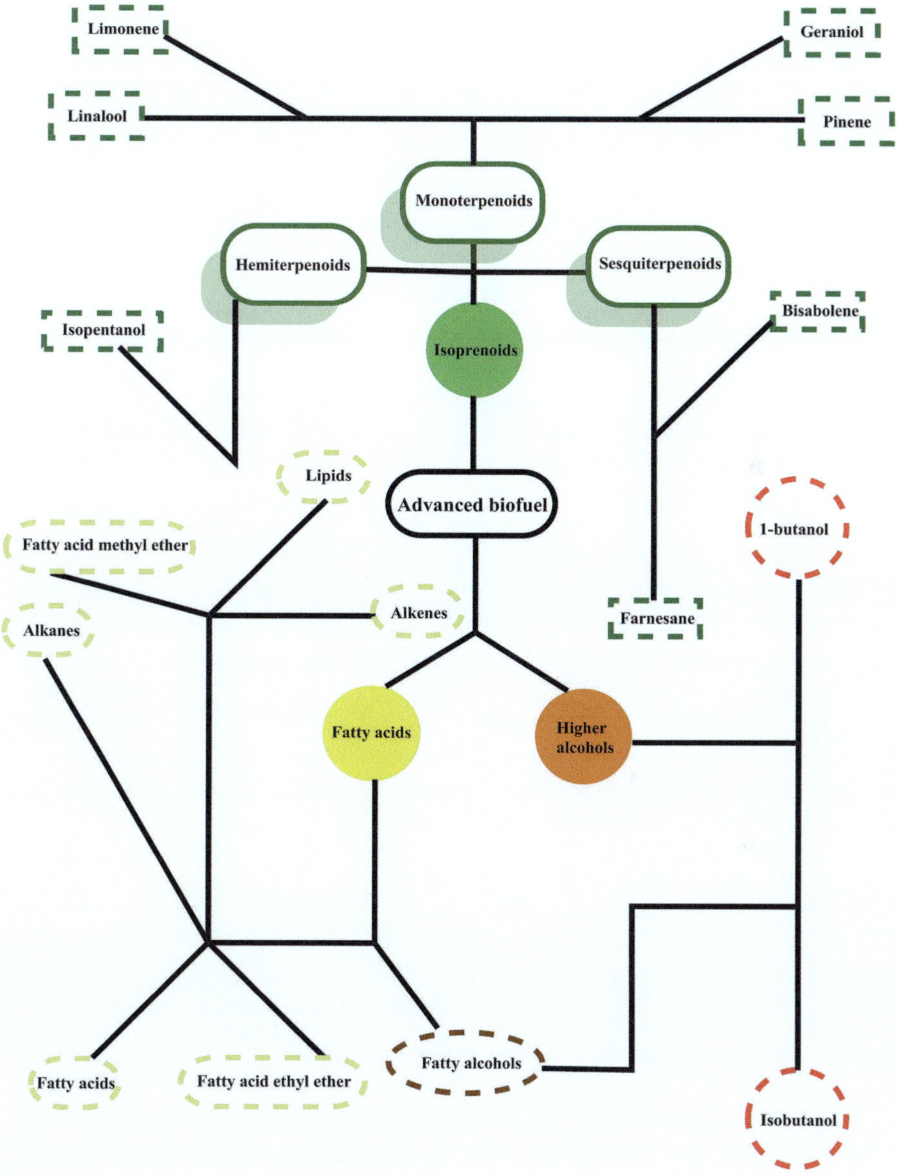

Fig. 9.1 Overview of advanced biofuels

Y. lipolytica is low expression of sugar-metabolizing genes for growth on biomass-derived sugars. Thus, Schwartz and Löbs (2018) developed a transcription-controlling gene through CRISPRa system by activation of two extracellular β-glucosidases (singularly and multiplexed) as an innovative path of *Y. lipolytica* growth promotion on biomass-derived sugar such as cellobiose.

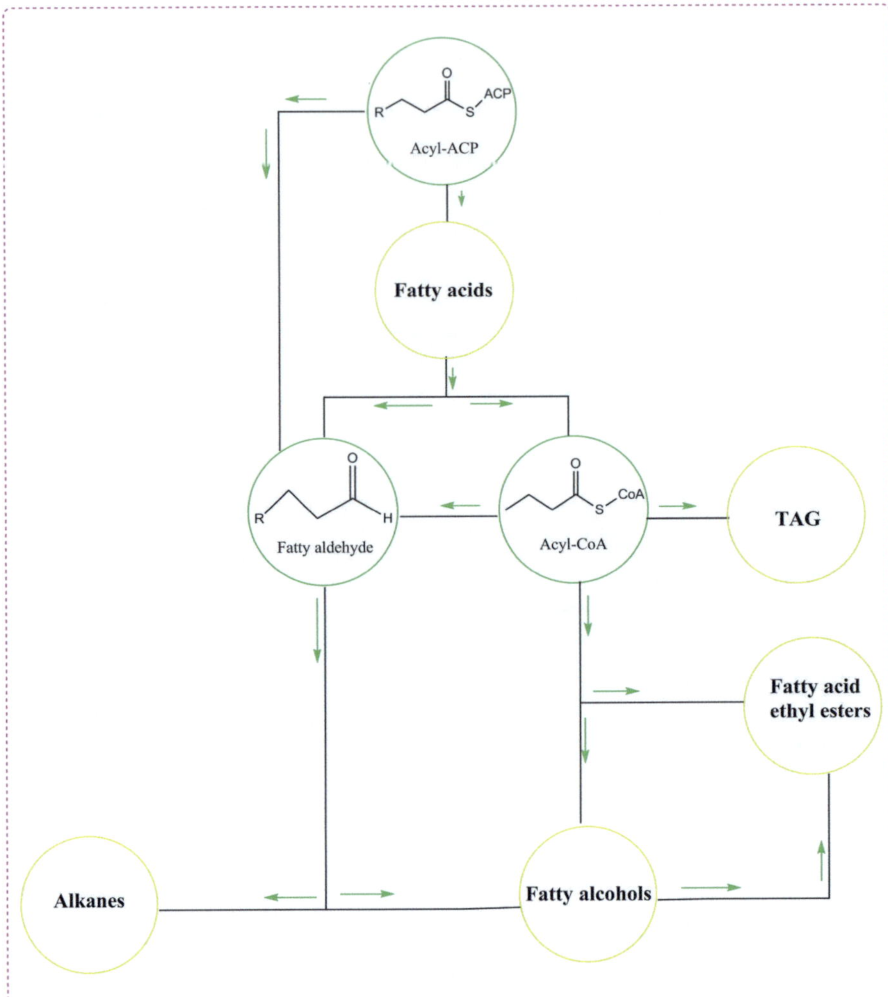

Fig. 9.2 Schematic of fatty acid biofuel biosynthetic pathways

There are two rate-limiting steps in the biosynthesis of fatty acids; they are catalyzed *via* acetyl-CoA carboxylase (ACC1), which improves the malonyl-CoA and fatty acyl-CoA pools, and also plasmid-based over-expression of both FAS I and FAS II converting malonyl-CoA to fatty acids. Friedlander et al. (2016) obtained 85 g•L^{-1} of lipid through fed-batch glucose fermentation. In another research, the cellular metabolism of *Saccharomyces cerevisiae* was reoriented to generate a high amount of TAGs utilizing over-expression of *PAH1* (phosphatidic acid phosphohydrolase), *DGA1* (diacylglycerol acyltransferase), and *ACC1*∗∗ (acetyl-CoA carboxylase) followed by deletion of *TGL3/4/5, ARE1, POX1, GUT2, PXA1*, and *FAA2* (Ferreira et al. 2018). In this way, the strain RF11 produced the highest TAG titers 1.76 g•L^{-1} with 27.4% of maximum theoretical yield. Teixeira et al. (2018) by lipid droplet dynamics strategy improves up to 138% TAG content in the *S. cerevisiae* by expression of *PLIN3* and *FIT2* genes and deletion of *ERD1* and *PMR1* genes that implemented in formerly engineered strains with ability to convey high flux of fatty acid biosynthesis and conversion of acyl-CoA into TAG.

Impairment methods of glycerol metabolism make 2033.8 mg•L^{-1} of free fatty acids (FFA) in *Y. lipolytica* (Yuzbasheva et al. (2017a). For the first time, Shi et al. (2016a), by cDNA library screening method, identified fatty acid accumulation-related enzymes in *Y. lipolytica*, secondly, by expression of these enzymes in *S. cerevisiae* (Shi et al. 2016a), improve 2.5-fold production of FFAs. Yu et al. (2018) reported the highest FFA at a titer of 33.4 g•L^{-1} by compounding adaptive laboratory evolution (ALE) with rational design method in *S. cerevisiae*. The same microorganism was examined by flux-based modeling strategies, and the results revealed that enhanced fatty acid generation by 70% in yeast (Ghosh et al. 2016).

Leber et al. (2015) presented evidence that degradation pathway engineering could improve short-chain fatty acid (SCFA) titer to 119-fold. Biosensor is designed in accordance with multi-copy yeast plasmid which embodies SMCFA-responsive PDR12 promoter for high-throughput screening of highly capable microbial strains to synthesize short- and medium-chain fatty acids (SMCFAs) through selection of hexanoic, heptanoic, and octanoic acid (Baumann et al. 2018). Mukherjee, Bhattacharyya, and Peralta-Yahya (2015) were the first to introduce G-protein coupled receptors (GPCRs) biosensors in yeast for medium-chain fatty acids. The role of elongase enzymes on enhancing MCFAs in *Y. lipolytica* was investigated by Rigouin et al. (2018). In comparison with *Y. lipolytica*, fatty acid metabolism pathway (FAS I) system from *Mycobacterium vaccae* was shown to over-express in *S. cerevisiae* for the first time (Yu et al. 2017), and tuning endogenous elongases which produced 83.5 mg•L^{-1} of docosanol.

Transesterification of lipids resulted in the generation of fatty acid biodiesels such as fatty acid methyl esters (FAMEs) and fatty acid ethyl esters (FAEEs) (Meadows et al. 2017; Yan et al. 2017). Markham and Alper (2018a) detected an ability of *Y. lipolytica* in industrial generation of fatty acid biodiesel in the form of FAMEs through cyclopropanation of unsaturated fatty esters. In their study, cyclopropane fatty acid synthase from *E. coli* was introduced into *Y. lipolytica*, which produced 3 g•L^{-1}of C19 cyclopropanated fatty acid titers in the bioreactor fermentation. In a study by Qiao et al. (2017), the efficiency and yield of FAMEs reached 0.269 (g•g^{-1}), 98.9 (g•L^{-1}), and 1.3 (g•L^{-1}•h^{-1}) titer by redox engineering of *Y. lipolytica*.

Acetyl-CoA and FFAs have major contribution to the biosynthesis of fatty alcohols; considerable efforts are needed for industrialization of this biosynthetic pathway to improve the yields of products as accumulation of precursors or intermediates and unbalanced fluxes are critical bottlenecks in this pathway. Two strategies have been used by Teixeira et al. (2017) for dynamic regulation of fatty alcohol pathway, which can be employed to resolve the above problems. First, their fatty alcohols were generated by FFAs only aiming at knocking out fatty acyl-CoA oxidase (POX1), fatty aldehyde dehydrogenase (HFD1), and fatty acyl-CoA synthetase (FAA1/FAA4) genes for inhibition β-oxidation pathway, fatty acyl-CoA, fatty acids formation as well as the expression of MmCAR from *M. marinum* and alcohol dehydrogenase (ADH5). However, merely 20% of fatty acids were converted into fatty alcohols, leading them to use techniques through application of both FFA and acyl-CoA pathways, respectively (Fig. 9.3, a). Additionally, they expressed genes in a previous work and also they expressed FaCoAR gene from *Marinobacter aquaeolei* to directly convert acyl-CoA to fatty alcohol. Lastly, final fatty alcohol titer amounted to 43.6 mg•L^{-1} after designing strong promoters for pathway regulation. Xu et al. (2016) reached to 2.15 g•L^{-1} fatty alcohol and introduced the constitution of acyl-CoAs from cytosolic activation of FFA as a critical step for fatty alcohol production. An amount of 6.33 g•L^{-1} of fatty alcohol titer in the fed-batch fermentation was obtained in *E. coli* by inducing fatty acid starvation in MGL2 strain by expressing fatty acyl reductase (*FAR*) and elimination of acyl-ACP thioesterases (tesA, tesB, and tesC), ldhA, pta, and ackA genes to block competing pathways (Liu et al. 2016) (Fig. 9.3, b). In addition, fatty alcohol production increased 63-fold for the first time via modifying the tri-module; deleting fatty alcohol degradation pathway, enhancing *TaFAR1* expression, and fatty acyl-CoA supply (Wang et al. 2016b). d'Espaux et al. (2017) produced 6.0 g•L^{-1} fatty alcohols titer of *S. cerevisiae* in fed-batch fermentation over-expressed *MmFAR1*, *ACC1*, *OLE1* and also deleted *DGA1*, *HFD1*, *ADH6*, and *GDH1* genes in the selected strains after screening of strains.

Various biosynthetic pathways have been discovered naturally for de novo microbial synthesis of alkanes and alkenes through decarbonylation of fatty aldehydes, fatty acid decarboxylation, head-to-head hydrocarbon biosynthesis, and polyketide synthase (PKS) pathway (Kang and Nielsen 2017) (Fig. 9.3, d). Nonetheless, the low titer, rate, and yield (TRY) continue to be challenges to produce alkanes/alkenes at industrial scale. Recently, developments made in engineered microbial strains suggest approaches to improve TRY extraordinarily using enzyme, precursor, and cofactor engineering. Zhou et al. (2018), for instance, have engineered a eukaryotic cell factory and noticed a tenfold improvement 1-alkene titer to 35.3 mg•L^{-1} in *S. cerevisiae* by transporter and cofactor engineering as well as dynamic enzyme control approach (Fig. 9.3, e). In *S. cerevisiae*, aldehyde intermediates have been reported to be a serious obstacle for the synthesis of a more productive medium-chain alkane. Thus, Zhu et al. (2017) after engineering fungal type I FASs in the one-step produced 3.35 mg•L^{-1} (1.25 mg^{-1}•OD^{-1}) alkenes titer by UndA decarboxylase. Similar investigations have tried enormously to characterize efficient aldehyde decarbonylases as a serious drawback in the biosynthesis of alkane. A strategy has been proposed to improve alkane production in yeast cell factories through evaluation of ADs from cyanobacteria (ADOs), an insect (*Drosophila melanogaster* CYP4G1), and a plant (*Arabidopsis* CER1) (Kang et al. 2017) (Fig. 9.3, c). Results indicate that cyanobacteria ADs are perfect AD enzymes. Then, under GAL promoters, alkane production enhanced to 0.20 mg•L^{-1}•$OD600^{-1}$

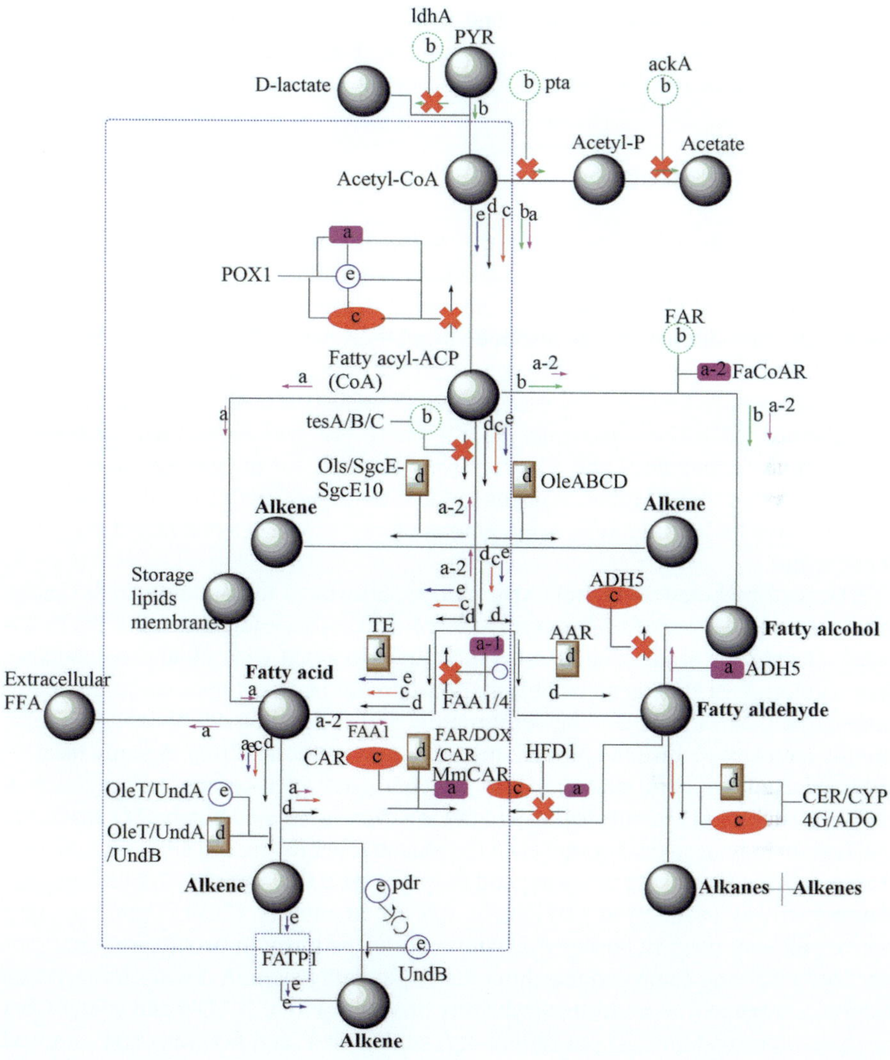

Fig. 9.3 Fatty acid biofuel biosynthetic pathways and enzymes. Alkane/alkene and fatty alcohol production pathway engineered in microorganisms. (**a**, Gonçalves Teixeira et al. 2017); **b**, Liu et al. 2016); **c**, Kang et al. 2017); **d**, Kang and Nielsen 2017; **e**, Zhou et al. 2018). Enzyme names are in (a) pink, (b) green, (c) red, (d) black, and (e) blue lines and shapes: Ols, olefin synthase; OleA, thiolase; OleD, short-chain dehydrogenase/reductase; OleC, AMP-dependent ligase/synthase; OleT, cytochrome P450 enzyme; OleB, α/β-hydrolase; UndA/B, aldehyde decarboxylase; FAA1/FAA4, fatty acyl-CoA synthetase; AAR, acylACP reductase; CER/CYP4G, aldehyde decarbonylase; FAR, fatty acid reductase; ADO, aldehyde deformylating oxygenase; FATP1, long-chain fatty acid transporter protein 1; DOX, fatty acid α-dioxygenase; CAR, carboxylic acid reductase; POX1, fatty acyl-CoA oxidase; pdr, putidaredoxin-putidaredoxinreductase (pdx/pdr); HFD1, aldehyde dehydrogenase; ADH5, alcohol dehydrogenase; ldhA, lactic dehydrogenase; pta, phosphate acetyl-transferase; ackA, acetate kinase; MmCAR, carboxylic acid reductase. The unsought genes in the biosynthetic pathways are removed as marked with "X"

by dynamic expression of *CwADO* and *SeADO* from *Crocosphaera watsonii* and *Thermosynechococcus elongates*, respectively. Furthermore, Cao et al. (2016) reported an alkane titer of 1.31 g•L^{-1} in *E. coli*.

9.3 Higher Alcohols

Presently, the ethanol generated by the fermentation process is advantageous to other biofuels due to its prolonged availability (Choi et al. 2014). Ethanol, however, comes with several physicochemical challenges drawing attention toward the use of higher alcohol alternatives to gasoline (Immethun et al. 2016). Otherwise stated, some abilities of higher alcohols such as higher energy density, lower vapor pressure, and lower hygroscopicity render them popular compared to ethanol (Atsumi and Connor 2010; Lan and Liao 2013). There are two biosynthetic pathways, namely, non-fermentative and fermentative pathways, for the generation of higher alcohols. A variety of problems occur in the biosynthetic pathways of higher alcohols affecting their production, necessitating plenty of research conducted to resolve these issues.

The acetone-butanol-ethanol (ABE) native pathway of fermentation in the bacterium *Clostridium acetobutilicum* (Fig. 9.4a) (Lütke-Eversloh and Bahl 2011) has been selected for production of n-butanol but associated with some disadvantages like regulation of the balance between downstream and upstream enzymes, maintaining the balance between the consumption and production of cofactor, and forcing the pathway to raise the product flux (Bond-Watts et al. 2011). A study used 20 g•L^{-1} of crude glycerol and the yield was 6.9 g•L^{-1} of n-butanol using different methods such as (1) tricarboxylic acid cycle suppression moderately; (2) anaerobic catabolism enhancement for glycerol; (3) gluconeogenic flux direction into the oxidative pentose phosphate pathway; and (4) glycolytic flux forcing *via* pyruvate oxidation pathway (Saini et al. 2017) (Fig. 9.4, b). Shen et al. (2011) improved ABE native pathway titers in hosts by driving force of pathway in two stages, viz., production of driving force and matching the target pathway with driving force. As an example, n-butanol titers were elevated by driving force of NADH and acetyl-CoA in some investigations. To put differently, adhE, ldhA, and frd were first removed for NADH and acetyl-CoA driving force, and then acetyl-CoA acetyltransferase (AtoB) from *E. coli* was replaced with *Clostridium* acetoacetyl-CoA thiolase (Thl) to avoid high changes in Gibbs energy due to the production of acetoacetyl-CoA from acetyl-CoA. Thereafter, pta encoding phosphate acetyltransferase was deleted resulting in lowering ATP synthesis and increasing acetyl-CoA availability matched with AtoB. Moreover, trans-enoyl-CoA reductase (Ter) irreversible reaction was employed to couple these driving forces to one another. The study reported 30 g•L^{-1} titer in fermenter.

The modified Ehrlich pathway was employed in the non-fermentative pathway, known as amino acid biosynthesis/Ehrlich pathway as keto acid intermediates, precursors in biosynthetic pathways of amino acid, resulting in alcohol production through the integration of a keto acid decarboxylase and an alcohol dehydrogenase.

The reverse β-oxidation and amino acid oxidation/Ehrlich pathways have been suggested to produce the longer chain alcohols (Immethun et al. 2016).

The combination of threonine-based keto-acid pathway and the citramalate pathway in *S. cerevisiae*, their optimization, and the expression of *KDC*, *ADH*, *LEU1* (two copies), *LEU4*, *LEU2* (two copies), *LEU5*, *CimA*, *NFS1*, *ADH7*, and *ARO10* in a bioreactor and an anaerobic glass, respectively, resulted in the production of 1.05 g•L^{-1} and 835 mg•L^{-1} of 1-butanol titers (Shi et al. 2016b) (Fig. 9.5, d). Si et al. (2014) also attained a 1-butanol titer of 242.8 mg•L^{-1} through removal of adh1 gene and the over-expression of enzymes, unactivated competition enzyme. Similarly, deletion of activated endogenous pathway gene as *ADH1* as well as the expression of heterologous ABE produced >2 g•L^{-1} of butanol titers (Swidah et al. 2018).

To enhance isobutanol production, the genes ILV5, ILV3, and ILV2 were overexpressed in the mitochondrial biosynthetic network of valine in *S. cerevisiae* under aerobic condition yielding 4.12 mg per g glucose (Chen et al. 2011) (Fig. 9.5, a). In a study by Matsuda et al. (2013), the production of isobutanol in two strains was respectively increased to 1.62 ± 0.11 g•L^{-1} and 1.61 ± 0.03 g•L^{-1} through competing pathways elimination methods and resolving cofactor imbalance (Matsuda et al. 2013) (Fig. 9.5b). Multiple membrane bottlenecks related to transportation of intermediates can reduce the yield of production and intake the intermediates through the competition. These issues were solved through mitochondrial compartmentalization to make alpha-ketoisovalerate (α-KIV) available, decline the absence of α-KIV in the competition reaction, and resolve α-KIV transfer blockages. To this end, the expression of *Llv2/3/5* and the overexpression of downstream *α-KDC* and *ADH* enzymes in mitochondria. Eventually, isobutanol was produced with titers of 635 ± 23 (mg•L^{-1}) (Avalos et al. 2013) (Fig. 9.5, c). As the products are inadequate in both *S. cerevisiae* and *E. coli*, such industrial organisms as *Pichia pastoris* were employed to produce higher alcohols, leading to the achievement of 2.22 g•L^{-1} of isobutanol titer (Siripong et al. 2018).

9.4 Isoprenoid-Derived Biofuel

According to Gupta and Phulara (2015), isoprenoids as valuable compounds have diverse physicochemical features and are synthesized through dimethylallyl pyrophosphate (DMAPP) and isopentenyl pyrophosphate (IPP) intermediates in the methylerythritol phosphate (MEP) pathway in prokaryotes and green algae and mevalonate (MVA) pathway in eukaryotes. Figure 9.6 demonstrates an overview of isoprenoid biofuel biosynthetic pathways. Numerous approaches of synthetic biology and metabolic engineering have been employed to produce different biofuels derived from isoprenoids, such as hemiterpenoids (e.g., isopentenol (George et al. 2015) (Fig. 9.7, e)), monoterpenes (e.g., geraniol (Zhao et al. 2016) (Fig. 9.7, c), pinene (Sarria et al. 2014; Tashiro et al. 2016; Niu et al. 2018) (Fig. 9.7, d), linalool (Rico et al. 2010; Cao et al. 2017), and limonene (Eng et al. 2018)), and sesquiterpenes (e.g., farnesol (Wang et al. 2016a), farnesene (Wang et al. 2011; Zhu et al. 2014; van Rossum et al. 2016; Yang et al. 2016) (Fig. 9.7, a), bisabolene (Kirby

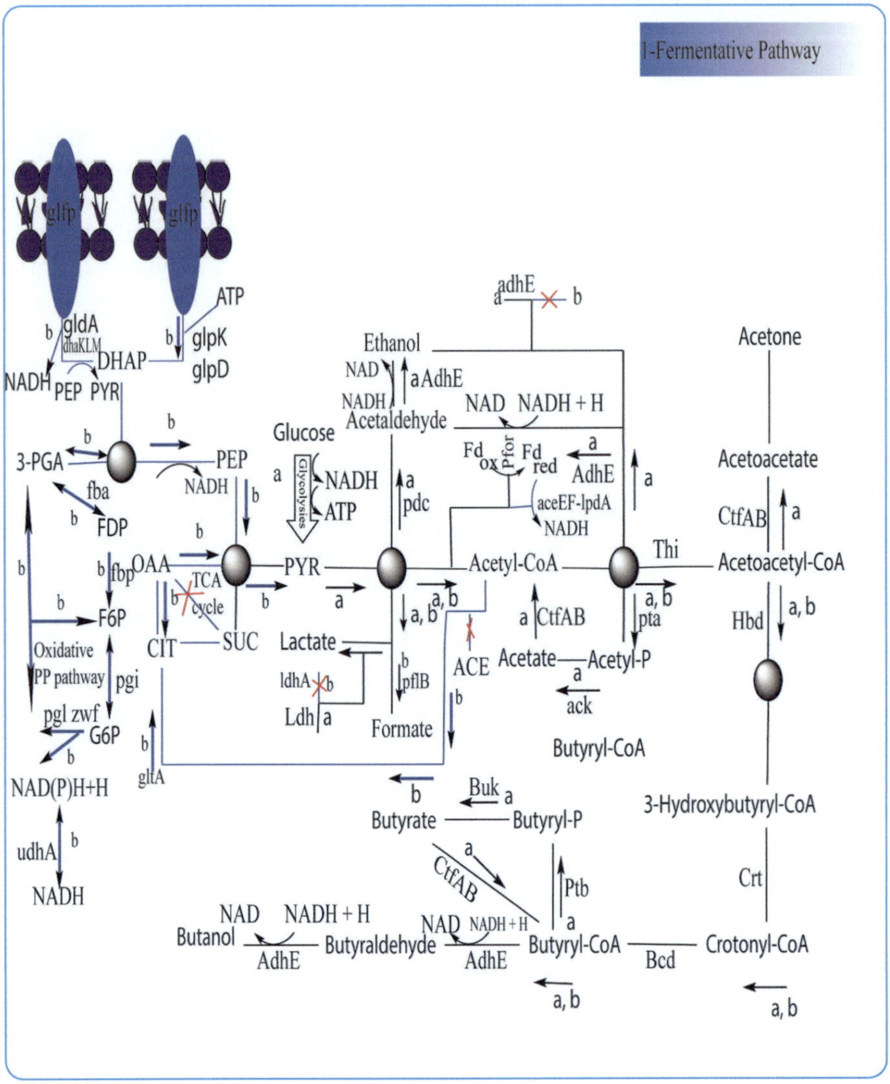

Fig. 9.4 Higher alcohol fermentative pathways. (a, Lütke-Eversloh and Bahl 2011; b, Saini et al. 2017). Intermediates and enzyme names are in lines: CIT, citrate; OAA, oxaloacetate; fba, fructose bisphosphate aldolase; Crt, crotonase; DHAP, dihydroxyacetone phosphate; ter, trans-enoyl-CoA reductase; 3-PGA, 3-phosphoglyceraldehyde; FDP, fructose 1,6-bisphosphate; PYR, pyruvate; SUC, succinate; PEP, phosphoenolpyruvate; crt, crotonase; frdABCD, fumarate reductase; adhE, aldehyde-alcohol dehydrogenase; Fd, ferredoxin; Etf, electron transfer flavoprotein; adhE2, butyraldehyde-butanol dehydrogenase; gldA, glycerol dehydrogenase; zwf, glucose-6-phosphate dehydrogenase; gltA, citrate synthase; Bcd, butyryl-CoA dehydrogenase; G6P, glucose-6-phosphate; Ack, acetate kinase; Ldh, lactate dehydrogenase; Pdc, pyruvate decarboxylase; phaA, acetoacetyl-CoA thiolase; Pfor, pyruvate:ferredoxin oxidoreductase; fbp, fructose 1,6-bisphosphatase; Pta, phosphotransacetylase; Ptb, phosphotransbutyrylase; udhA, transhydrogenase; glpF, glycerol facilitator; Adc, acetoacetate decarboxylase; F6P, fructose-6-phosphate; hbd, 3-hydroxybutyryl-

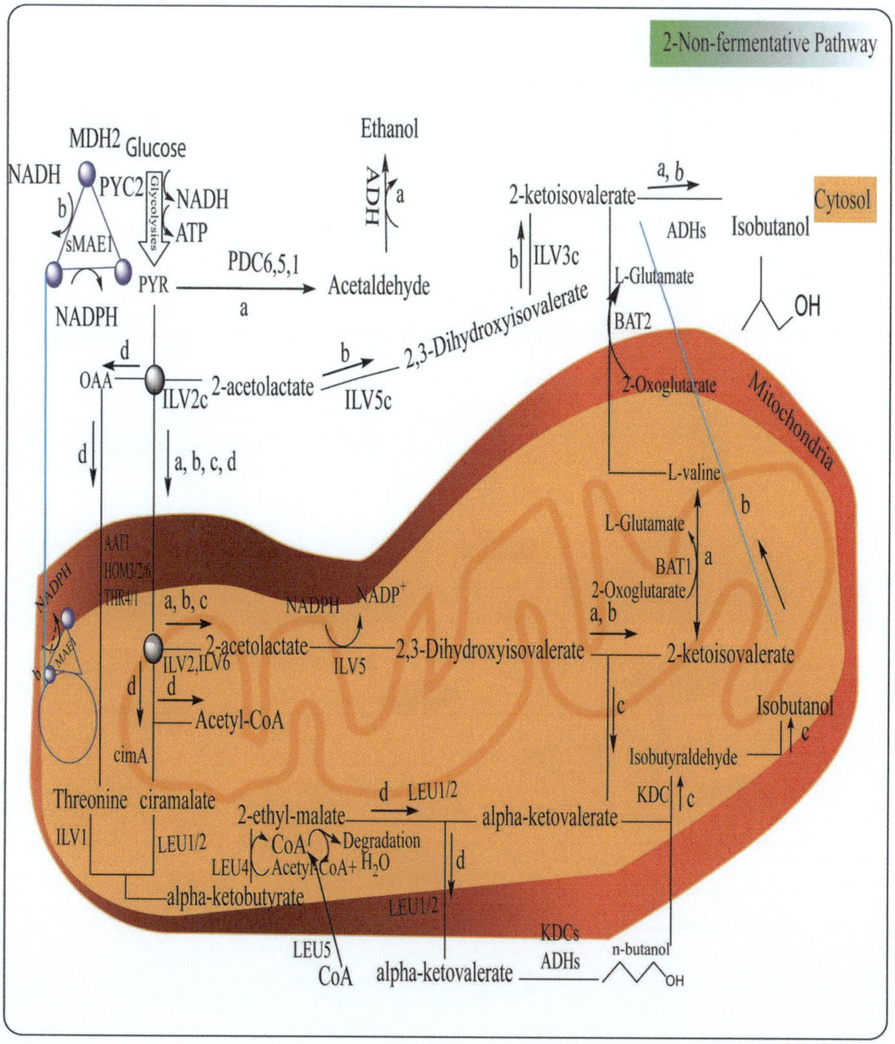

Fig. 9.5 Higher alcohol non-fermentative pathways. (a, Chen et al. 2011; b, Matsuda et al. 2013; c, Avalos et al. 2013; d, Shi et al. 2016b). Enzyme names are in lines: BAT1/2, aminotransferase; α-KDC, α-ketoacid decarboxylase; PDC6,5,1, pyruvate decarboxylases; ADH, alcohol dehydrogenase; ILV2/6, synthase; ILV3, dehydratase; ILV5, reductoisomerase; ILV3c, ILV2c, and ILV5c, valine cytosolic pathway; CimA, citramalate synthase; PYC2, MDH2, and sMAE1, transhydrogenase-like shunt; LEU4, LEU2, and LEU1, synthetase

Fig. 9.4 (continued) CoA dehydrogenase; dhaKLM, dihydroxyacetone kinase; SADH, secondary alcohol dehydrogenase; Thi, thiolase, SUC, succinate; ldhA, lactate dehydrogenase; Buk, butyrate kinase; glpK, glycerol kinase; pflB, pyruvate-formate lyase; glpD, glycerol-3-phosphate dehydrogenase; CtfAB, acetoacetyl-CoA:acyl-CoA transferase; pgl, lactonase; pgi, phosphoglucose isomerase; aceEFlpdA*, pyruvate dehydrogenase complex. The unsought genes in the pathways have been eliminated as indicated by "X"

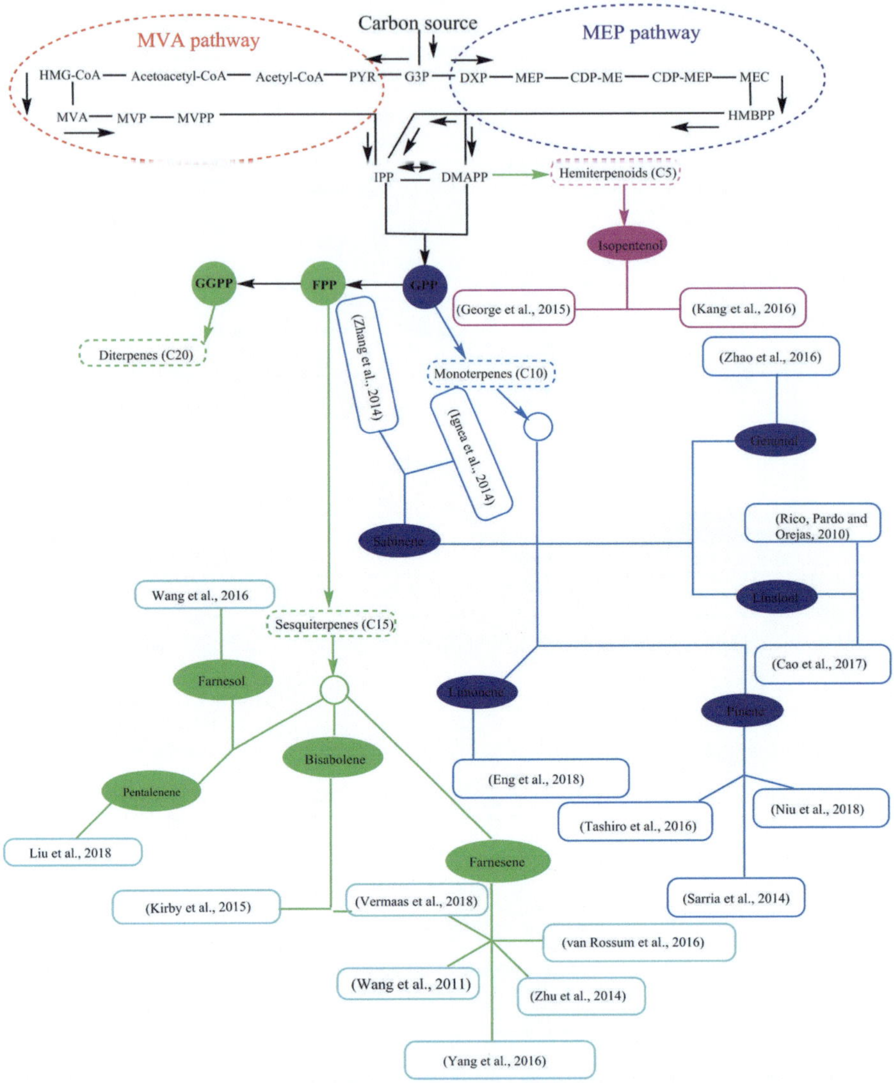

Fig. 9.6 Schematic of isoprenoid biofuel biosynthetic pathways. Pathway intermediates: MVPP, mevalonate pyrophosphate; MVP, mevalonate 5-phosphate; HMBPP, hydroxymethylbutenyl-4-diphosphate; GGPP, geranylgeranyl diphosphate; IPP, isopentenyl pyrophosphate; CDP-MEP, 4-diphosphocytidyl-2-C-methylerythritol-2-phosphate; MVA, mevalonate; CDP-ME, 4-diphosph ocytidyl-2-C-methylerythritol; FPP, farnesyl diphosphate; HMG-CoA, hydroxymethylglutaryl-CoA; MEC, 2-C-methylerythritol-2,4-cyclodiphosphate; MEP, 2-C-methyl-D-erythritol-4-phosphate; GPP, geranyl diphosphate; PYR, pyruvate; DXP, 1-deoxy-D-xylulose-5-phosphate; DMAPP, dimethylallyl pyrophosphate

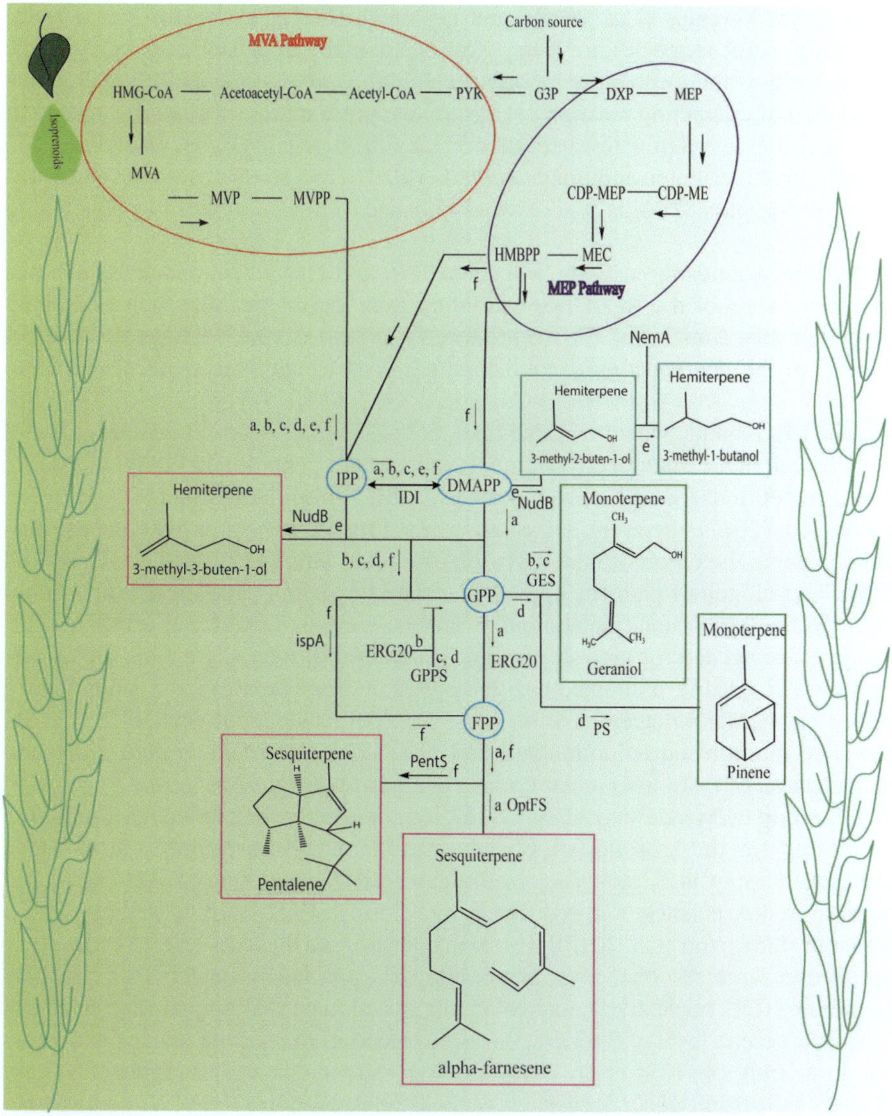

Fig. 9.7 Isoprenoid biofuel biosynthetic pathways and enzymes. (a, Yang et al. 2016; b, Lin et al. 2018; c, Zhao et al. 2016; d, Sarria et al. 2014; e, George et al. 2015; f, Liu et al. 2018). Pathway intermediates: Mev-PP, mevalonate pyrophosphate; MevP, mevalonate-5-phosphate; HMG-CoA, hydroxymethylglutaryl-CoA; GPP, geranyl diphosphate; AA-CoA, acetoacetyl-CoA; DMAPP, dimethylallyl pyrophosphate; IPP, isopentenyl pyrophosphate; A-CoA, acetyl-CoA. Enzyme names are in lines and shapes: atoB, thiolase; IDI, IPP isomerase; ispA, synthase; GPPS1/2, geranyl diphosphate synthase; ERG20, farnesyl diphosphate synthase; PMK, phosphomevalonate kinase; MK, mevalonate kinase; HMGR/tHMG, HMG-CoA reductase; GES, synthase; HMGS,HMG-CoA synthase; FPP, farnesyl diphosphate; PS, pinene synthase; NemA, reductase; NudB, phosphatase; PMD, diphosphomevalonate decarboxylase

et al. 2015; Vermaas et al. 2018), and pentalene (Liu et al. 2018) (Fig. 9.7, f)). Moreover, some strategies are there to eliminate many drawbacks and enhance the yields, such as the localization of specific reactions, the knock-out or knock-down regulation of competing reactions, the complete presentation of an exogenous pathway, and alleviation of pathway bottlenecks by the balancing of enzyme expression (Ward, Chatzivasileiou, and Stephanopoulos 2018). In a study of Kang et al. (2016) a new IPP-bypass MVA pathway was established to increase titer of isopentanol in E. coli.

Monoterpenoids farnesyl diphosphate (FPP) and HMG-CoA reductase are also regulatory sites of the MVA pathway other than hemiterpenoids such as isopentenol. For this, Amiri et al. (2016) produced a strain to express LIS gene from lavender *(Lavandula angustifolia)* for the induction of the ability to synthesize linalool in *S. cerevisiae*. Afterward, HMG-CoA reductase (tHMG1) over-expressed and suppressed ERG9 gene, which elevated linalool by 95 $\mu g{\bullet}L^{-1}$, though toxicity was still a bottleneck in the linalool production. Basal monoterpenoids and FPP biosynthesis have competitions for GPP consumption in the MVA biosynthetic pathway. Mendez-Perez et al. (2017), therefore, increased availability of GPP for monoterpenoid production using the chromosomal mutation of *E. coli* native FPP synthase (*IspA*) and MVA heterologously pathway approaches, thereby yielding 505 $mg{\bullet}L^{-1}$ of linalool titer. On the other hand, metabolomics analysis revealed that basal levels of GPP and FPP are needed for high-throughput generation of monoterpenoids. In a study by Lin et al. (2018), a coupled enzyme-based fluorogenic assay was suggested to screen the geraniol detection, so that the geraniol production resulted in 1.2-fold elevation through saturation mutagenesis of *GES* of *Castellaniela defragrans* and the detection of F418 to Q mutation (Lin et al. 2018) (Fig. 9.7b).

In a study by Meadows et al. in (2016), four non-native metabolic reactions were introduced for the biosynthesis of the cytosolic acetyl coenzyme A to enhance farnesene supply in *S. cerevisiae* with decreased consumption of ATP, increased pathway redox balance, and reduced carbon loss to CO2-emitting reactions. In a different study (You et al. 2017), IPP toxic intermediate and cost-effective farnesene production were resolved after expression of ispA and isopentenyl diphosphate isomerase (IDI), respectively, and balancing mevalonate (MVA) pathway, thereby a final β-farnesene level of 2.83 $g{\bullet}L^{-1}$ was produced in *E. coli* in a lab-scale bioreactor. In a study of (Kim et al. 2017) the use of dynamic sensor-regulator system (LuxI/R QS) from *Vibrio fischeri* was suggested to optimize the MVA pathway for the enhancement of bisabolene synthesis, as alternative to the use of inducer promoter as IPTG. A report by Alonso-Gutierrez et al. (2017) also indicates a new method for promoting bisabolene production via the chromosomal metabolic engineering of MVA pathway. For this purpose, they engineered sucrose consumption pathway in *E. coli* and developed a CRISPR-Cas9 system to increase transcriptional rates and enzyme expression in this biosynthetic pathway. However, toxicity was the major challenge of this approach as it hinders extensive additional research in this field.

9.5 Conclusion

Recent years have been associated with the use of microorganisms capable of producing high-energy fuels in order to solve the lack of fossil fuels by environmentally sustainable and economically efficient bioprocess. However, the presence of many challenges in microbial biosynthetic pathways hinders addressing these issues, which require extensive application of optimization techniques (e.g., CRISPR methods) and also biosensors to regulate and facilitate biofuel network in the microorganisms. Additionally, appropriate commercialization of biofuels is possible by finding novel microorganisms that possess high capacity in the production of biofuels.

References

Abdel-Mawgoud AM et al (2018) Metabolic engineering in the host Yarrowia lipolytica. Metabolic engineering. Elsevier Inc. https://doi.org/10.1016/j.ymben.2018.07.016

Alonso-Gutierrez J et al (2017) Toward industrial production of isoprenoids in *Escherichia coli*: lessons learned from CRISPR-Cas9 based optimization of a chromosomally integrated mevalonate pathway. Biotechnol Bioeng 115:1000–1013. https://doi.org/10.1002/bit.26530

Amiri Pet al (2016) Metabolic engineering of *Saccharomyces cerevisiae* for linalool production. Biotechnol Lett Springer Netherlands 38(3):503–508. https://doi.org/10.1007/s10529-015-2000-4

Atsumi S, Connor MR (2010) Synthetic biology guides biofuel production. J Biomed Biotechnol 2010:9. https://doi.org/10.1155/2010/541698

Avalos JL, Fink GR, Stephanopoulos G (2013) Compartmentalization of metabolic pathways in yeast mitochondria improves the production of branched-chain alcohols. Nat Biotechnol 31(4):335–341. https://doi.org/10.1038/nbt.2509

Baumann L et al (2018) A yeast-based biosensor for screening of short- and medium-chain fatty acid production. ACS Synth Biol 7(11):2640–2646. https://doi.org/10.1021/acssynbio.8b00309

Bond-Watts BB, Bellerose RJ, Chang MCY (2011) Enzyme mechanism as a kinetic control element for designing synthetic biofuel pathways. Nat Chem Biol Nature Publishing Group 7(4):222–227. https://doi.org/10.1038/nchembio.537

Cao YX et al (2016) Heterologous biosynthesis and manipulation of alkanes in *Escherichia coli*. Metab Eng Elsevier 38:19–28. https://doi.org/10.1016/j.ymben.2016.06.002

Cao X et al (2017) Enhancing linalool production by engineering oleaginous yeast Yarrowia lipolytica. Bioresour Technol 245(Pt B):1641–1644. https://doi.org/10.1016/j.biortech.2017.06.105

Chen X et al (2011) Increased isobutanol production in Saccharomyces cerevisiae by overexpression of genes in valine metabolism. Biotechnol Biofuels 4:1–12. https://doi.org/10.1186/1754-6834-4-21

Choi YJ et al (2014) Metabolic engineering of microorganisms for the production of higher alcohols. mBio5(5):1–10. https://doi.org/10.1128/mBio.01524-14.Copyright

d'Espaux L et al (2017) Engineering high-level production of fatty alcohols by Saccharomyces cerevisiae from lignocellulosic feedstocks. Metab Eng Elsevier Inc 42:115–125. https://doi.org/10.1016/j.ymben.2017.06.004

Ekas H, Deaner M, Alper HS (2019) Recent advancements in fungal-derived fuel and chemical production and commercialization. Curr Opin Biotechnol Elsevier Ltd 57:1–9. https://doi.org/10.1016/j.copbio.2018.08.014

Eng T et al (2018) Restoration of biofuel production levels and increased tolerance under ionic liquid stress is enabled by a mutation in the essential *Escherichia coli* gene cydC. Microb Cell Fact BioMed Central 17(1):159. https://doi.org/10.1186/s12934-018-1006-8

Ferreira R, Teixeira PG, Gossing M, David F, Siewers V, Nielsen J (2018) Metabolic engineering of Saccharomyces cerevisiae for overproduction of triacylglycerols. Metab Eng Commun 6:22–27

Friedlander J et al (2016) Engineering of a high lipid producing Yarrowia lipolytica strain. Biotechnol Biofuels 9:77. https://doi.org/10.1186/s13068-016-0492-3

George K et al (2015) Metabolic engineering for the high-yield production of isoprenoid-based C 5 alcohols in *E. coli*. Sci Rep 5:11128. https://doi.org/10.1038/srep11128

Ghosh A et al (2016) (13)C metabolic flux analysis for systematic metabolic engineering of *S. cerevisiae* for overproduction of fatty acids. Front Bioeng Biotechnol 4:76. https://doi.org/10.3389/fbioe.2016.00076

Gupta P, Phulara SC (2015) Metabolic engineering for isoprenoid-based biofuel production. J Appl Microbiol 119(3):605–619. https://doi.org/10.1111/jam.12871

Immethun CM et al (2016) Chapter 1–engineering central metabolism for production of higher alcohol-based biofuels. In: Eckert CA, Trinh CT (eds) Biotechnology for biofuel production and optimization. Elsevier, Amsterdam, pp 1–34. https://doi.org/10.1016/B978-0-444-63475-7.00001-7

Jiang W, Gu P, Zhang F (2018) Steps towards "drop-in" biofuels: focusing on metabolic pathways. Curr Opin Biotechnol 53:26–32. https://doi.org/10.1016/j.copbio.2017.10.010

Kang M-K, Nielsen J (2017) Biobased production of alkanes and alkenes through metabolic engineering of microorganisms. J Ind Microbiol Biotechnol 44(4):613–622. https://doi.org/10.1007/s10295-016-1814-y

Kang A et al (2016) Isopentenyl diphosphate (IPP)-bypass mevalonate pathways for isopentenol production.Metab Eng Elsevier 34:25–35. https://doi.org/10.1016/j.ymben.2015.12.002

Kang M-K et al (2017) Functional screening of aldehyde decarbonylases for long-chain alkane production by Saccharomyces cerevisiae. Microb Cell Factories 16:1–8. https://doi.org/10.1186/s12934-017-0683-z

Kim EM et al (2017) Autonomous control of metabolic state by a quorum sensing (QS)-mediated regulator for bisabolene production in engineered *E. coli*. Metab Eng Elsevier Inc.44:325–336. https://doi.org/10.1016/j.ymben.2017.11.004

Kirby J et al (2015) Enhancing Terpene yield from sugars via novel routes to 1-deoxy-D-xylulose 5-phosphate. Appl Environ Microbiol 81(1):130–138. https://doi.org/10.1128/AEM.02920-14

Lan EI, Liao JC (2013) Microbial synthesis of n-butanol, isobutanol, and other higher alcohols from diverse resources. Bioresour Technol 135:339–349. https://doi.org/10.1016/j.biortech.2012.09.104

Lazar Z, Liu N, Stephanopoulos G (2018) Holistic approaches in lipid production by Yarrowia lipolytica. Trends Biotechnol Elsevier 36(11):1157–1170. https://doi.org/10.1016/j.tibtech.2018.06.007

Leber C et al (2015) Disrupted short chain specific β-oxidation and improved synthase expression increase synthesis of short chain fatty acids in *Saccharomyces cerevisiae*. Biotechnol Bioeng 113(4):895–900. https://doi.org/10.1002/bit.25839

Lin J-L et al (2018) An enzyme-coupled assay enables rapid protein engineering for geraniol production in yeast. Biochem Eng J Elsevier 139:95–100. https://doi.org/10.1016/J.BEJ.2018.08.011

Liu Y et al (2016) High production of fatty alcohols in *Escherichia coli* with fatty acid starvation. Microb Cell Fact BioMed Central 15(1):1–10. https://doi.org/10.1186/s12934-016-0524-5

Liu C-L et al (2018) Renewable production of high density jet fuel precursor sesquiterpenes from *Escherichia coli.* Biotechnol Biofuels BioMed Central 11(1):285. https://doi.org/10.1186/s13068-018-1272-z

López-Lara IM, Soto MJ (2018) Fatty acid synthesis and regulation. In: Geiger O (ed) Biogenesis of fatty acids, lipids and membranes. Springer, Cham, pp 1–17. https://doi.org/10.1007/978-3-319-43676-0_26-1

Lütke-Eversloh T, Bahl H (2011) Metabolic engineering of Clostridium acetobutylicum: recent advances to improve butanol production. Curr Opin Biotechnol Elsevier Current Trends 22(5):634–647. https://doi.org/10.1016/J.COPBIO.2011.01.011

Marella ER et al (2018) Engineering microbial fatty acid metabolism for biofuels and biochemicals. Curr Opin Biotechnol 50(Table 1):39–46. https://doi.org/10.1016/j.copbio.2017.10.002

Markham KA, Alper HS (2018a) Engineering Yarrowia lipolytica for the production of cyclopropanated fatty acids. J Ind Microbiol Biotechnol 45(10):881–888. https://doi.org/10.1007/s10295-018-2067-8

Markham KA, Alper HS (2018b) Synthetic biology expands the industrial potential of Yarrowia lipolytica. Trends Biotechnol Elsevier Ltd 36(10):1085–1095. https://doi.org/10.1016/j.tibtech.2018.05.004

Matsuda F et al (2013) Increased isobutanol production in *Saccharomyces cerevisiae* by eliminating competing pathways and resolving cofactor imbalance. Microb Cell Factories 12:119. https://doi.org/10.1186/1475-2859-12-119

Meadows AL et al (2016) Rewriting yeast central carbon metabolism for industrial isoprenoid production. Nature Nature Publishing Group537(7622):694–697. https://doi.org/10.1038/nature19769

Meadows CW, Kang A, Lee TS (2017) Metabolic engineering for advanced biofuels production and recent advances toward commercialization. Biotechnol J 13(1):1600433. https://doi.org/10.1002/biot.201600433

Mendez-Perez D et al (2017) Production of jet fuel precursor monoterpenoids from engineered *Escherichia coli*. Biotechnol Bioeng 114(8):1703–1712. https://doi.org/10.1002/bit.26296

Mukherjee K, Bhattacharyya S, Peralta-Yahya P (2015) GPCR-based chemical biosensors for medium-chain fatty acids. ACS Synth Biol American Chemical Society4(12):1261–1269. https://doi.org/10.1021/sb500365m

Nielsen J, Keasling JD (2016) Engineering cellular metabolism. Cell Elsevier Ltd 164(6):1185–1197. https://doi.org/10.1016/j.cell.2016.02.004

Niu F et al (2018) Enhancing production of Pinene in Escherichia coli by using a combination of tolerance, evolution, and modular co-culture engineering. Front Microbiol 9:1–14. https://doi.org/10.3389/fmicb.2018.01623

Peralta-Yahya PP, Keasling JD (2010) Advanced biofuel production in microbes. Biotechnol J 5(2):147–162. https://doi.org/10.1002/biot.200900220

Pfleger BF, Gossing M, Nielsen J (2015) Metabolic engineering strategies for microbial synthesis of oleochemicals. Metab Eng Elsevier 29:1–11. https://doi.org/10.1016/j.ymben.2015.01.009

Qiao K et al (2017) Lipid production in Yarrowia lipolytica is maximized by engineering cytosolic redox metabolism. Nat Biotechnol Nature Publishing Group 35(2):173–177. https://doi.org/10.1038/nbt.3763

Rico J, Pardo E, Orejas M (2010) Enhanced production of a plant monoterpene by overexpression of the3-hydroxy-3-methylglutaryl coenzyme A reductase catalytic domain in *Saccharomyces cerevisiae*. Appl Environ Microbiol 76(19):6449–6454. https://doi.org/10.1128/AEM.02987-09

RigouinCet al (2018) Increasing medium chain fatty acids production in Yarrowia lipolytica by metabolic engineering. Microb Cell Fact BioMed Central 17(1):142. https://doi.org/10.1186/s12934-018-0989-5

Runguphan W, Keasling JD (2014) Metabolic engineering of *Saccharomyces cerevisiae* for production of fatty acid-derived biofuels and chemicals. Metab Eng 21:103–113. https://doi.org/10.1016/j.ymben.2013.07.003

Saini M et al (2017) Metabolic engineering of *Escherichia coli* for production of n-butanol from crude glycerol. Biotechnol Biofuels 10:173. https://doi.org/10.1186/s13068-017-0857-2

Sànchez i Nogué V et al (2018) Integrated diesel production from lignocellulosic sugars via oleaginous yeast. Green Chem 20:4349–4365. https://doi.org/10.1039/C8GC01905C

Sarria S et al (2014) Microbial synthesis of Pinene. ACS Synth Biol 3:466–475. https://doi.org/10.1021/sb4001382

Schwartz C, Löbs A (2018) Multiplexed CRISPR activation of cryptic sugar metabolism enables Yarrowia lipolytica growth on cellobiose, pp 1–21

Shen CR et al (2011) Driving forces enable high-titer anaerobic 1-butanol synthesis in *Escherichia coli*. Appl Environ Microbiol 77(9):2905–2915. https://doi.org/10.1128/AEM.03034-10

Shi S, Zhao H (2017) Metabolic engineering of oleaginous yeasts for production of fuels and chemicals. Front Microbiol 8:2185. https://doi.org/10.3389/fmicb.2017.02185

Shi S, Ji H et al (2016a) Improved production of fatty acids by *Saccharomyces cerevisiae* through screening a cDNA library from the oleaginous yeast *Yarrowia lipolytica*. FEMS Yeast Research. Oxford University Press 16(1):fov108. https://doi.org/10.1093/femsyr/fov108

Shi S, Si T et al (2016b) Metabolic engineering of a synergistic pathway for n-butanol production in *Saccharomyces cerevisiae*. Sci Rep 6:25675. https://doi.org/10.1038/srep25675

Si T et al (2014) Utilizing an endogenous pathway for 1-butanol production in *Saccharomyces cerevisiae*.Metab Eng Elsevier22:60–68. https://doi.org/10.1016/j.ymben.2014.01.002

Siripong Wet al (2018) Metabolic engineering of Pichia pastoris for production of isobutanol and isobutyl acetate. Biotechnol Biofuels BioMed Central 11(1):1–16. https://doi.org/10.1186/s13068-017-1003-x

Stephanopoulos G, Gill RT (2001) After a decade of progress, an expanded role for metabolic engineering. In: Nielsen J et al (eds) Metabolic engineering. Springer, Berlin/Heidelberg, pp 1–8. https://doi.org/10.1007/3-540-45300-8_1

Stephanopoulos G, Vallino JJ (1991) Network rigidity and metabolic engineering in metabolite overproduction. Science 252(5013):1675 LP–1671681. Available at: http://science.sciencemag.org/content/252/5013/1675.abstract

Swidah R et al (2018) N-Butanol production in *S. cerevisiae*: co-ordinate use of endogenous and exogenous pathways. Appl Microbiol Biotechnol 102:9857–9866

Tai M, Stephanopoulos GN (2015) Metabolic engineering: enabling technology for biofuels production. In: Advances in bioenergy. Wiley, Hoboken, pp 1–10. https://doi.org/10.1002/9781118957844.ch1

Tashiro M et al (2016) Bacterial production of Pinene by a laboratory-evolved Pinene-synthase. ACS Synth Biol 5(9):1011–1020. https://doi.org/10.1021/acssynbio.6b00140

Teixeira PG et al (2017) Dynamic regulation of fatty acid pools for improved production of fatty alcohols in *Saccharomyces cerevisiae*. Microb Cell Factories 16(1):45. https://doi.org/10.1186/s12934-017-0663-3

Teixeira PG et al (2018) Engineering lipid droplet assembly mechanisms for improved triacylglycerol accumulation in Saccharomyces cerevisiae. FEMS Yeast Res. Oxford University Press 18(6). https://doi.org/10.1093/femsyr/foy060

van Rossum H et al (2016) Engineering cytosolic acetyl-coenzyme a supply in *Saccharomyces cerevisiae*: pathway stoichiometry, free-energy conservation and redox-cofactor balancing. Metab Eng 36:99–115. https://doi.org/10.1016/j.ymben.2016.03.006

Vermaas JV et al (2018) Membrane permeability of Terpenoids explored with molecular simulation. J Phys Chem B 122(45):10349–10361. https://doi.org/10.1021/acs.jpcb.8b08688

Wang C et al (2011) Metabolic engineering of Escherichia coli for α-farnesene production. Metab Eng 13(6):648–655. https://doi.org/10.1016/j.ymben.2011.08.001

Wang C et al (2016a) Farnesol production in Escherichia coli through the construction of a farnesol biosynthesis pathway – application of PgpB and YbjG phosphatases. Biotechnol J 11(10):1291–1297. https://doi.org/10.1002/biot.201600250

Wang G et al (2016b) Exploring fatty alcohol-producing capability of Yarrowia lipolytica. Biotechnol Biofuels 9:107. https://doi.org/10.1186/s13068-016-0512-3

Ward VCA, Chatzivasileiou AO, Stephanopoulos G (2018) Metabolic engineering of Escherichia coli for the production of isoprenoids. FEMS Microbiol Lett 365(10):fny079. https://doi.org/10.1093/femsle/fny079

Xu P et al (2016) Engineering *Yarrowia lipolytica* as a platform for synthesis of drop-in transportation fuels and oleochemicals. Proc Natl Acad Sci 113(39):10848–10853. https://doi.org/10.1073/pnas.1607295113

Xu P, Qiao K, Stephanopoulos G (2017) Engineering oxidative stress defense pathways to build a robust lipid production platform in *Yarrowia lipolytica*. Biotechnol Bioeng 114(7):1521–1530. https://doi.org/10.1002/bit.26285

Yan J et al (2017) Harnessing biodiesel-producing microbes: from genetic engineering of lipase to metabolic engineering of fatty acid biosynthetic pathway. Crit Rev Biotechnol 37(1):26–36. https://doi.org/10.3109/07388551.2015.1104531

Yang X et al (2016) Heterologous production of α-farnesene in metabolically engineered strains of Yarrowia lipolytica. Bioresour Technol 216:1040–1048. https://doi.org/10.1016/j.biortech.2016.06.028

You S et al (2017) Utilization of biodiesel by-product as substrate for high-production of β-farnesene via relatively balanced mevalonate pathway in Escherichia coli. Bioresour Technol 243:228–236. https://doi.org/10.1016/j.biortech.2017.06.058

Yu Tet al (2017) Metabolic engineering of Saccharomyces cerevisiae for production of very long chain fatty acid-derived chemicals.Nat Commun. Nature Publishing Group 8:15587. https://doi.org/10.1038/ncomms15587

Yu T et al (2018) Reprogramming yeast metabolism from alcoholic fermentation to Lipogenesis. Cell Elsevier Inc 174(6):1549–1558.e14. https://doi.org/10.1016/j.cell.2018.07.013

Yuzbasheva E et al (2017a) A metabolic engineering strategy for producing free fatty acids by the Yarrowia lipolytica yeast based on impairment of glycerol metabolism: impairment of glycerol metabolism for FFA production. Biotechnol Bioeng 115:433–443. https://doi.org/10.1002/bit.26402

Yuzbasheva EY et al (2017b) Co-expression of glucose-6-phosphate dehydrogenase and acyl-CoA binding protein enhances lipid accumulation in the yeast Yarrowia lipolytica. New Biotechnol 39.(Pt A):18–21. https://doi.org/10.1016/j.nbt.2017.05.008

Zhao J et al (2016) Improving monoterpene geraniol production through geranyl diphosphate synthesis regulation in *Saccharomyces cerevisiae*. Appl Microbiol Biotechnol 100(10):4561–4571. https://doi.org/10.1007/s00253-016-7375-1

Zhou YJ et al (2018) Engineering 1-alkene biosynthesis and secretion by dynamic regulation in yeast. ACS Synth Biol Am Chem Soc 7(2):584–590. https://doi.org/10.1021/acssynbio.7b00338

Zhu F et al (2014) In vitro reconstitution of Mevalonate pathway and targeted engineering of farnesene overproduction in *Escherichia coli*. Biotechnol Bioeng 111:1396–1405. https://doi.org/10.1002/bit.25198

Zhu Z et al (2017) Enabling the synthesis of medium chain alkanes and 1-alkenes in yeast. Metab Eng Academic Press 44:81–88. https://doi.org/10.1016/J.YMBEN.2017.09.007

Nanomaterials and Its Application in Biofuel Production

10

Satish Kumar Sharma, Aradhana Kumari, Vinay Dwivedi, Pankaj Kumar Rai, and Monika Gupta

Abstract

Today's life is highly energy dependent. Is it possible to meet our energy demand in the future by rapidly depleting energy resources? The answer is no. The alarming hike in crude oil and its final products prices evidently shows the terrible future conditions. This situation urges individuals to explore the new way or techniques of finding nonconventional energy sources that will be able to meet growing demands of the fuel in the future. Biofuel can be assumed to be one of the alternative energy sources of having low carbon footprint. With the help of nanotechnology and its constituent areas, it is feasible to create an effective contribution to the biofuel manufacturing sector. To increase the manufacturing rate of biofuels, it is technically feasible to use nanomaterial bound microbial enzymes and nanocatalysts. Some kinds of nanomaterial additives are presently used for long-term stability and enhanced biofuel/biogas manufacturing yields at low input expenses. Developments of nanomaterials technology could also enhance the efficacy of extracting lipids from biosources.

S. K. Sharma · A. Kumari
Jawaharlal Nehru Krishi Vishwa Vidyalaya, College of Agriculture, Ganj Basoda, Vidisha, Madhya Pradesh, India

V. Dwivedi
Department of Biotechnology, Naraina Vidyapeeth Engineering & Management Institute, Kanpur, India

P. K. Rai
Departments of Biotechnology, Invertis University, Bareilly, India

M. Gupta (✉)
Departments of Biotechnology, Invertis University, Bareilly, India

Amity Institute of Biotechnology, Amity University Madhya Pradesh, Gwalior, Madhya Pradesh, India

N. Srivastava et al. (eds.), *Substrate Analysis for Effective Biofuels Production*, Clean Energy Production Technologies, https://doi.org/10.1007/978-981-32-9607-7_10

Keywords
Biofuel · Nanotechnology · Nanomaterials · Energy resources

10.1 Introduction

Today's life is highly energy dependent. The world is currently facing the problem of rapidly depleting energy resources which can be clearly visualized from the increased cost of crude oil and its products after refinements. Therefore, the increasing energy demand pushes scientists to search for various alternative energy sources to meet the demand. To date, most of the world's 90 percent demand for energy is being supplied in forms of nonrenewable/conventional electricity sources such as coal, fossil fuels, natural gas and crude oil. This rapid consumption of the above-mentioned nonrenewable fossil fuel reserves of the world will lead to the decrease in fossil fuel reserves in Nature by 2050 (Hussein 2015). Besides price escalation, these are other various factors related to fossil fuels which contribute considerably to environmental pollution and ecological devastation, causing depletion of ozone layer, global warming, etc. These factors act as driving force for the fuel industry to move towards sustainable sources of alternative energy, i.e. renewable energy sources. Hence, biofuel is an alternate source of energy derived from organic material (Rai et al. 2018). Biofuel is commonly advocated as a cost-effective and environmentally benign alternative to petroleum and other fossil fuels and may be known as bioethanol, biohydrogen, biodiesel, etc. Tremendous efforts are going on to increase the production of biofuel, and various techniques are being constantly researched. Nanotechnology seemed to be one of the promising techniques which can contribute to a large extent in the efficient biofuel production industry.

The renewable energy is derived from biological carbon fixation or during photo carbon assimilation. It is produced from organic product and wastes. It included fuel derived from liquefaction or biomass conversion process through solid biomass. Biodiesel is esters based originated biofuel and utilizes domestically available renewable resources (Puri et al. 2012) and is biodegradable and nontoxic in nature (Pugh et al. 2011). It can be produced from various feedstocks such as vegetable oils and biomass (Kralova and Sjöblom 2010). Anaerobic digestion of agricultural residues (Karellas et al. 2010), animal manures (Bidart et al. 2014), organic food wastes (Zhang et al. 2016), sewage sludge (Cao and Pawowski 2012) and various other energy crops (Lönnqvist et al. 2013; Rai 2016) are used as renewable source of energy. Product of soybean, groundnut and Jatropha can be used efficiently for biogas production by the use of nanomaterial techniques. In 1959, Feynman introduced the term 'nanotechnology', which refers to the manufacture and use of nanometre-scale materials (materials with at least one dimension less than 100 nm) (Jiang et al. 2013).

Nanotechnology is one of the big areas of science that show promising opportunities for turning original research into a successful development (Fig. 10.1). The

Fig. 10.1 Different disciplines and nanotechnology. (Adapted and modified from Kumar et al. 2014)

macromolecular, atomic and molecular scale of 'nanoscience' has also been researched in the material manipulation. The characteristics of these individuals' nanoparticles on a large scale have varied considerably. The nanomaterials have distinctive mechanical, chemical, optical, magnetic and electronic characteristics compared to their massive scale counterparts (Biswas et al. 2012). It also includes a wide variety of techniques in a wider context across a number of streams such as engineering, medical, health, defence, power, biotechnology, data technology, agriculture, water, food and environment. Nanomaterials play an important role in the field of energy owing to certain specific properties such as distinctive structure, high energy storage capability, comparatively large surface area and excellent lighting and heating effectiveness (Serrano et al. 2009; Ansari and Husain 2012). A particle is described in nanotechnology as a tiny item that acts as a whole in terms of its transportation and characteristics. The particles cover the size from 100 to 2500 nanometres in terms of diameter. On the other hand, between 1 and 100 nanometres ultrafine particles come. Nanoparticles exhibit the characteristic properties which are quite different from those seen in bulk materials or microscopic particles. The properties of nanofibers or nanoparticles can be used to create a large surface area and to create an abiotic/biotic interface between the devices. Dimensionally, nanomaterials can be compared to biological macromolecules like enzymes or nucleic acids (Stark 2011; Verma et al. 2013).

10.2 The Rise of Nanotechnology

Nanotechnology is allowing the improvement of devices in a scale ranging from one to a few hundred nanometres which are able to achieve simple tasks such as computing or actuation, data storing, sensing and the *nanoscale (1x nanometre = 1x* 10^{-9} *metre)*. The biological, chemical and physical properties of novel nanomaterials and nanoparticles vary in valuable ways, and fundamental properties vary from the properties of those at microscopic level, therefore demonstrating the transition of technology from macroscale to nanoscale (Fig. 10.1).

10.3 Characterization of Nanomaterials

The synthesis of nanoparticles is essential for understanding the nanoparticles properties and controlling them. Nanoparticles characters are characterized using different techniques. Common techniques such as ultraviolet-visible spectroscopy, powder x-ray diffractometry (XRD), Fourier-transform infrared spectroscopy (FTIR), atomic force microscopy (AFM), dynamic light scattering (DLS), electron microscopy (SEM, TEM), photoluminescence spectra, matrix-assisted laser-desorption time-of-flight mass spectrometry (MALDI-TOF) and x-ray photoelectron spectroscopy (XPS).

10.4 Properties of Nanomaterials

Nanoparticles are said to effectively make a bridge between an atom and bulk material which is of great scientific interest. Bulk content should exhibit constant physical properties regardless of its size, but this is often not the case at the nanoscale. The dependent nature and properties are also seen, such as quantum imprisonment in semiconductor molecules, super magnetism and plasmon resonance in surface magnetic materials in some metal particles. The nature and characteristics of the materials change as their shape gets close to the nanoscale and as quickly as the proportion of atoms on the exterior surface of a substance becomes significant. The nanoparticles' interesting and sometimes unexpected characteristics are due in part to the exterior surface elements of the material being dominated by properties, i.e. bulk properties instead.

Nanoparticles display many special properties relative to bulk content. Copper nanoparticles smaller than 50 nanometres are considered super hard ingredients and do not exhibit ductility and malleability similar to bulk copper. Changes in properties are not always desirable. Smaller ferroelectric material less than 10 nanometres can switch the temperature of the room using its thermal energy to switch its magnetic direction making them useless for storage. Suspension of nanoparticles is possible because the interaction between the surfaces of the particle with the solvent is strong enough to remove the difference in density. As a result, the liquid usually contains materials to be drowned or floating. Nanoparticles often have visual characteristics that are unpredictable, because they are tiny enough to restrict and generate quantum effects. For instance, dark red appears to be black in gold nanoparticles solutions. The nanoparticles attain a very high surface-to-volume ratio. It basically gives a tremendous driving force for the spread. Especially at high temperatures, sintering may occur for short periods of time compared to large particles. This principle does not affect the density of the final product, although the difficulties of flow and trend of nanoparticles make matters complicated. Large surface area-to-volume ratio also reduces the transient melting temperature of nanoparticles. In addition, nanoparticles have been provided to some extraordinary properties. Zinc oxide nanoparticles have been found to have better UV blocking properties than their bulk options. This is one of the reasons that it is often used in screen lotion. Nanoparticles have also been linked to textile fibre to make functional and smart fabrics.

10.5 Synthesis of Nanomaterials

Synthesis of nanometre dimensions particles with narrow shapes distribution and accuracy is one of the biggest challenges. Various approaches have been used to obtain nanostructure material. As a principle, any method capable of producing polymeric content of very rich grains can be used in the production of nanocrystalline materials. There are currently two approaches. First is reducing the size, also known as the top-down approach, as well as the self-assembly or bottom-up approach to obtain nanosized material. Secondly, large materials are reduced to small dimensions of the order of a nanometre, and in the down-up approach, the material is produced up to the nanometre size. Nanomaterials can be synthesized chemically as well as by means of physical methods. Physical methods used for nanomaterials synthesis are consolidation, monomer gas aggregation, sputtering, inert gas evaporation, lithography, ion beam method and ball milling. Chemical methods are chemical precipitation and capping, microemulsion, sol-gel method, condensed phase synthesis, electrochemical deposition and reduction technique.

10.6 Biodiesel as a Biofuel

Biodiesel is a highly inflammable biofuel which usually burns with more or less no ash. Biodiesel is ester-based biofuel, and it is retrieved or obtained from renewable biomass resources, like soybean, groundnut, neem and Jatropha, efficiently (Fig. 10.2). Chemically, biodiesel comprises long-chain fatty esters of organic compound methyl or ethyl esters. It is suitable to use in vehicles as similar of

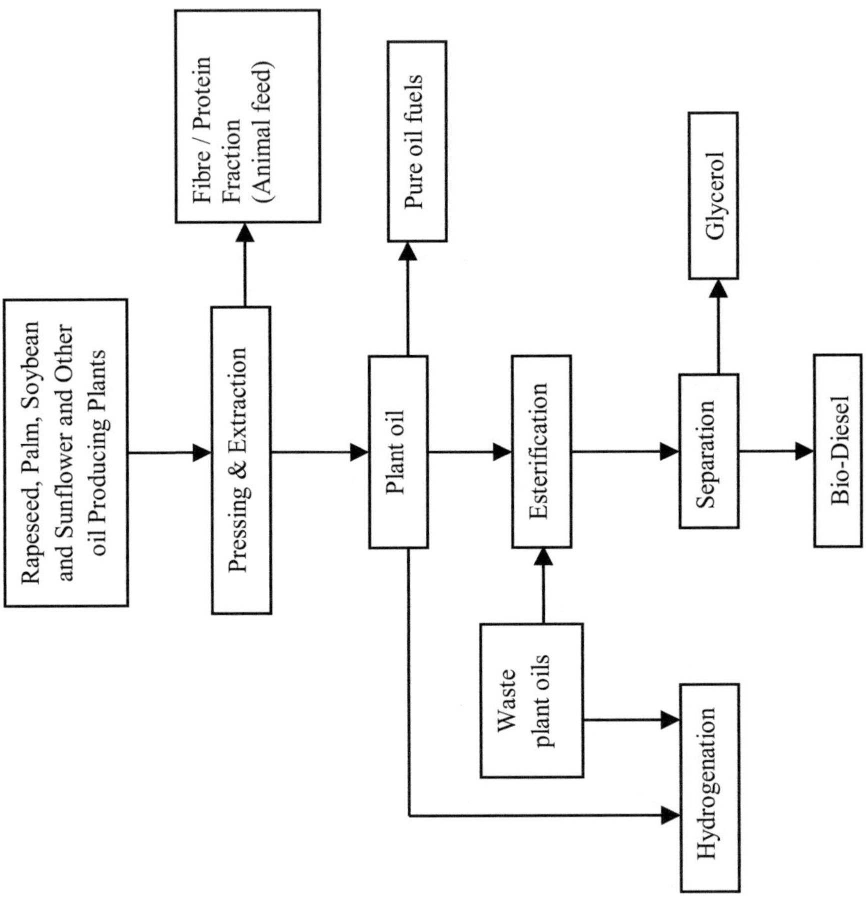

Fig. 10.2 Biodiesel production chart

petroleum product diesel, so it is called biofuel or biodiesel. Certain properties of biodiesel such as air/fuel ratio, density and heat of vaporization are similar to fossil diesel and are also known to reduce carbon monoxide, sulphur oxide and smoke emissions. The oils are transesterified with methanol in the presence of a suitable alkaline catalyst, mostly NaOH in the usual method of biofuel preparation from vegetable/plant originated oils (Palaniappan 2017). The major problems associated with this process include saponification, deactivation of the catalyst and low reaction rate (Palaniappan 2017). In order to overcome the abovementioned hiccups, Ti-incorporated SBA-15 (Santa Barbara Amorphous) mesoporous silica was found to be a highly efficient and recyclable solid acid catalyst to produce biofuels from vegetable oils (Chen et al. 2013). During material-based biodiesel synthesis, nanotechnology enables high strength, durable, sensorial and active materials in developments. Using chemical vapour deposition (CVD), nanolayers of plant oils during esterification are deposited which is the base material used to make bioethanol. Further small, nano and micro sensors integrated into making final product as biodiesel. In present era, we know that conventional petroleum recourses are going to be diminished. Hence, biofuel can be a good option for use as it is eco-friendly and economically feasible. Some edible oils may also be used for preparation of biodiesel, e.g. sunflower, soybean and groundnut, through nanotechnology research beside other nonedible oil – Jatropha, Karanj, (*Pongamia pinnata*) and Nagchampa.

10.7 Present Status and Scenario of Biofuel in India

In the present scenario, ever-rising energy demand, technological advancements, greenhouse gases (GHGs) emissions, diversified energy supply sources, etc., are some of the reasons for promoting biofuel production and future use of the biodiesel as a green and clean energy source across the world. In order to access the clean energy at an affordable price, it is a prerequisite that the biofuel is generated from cheap raw material. For promoting biofuels in India, diversifying energy sources, augmenting domestic supply sources, reducing local emissions and improving the livelihood and income generation opportunities in rural areas especially in the semi-arid regions are additional factors for promoting biofuels. In India, Jatropha oil blended with sugarcane-based ethanol is used as a substituted of petrol. Seed oils from other sources are used as a substitute for diesel. The biodiesel production effort are focused on using nonedible oils from plants (*Jatropha curcas, Pongamia pinnata* and other treeborne oilseeds) and animal fats like fish oil.

10.8 Feasibility Scenario of Biodiesel Production

Based on the survey reports on Rajasthan and Orissa states in general, it would be better to say that the promotion of biofuel/biodiesel is a very complex issue in particular and is solely dependent on various affecting factors like knowledge, infrastructure support, policy issues, inter-sectoral promotion and technology transfer beside local circumstances such as climatic, agronomic, mechanical, economic and social issues.

10.9 Land and Climate for Biofuel

Mainly, wasteland is suitable for the cultivation of biofuel crops, viz., *Jatropha, Karanj (Pongamia pinnata)* in Rajasthan and Orissa; this includes dunes prone areas, deserts tract, sand and red desert soils in western part of the country and the red soils and alluvial in depressions of foot hills in the southern parts of India Tamil Nadu and Kerala. Available wasteland is suitable for the cultivation of biofuel crops due to degradation, undulating terrain erosion and low fertility. These types of soil offer a wide range of variation from lateritic, red to yellow and mixed red to black soils. These soils are either acidic or alkaline in nature, thereby limiting the possibility of raising any normal agricultural crops. The texture varies from sandy, coarse sand and sandy loam to gravel; such type of land is suited for growing these green energy crops. The average annual rainfall (AAR) generally varies as of 100 mm is received (in Rajasthan). In Orissa fairly good rainfall of around 1500 mm is received including summer. These climatic conditions are favorable for the Jatropha plantation for which the availability of water during the early stages of growth plays a critical role. Also the fruiting and the seed yield of the plant are highly dependent on the availability of water (rain or irrigation) during critical growth stages. As far as temperature is concerned, a wide range of constant variation in temperature from −2 °C (in north western plain) to 45 °C in south eastern region of Rajasthan is not considered suitable for the growth of Jatropha plantation. Similarly, the temperature observed as 12–40 °C experienced in all the agroclimatic regions of Orissa is ideal for the growth of Jatropha plantation.

10.10 Significance of Biofuels

As far as biodiesel is concerned, it is an eco-friendly green fuel that has almost no sulphur and no aromatic compounds (carbon benzene, phenol, haloarene, etc.) and about 10% built in oxygen. It is a fatty acid or methyl ester-based fuel. Biodiesel needs no separate infrastructure for storage; dispensing and localized existing tank of conventional diesel can be used. Biodiesel is safe to handle, and its flash point is higher than those of conventional diesel. So there is a remarkable increased consumption observed in the mixed use of biodiesel with conventional fuel also. Biodiesel is made of virgin or used vegetable oil (both edible and nonedible) and animal fats. Biodiesel can be operated in CI engines like petrol diesel conventional fuels. Also, it can be blended in a ratio. Some of advantages over conventional fuels are:

(a) Reduction in carbon monoxide gas up to 35–50%.
(b) Able to produce same mileage.

(c) It can be blended with conventional fuel up to any ratio.
(d) CO_2 emission 57% lower.
(e) It is nontoxic in comparison to conventional diesel fuel and gets more lubricant efficiency by 60–65%.

10.11 Nanotechnique-Based Process of Making Biofuel

Biodiesel is prepared by chemical reactions in which nanomaterial's technology is involved in a process called transesterification. In this reaction, vegetable oil reacts with ethyl or methyl alcohol in the presence of a catalyst NaOH/KOH. Then, this reaction gives one molecule of triglyceride having three pairs of –OH group. Temperature is maintained at 60 °C, and a product of three molecules methyl ester (biodiesel) is obtained with one molecule of glycerol. After completion of 24 h, this product is divided into two layers. Upper layer is ester and lower is glycerol which is well separated out. Methyl or ethyl ester as a fuel and coproduct glycerol can be used in preparation of soap, candle, glycerin, cosmetics and plastic materials (Fig. 10.2).

10.12 Nanoparticles and Performance of Biofuel

A new study showed that addition of alumina nanoparticles can improve the performance and combustion of biodiesel, while producing fewer emissions (Basha and Anand 2011). This was mainly attributed to the high surface-to-volume ratio of nanoparticles which increases their efficiency as chemical catalysts, thus increasing fuel combustion. The presence of these particles enhances fuel-air mixing properly in the fuel which leads to complete burning or combustion of biofuel in IC engines. This ultimately leads to the better thermal as well as mechanical efficiency of internal combustion engines. In comparison to regular biofuel, the nanoparticle-spiked fuels create less smoke and significantly lower the production of nitrogen oxide and carbon monoxide gases which also prevent photo smog in cold winter regions, save energy and are environmental friendly.

10.13 Conclusion

This chapter provides a general overview of the biofuels production as well as discusses the range of nanomaterials that have been considered as potential candidates for better biofuel manufacturing. Nanomaterials are commonly used in the manufacturing of biodiesel as catalysts. Consequently, it is obvious that nanotechnology can be efficiently used in making biofuels in view of the green technology enhancement and further reduces environmental hazardous issues like both the impact and limits of the usage of fossil fuels in general.

References

Ansari SA, Husain Q (2012) Potential applications of enzymes immobilized on/in nano materials: a review. Biotechnol Adv 30:512–523

Basha JS, Anand RB (2011) Role of nanoadditive blended biodiesel emulsion fuel on the working characteristics of a diesel engine. J Renewable Sustain Energy 3:023106

Bidart C, Fröhling M, Schultmann F (2014) Livestock manure and crop residue for energy generation: macro-assessment at a national scale. Renew Sust Energ Rev 38:537–550

Biswas A, Bayer IS, Biris AS, Wang T, Dervishi E, Faupel F (2012) Advances in top-down and bottom-up surface nanofabrication: techniques, applications & future prospects. Adv Colloid Interf Sci 170:2–27

Cao Y, Pawowski A (2012) Sewage sludge-to energy approaches based on anaerobic digestion and pyrolysis: brief overview and energy efficiency assessment. Renew Sust Energ Rev 16:1657–1665

Chen SY, Mochizuki T, Abe Y, Toba M, Yoshimura Y (2013) Production of high-quality biodiesel fuels from various vegetable oils over Ti-incorporated SBA-15 mesoporous silica. Catal Commun 41:136–139

Hussein AK (2015) Applications of nanotechnology in renewable energies—a comprehensive overview and understanding. Renew Sust Energ Rev 42:460–476

Jiang S, Win KY, Liu S, Teng CP, Zheng Y, Han MY (2013) Surface-functionalized nanoparticles for biosensing and imaging-guided therapeutics. Nanoscale 5:3127–3148

Karellas S, Boukis I, Kontopoulos G (2010) Development of an investment decision tool for biogas production from agricultural waste. Renew Sust Energ Rev 14:1273–1282

Kralova I, Sjöblom J (2010) Biofuels-renewable energy sources: a review. J Dispers Sci Technol 31:409–425

Kumar S, Pant G, Sharma V, Bisht P (2014) Nanotechnology in computers. Int J Inf Comput Technol 4:0974–2239

Lönnqvist T, Silveira S, Sanches-Pereira A (2013) Swedish resource potential from residues and energy crops to enhance biogas generation. Renew Sust Energ Rev 21:298–314

Palaniappan K (2017) An overview of applications of nanotechnology in biofuel production. World Appl Sci J 35:1305–1311

Pugh S, McKenna R, Moolick R, Nielsen DR (2011) Advances and opportunities at the interface between microbial bioenergy and nanotechnology. Can J Chem Eng 89:2–12

Puri M, Abraham RE, Barrow CJ (2012) Biofuel production: prospects, challenges and feedstock in Australia. Renew Sust Energ Rev 16:6022–6031

Rai PK (2016) Recent advances in substrate utilization for fermentative hydrogen production. J Appl Biol Biotechnol 4:59–67

Rai PK, Kadier A, Kumar M, Singh SP (2018) Chapter 7: Utilization of microalgal biomass as a source of bioenergy. In: Tripathi K, Kumar N, Abraham G (eds) Role of photosynthetic microbes in agriculture and industry. Nova Science Publishers, New York

Serrano E, Rus G, García-Martínez J (2009) Nanotechnology for sustainable energy. Renew Sust Energ Rev 13:2373–2384

Stark WJ (2011) Nanoparticles in biological systems. Angew Chem Int Ed 50:1242–1258

Verma ML, Barrow CJ, Puri M (2013) Nanobiotechnology as a novel paradigm for enzyme immobilization and stabilisation with potential applications in biodiesel production. Appl Microbiol Biotechnol 97:23–39

Zhang Z, O'Hara IM, Mundree S, Gao B, Ball AS, Zhu N, Bai Z, Jin B (2016) Biofuels from food processing wastes. Curr Opin Biotechnol 38:97–105

MIX
Papier aus verantwortungsvollen Quellen
Paper from responsible sources
FSC® C105338

If you have any concerns about our products,
you can contact us on
ProductSafety@springernature.com

In case Publisher is established outside the EU,
the EU authorized representative is:
Springer Nature Customer Service Center GmbH
Europaplatz 3, 69115 Heidelberg, Germany

Printed by Libri Plureos GmbH
in Hamburg, Germany